Satellite Soil Moisture Retrieval

Satellite Soil Moisture Retrieval

Techniques and Applications

Edited by

Prashant K. Srivastava
NASA Goddard Space Flight Center, USA
and Institute of Environment and Sustainable Development,
Banaras Hindu University, India

George P. Petropoulos
Geography & Earth Sciences
Aberystwyth University
Aberystwyth, SY23 3DB
Wales, UK

Yann H. Kerr
CESBIO
Toulouse, France

ELSEVIER　　AMSTERDAM • BOSTON • HEIDELBERG • LONDON • NEW YORK • OXFORD
PARIS • SAN DIEGO • SAN FRANCISCO • SINGAPORE • SYDNEY • TOKYO

Elsevier
Radarweg 29, PO Box 211, 1000 AE Amsterdam, Netherlands
The Boulevard, Langford Lane, Kidlington, Oxford OX5 1GB, UK
50 Hampshire Street, 5th Floor, Cambridge, MA 02139, USA

Notices
Knowledge and best practice in this field are constantly changing. As new research and
experience broaden our understanding, changes in research methods, professional practices,
or medical treatment may become necessary.

Practitioners and researchers must always rely on their own experience and knowledge in
evaluating and using any information, methods, compounds, or experiments described herein.
In using such information or methods they should be mindful of their own safety and the
safety of others, including parties for whom they have a professional responsibility.

To the fullest extent of the law, neither the Publisher nor the authors, contributors, or editors,
assume any liability for any injury and/or damage to persons or property as a matter of products
liability, negligence or otherwise, or from any use or operation of any methods, products,
instructions, or ideas contained in the material herein.

Library of Congress Cataloging-in-Publication Data
A catalog record for this book is available from the Library of Congress

British Library Cataloguing-in-Publication Data
A catalogue record for this book is available from the British Library

ISBN: 978-0-12-803388-3

For information on all Elsevier publications
visit our website at https://www.store.elsevier.com/

Working together
to grow libraries in
developing countries

www.elsevier.com • www.bookaid.org

I would like to dedicate this book to my parents, Bishwambhar N Srivastava and Nirmala Devi, as well as to my beloved wife Manika for their continuous support

—Prashant K. Srivastava

I dedicate this book to my parents, Panagiotis and Evgenia, for their continuing love and support in all the endeavors of my life

—George P. Petropoulos

I would like to dedicate this book to my wife and children for their patience and understanding as well as to my colleagues for their significant contributions

—Yann H. Kerr

Contents

Section II
Optical and Infrared Techniques & Synergies Between them

Section III
Microwave Soil Moisture Retrieval Techniques

11. Intercomparison of Soil Moisture Retrievals From In Situ, ASAR, and ECV SM Data Sets Over Different European Sites

B. Barrett, C. Pratola, A. Gruber and E. Dwyer

Section IV
Advanced Applications of Soil Moisture

12. Use of Satellite Soil Moisture Products for the Operational Mitigation of Landslides Risk in Central Italy

L. Brocca, L. Ciabatta, T. Moramarco, F. Ponziani,
N. Berni and W. Wagner

Section V
Future Challenges in Soil Moisture Retrieval and Applications

List of Contributors

Numbers in parenthesis indicate the pages on which the authors' contributions begin.

R. Akbar (187), Ming Hsieh Department of Electrical Engineering, University of Southern California, Los Angeles, CA

A. Al Bitar (3, 351), Centre d'Études Spatiales de la Biosphère, Toulouse, France

V. Anagnostopoulos (91), National Technical University of Athens; InfoCosmos Ltd, Athens, Greece

B. Barrett (209), University College Cork (UCC), Cork, Ireland

A. Berg (47), University of Guelph, Guelph, ON, Canada

N. Berni (231), Civil Protection Centre, Foligno, Italy

L. Brocca (231, 351), National Research Council; Research Institute for Geo-Hydrological Protection, National Research Council, Perugia, Italy

D. Chaparro (249), Universitat Politècnica de Catalunya, IEEC/UPC; Barcelona Expert Center, Institute of Marine Sciences, CSIC, Barcelona, Spain

L. Ciabatta (231), National Research Council, Perugia, Italy

Q. Dai (289), Nanjing Normal University, Nanjing, China

N. Das (187), NASA/Jet Propulsion Laboratory (JPL), Pasadena, CA

E. De Witte (379), Airbus Defence and Space Ltd, Stevenage, United Kingdom

E. Dwyer (209), EurOcean—European Centre for Information on Marine Science and Technology, Lisbon, Portugal

D. Entekhabi (187), Department of Civil and Environmental Engineering, Massachusetts Institute of Technology, Cambridge, MA

F. Gerard (379), Centre for Ecology & Hydrology, Wallingford, United Kingdom

N. Ghilain (309), Royal Meteorological Institute of Belgium, Brussels, Belgium

H. Griffiths (91), Department of Geography & Earth Sciences, Ceredigion, Wales, United Kingdom

A. Gruber (209), Vienna University of Technology, Vienna, Austria

D.K. Gupta (159, 333), Indian Institute of Technology (BHU), Varanasi, India

M. Gupta (29, 271, 333), NASA Goddard Space Flight Center, Greenbelt, MD, United States

D. Han (289), University of Bristol, Bristol, United Kingdom

C. Hodges (91), Geo Smart Decisions Ltd, Llanidloes, Powys, United Kingdom

M. Holzman (73), Consejo Nacional de Investigaciones Científicas y Técnicas – Instituto de Hidrología de Llanuras "Dr. Eduardo J. Usunoff", Rep. Italia 780, Azul, Buenos Aires, Argentina

G. Ireland (91), Department of Geography & Earth Sciences, Ceredigion, Wales, United Kingdom

T. Islam (29, 91, 159, 271, 289, 333), NASA Jet Propulsion Laboratory; California Institute of Technology, Pasadena, CA, United States

J.C. Jiménez-Muñoz (135), Universitat de València, Valencia, Spain

D. Kalivas (91), Agricultural University of Athens, Athens, Greece

Y.H. Kerr (3), Centre d'Études Spatiales de la Biosphère, Toulouse, France

C. Mattar (135), University of Chile, Santiago, RM, Chile

A. Mialon (3), Centre d'Études Spatiales de la Biosphère, Toulouse, France

V. Mironov (169), Kirensky Institute of Physics, Krasnoyarsk, Russia

M. Moghaddam (187), Ming Hsieh Department of Electrical Engineering, University of Southern California, Los Angeles, CA

T. Moramarco (231), National Research Council, Perugia, Italy

J. Muñoz-Sabater (351), European Centre for Medium Range Weather Forecasts, Reading, United Kingdom

V. Pandey (29), Banaras Hindu University, Banaras, India

A. Petrie (379), Centre for Ecology & Hydrology, Wallingford, United Kingdom

G.P. Petropoulos (91, 333), Department of Geography & Earth Sciences, Ceredigion, Wales, United Kingdom

M. Piles (109, 249), Universitat Politècnica de Catalunya, IEEC/UPC; Barcelona Expert Center, Institute of Marine Sciences, CSIC, Barcelona, Spain

F. Ponziani (231), Civil Protection Centre, Foligno, Italy

R. Prasad (159, 333), Indian Institute of Technology (BHU), Varanasi, India

C. Pratola (209), University College Cork (UCC), Cork, Ireland

P. Rahimzadeh-Bajgiran (47), University of Maine, Orono, ME, United States

R. Rivas (73), Comisión de Investigaciones Científicas de la provincia de Buenos Aires – Instituto de Hidrología de Llanuras "Dr. Eduardo J. Usunoff", Campus Universitario, Tandil, Buenos Aires, Argentina

A. Santamaría-Artigas (135), University of Chile, Santiago, RM, Chile

I. Savin (169), Kirensky Institute of Physics, Krasnoyarsk, Russia

S.K. Singh (333), University of Allahabad, Allahabad, India

N. Sánchez (109), University of Salamanca/CIALE, Villamayor, Spain

J.A. Sobrino (135), Universitat de València, Valencia, Spain

P.K. Srivastava (3, 29, 91, 159, 271, 333), NASA Goddard Space Flight Center, Greenbelt, MD, United States; Banaras Hindu University, Banaras, India

S. Suman (29), Banaras Hindu University, Banaras, India

E. Tebbs (379), King's College London, Department of Geography, Strand, London, United Kingdom

M. Vall-llossera (249), Universitat Politècnica de Catalunya, IEEC/UPC; Barcelona Expert Center, Institute of Marine Sciences, CSIC, Barcelona, Spain

W. Wagner (231), Vienna University of Technology, Vienna, Austria

W.Z. Wan Jaafar (333), University of Malaya, Kuala Lumpur, Malaysia

J.-P. Wigneron (3), INRA ISPA, Bordeaux, France

L. Zhuo (289), University of Bristol, Bristol, United Kingdom

Author Biographies

Prashant K. Srivastava is presently working in Hydrological Sciences, NASA Goddard Space Flight Center on SMAP satellite soil moisture retrieval algorithm development, instrumentation, and simulation for various applications, and affiliated with IESD, Banaras Hindu University as a faculty. He received his PhD degree from Department of Civil Engineering, University of Bristol, Bristol, UK. His primary research focuses on the use of optical and microwave satellite synergy with Weather Research and Forecasting for hydrological applications. He is also a working group partner with NASA JPL on SMAP soil moisture calibration and validation. He has been a recipient of several awards such as University of Maryland Fellowship, USA; Commonwealth Fellowship, UK; CSIR-UGC-JRF-NET (2005), CSIR-JRF-NET (2006), University Grant Commission (UGC)-NET (2006), and Ministry of Human Resource Development fellowships from India. He has 100+ publications in peer-reviewed journals, published 4 books with reputed publishing houses such as Springer, Taylor and Francis, and Elsevier and published several book chapters. He is also acting as editor of few journals and is currently a member of Indian Society of Geomatics, Indian Society of Remote Sensing, Indian Association of Hydrologists (IAH), International Society for Agrometeorology (INSAM), International Association of Hydrological Sciences (IAHS), and few others.

George P. Petropoulos is a senior lecturer in Remote Sensing & GIS at Aberystwyth University, UK. He completed his graduate studies (MSc, PhD) at the University of London in 2008, specializing in Earth Observation (EO) Modelling. His PhD was focusing specifically in the retrievals of energy fluxes and soil moisture from the synergy of optical EO data with simulation process models.

Petropoulos' research focuses on exploiting EO data alone or synergistically with land surface process models for computing key state variables of the Earth's energy and water budget, including energy fluxes and soil surface moisture. He is also conducting research on the application of remote sensing technology to land cover mapping and its changes occurred from either anthropogenic activities or geohazards (mainly floods, wildfires, frost). In this framework, he contributes to the development of open source software tools in EO modelling and develops and implements all-inclusive benchmarking approaches to either EO operational algorithms/products or surface process models, including advanced sensitivity analysis.

Petropoulos serves as a Council member of the Remote Sensing & Photogrammetric Society (RSPSoC), Editorial Board member on several international peer-reviewed scientific journals in EO and environmental modelling, and as a reviewer for various funding bodies (including the European Commission). He is editor/co-editor of 2 books, author/co-author of + *50* peer-reviewed journal articles and of +*90* international conferences. He has developed fruitful collaborations with key scientists in his area of specialization globally, and his research work so far has received international recognition via several noteworthy awards he has obtained.

Yann H. Kerr received the engineering degree from Ecole Nationale Supérieure de l'Aéronautique et de l'Espace, the MSc degree in electronics and electrical engineering from Glasgow University, Glasgow, Scotland, UK, and the PhD degree in Astrophysique Géophysique et Techniques Spatiales, Université Paul Sabatier, Toulouse, France. Affiliated to the French space agency (CNES) since 1980, he joined LERTS in 1985; for which he was director in 1993–94. He spent 19 months at JPL, Pasadena, in 1987–88. He has been working at CESBIO since 1995 as deputy director (1995–99) and director (2007–16).

His fields of interest are in the theory and techniques for microwave and thermal infrared remote sensing of the Earth, with emphasis on hydrology, water resources management, and vegetation monitoring.

He was an EOS principal investigator (interdisciplinary investigations), and PI and precursor of the use of the SCAT over land. In 1989, he started to work on the interferometric concept applied to passive microwave earth observation and was subsequently the science lead on the MIRAS for ESA. He is a member of the SMAP Science Team. In 1997, he first proposed the natural outcome of the previous MIRAS work with what was to become the SMOS Mission selected by ESA in 1999 with him as the SMOS mission lead-investigator and chair of the Science Advisory Group. He has organized all the SMOS workshops and was guest editor on two IEEE and one RSE special issues. He is also working on the SMOS-Next concept. He received the World Meteorological Organization 1st prize (Norbert Gerbier), the USDA Secretary's team award for excellence (Salsa Program), the GRSS certificate of recognition for leadership in development of the first synthetic aperture microwave radiometer in space and success of the SMOS mission, and the ESA team award.

Preface

Information on our planet's soil moisture is indispensable to a number of practical applications related to food security, society, and ecosystems. It plays an important role in the Earth's water cycle; it is a key variable in the water and energy exchanges that occur at the land-surface/atmosphere interface and conditions the evolution of weather and climate over continental regions. Therefore, globally, the monitoring of the soil moisture has developed into a very important and urgent research direction, especially towards water resource management, irrigation scheduling, improved weather forecasts, natural hazards mitigation analysis, predictions of agricultural productivity, crop insurance, climate predictions, ecological health and services, improved trafficability, groundwater recharge, water quality and quantity, etc. In this context, there is a growing need of monitoring and understanding the soil moisture retrieval techniques and its applications. This can be of crucial importance, particularly to regions on which the amount of water available is limited.

This book is motivated by the desire to solve the problem of soil moisture retrievals in a cost effective and timely way. The launch of advanced Earth Observation (EO) satellites in optical/IR and microwave domain has the potential to reshape the soil moisture retrieval system and can be used to monitor soil moisture regularly, accurately, and in real time. These satellites/instruments provide necessary data that can make up for the lack of on-the-ground monitoring of soil moisture around the world and because of substantial importance of soil moisture both ESA and NASA have launched fully dedicated soil moisture satellites known as SMOS and SMAP, respectively, to provide a better estimate soil moisture over the globe.

After the launch of many sophisticated satellites (such as SSM, AMSR E/2, SMOS, ASCAT, Aquarius, SMAP, Landsat series, MODIS, and many others), soil moisture retrieval techniques have gained considerable momentum among the earth and environmental science communities for solving and understanding various complex problems. In this essence, a comprehensive book is needed devoted to putting together a collection of the recent developments and rigorous applications of the soil moisture from optical/infrared and microwave satellites. In order to understand the retrieval system and applications, the book is designed to advance the scientific understanding, development, and application of soil moisture retrieval techniques, and its applications for various environmental problems. This book will promote the synergistic and multidisciplinary activities among scientists. Therefore, we would consider this a must read book that promotes the synergistic and multidisciplinary activities among scientists and users working in the field of hydrometeorological sciences and other related disciplines.

The book becomes possible because of extensive and valuable contributions from interdisciplinary experts from all over the world in the field of soil moisture retrievals and applications. In order to simplify the soil moisture retrieval techniques and applications for most of the students and researchers, this book focused on three working methodologies, *viz*, theory, abstraction, and applications as they are fundamental to all research programs. The book has been divided into five sections. Section I contains Introduction, Section II details the Optical and Infrared Techniques and Synergies Between Them, Section III provides several methodologies for Microwave Soil Moisture Retrieval while Sections IV and V deal with Advance Application of Soil Moisture and Future Challenges in Soil Moisture Retrieval and Applications.

Chapter 1 in Introduction section written by Kerr and team reviews various means of measuring soil moisture, satellite missions, retrieval system, and possible pitfalls. Chapter 2 by Srivastava et al. furnishes a brief description of available datasets for terrestrial soil moisture estimation. The Section II of the book contains chapters related to optical/IR methods. Chapter 3 of this section by Rahimzadeh-Bajgiran and Berg provides a brief history and concept of optical/TIR models for soil moisture estimation with case study of Canadian Prairies. Chapter 4 by Holzman and Rivas deals with the estimation of soil moisture using temperature vegetation dryness index. In Chapter 5, Petropoulos et al. interestingly described the soil moisture estimation using the "triangle" approach using temperature and vegetation index feature space. As finer resolution dataset are important for hydrological community, Chapter 6 by Piles and Sanchez provided a review of several approaches for soil moisture downscaling using the semiempirical model and its validation in Duero basin. Chapter 7 of this section written by Mattar et al. provided an overview of the different microwave soil moisture retrieval model with their theories and applications.

Section III of the book made available an overview of methodologies for microwave soil moisture retrieval. Chapter 8 provided by Gupta et al. briefly described diverse nonparametric models for the retrieval of soil moisture by using bistatic scatterometer datasets. In Chapter 9, Mironov and Savin provided the usefulness of dielectric model for microwave soil moisture retrieval while Akbar et al. in Chapter 10 provided the active and passive synergy for soil moisture retrieval and discuss the merits of active and passive systems in spatial resolution enhancement. Chapter 11 by Barrett et al. summarized the performance of different soil moisture products with seasonal and temporal analysis.

Section IV of this book deals with chapters related with the advanced applications of soil moisture. Chapter 12 of this section by Brocca et al. provided an overview of PRESSCA early warning system for landslide risk prediction using satellite soil moisture information. In Chapter 13, Chaparro et al. presented the role of soil moisture in wildfire prevention using the soil moisture from SMOS and SMAP satellites. In Chapter 14, Gupta et al. demonstrated the role of soil moisture for estimation of soil hydraulic parameters for an improved agricultural water management. Zhuo et al., in Chapter 15, presented the importance of satellite soil moisture in rainfall runoff modeling for an effective discharge prediction. Chapter 16 provided by Ghilain presented the usefulness of satellite soil moisture assimilation for improving the evapotranspiration on a daily and sub-daily scale. He further explained the gain and loss of accuracy in choosing a new source for soil moisture input and provided the validation of the prototype by comparison with in situ observations. Chapter 17 written by Srivastava et al. furnished the use of microwave soil moisture from SMOS for soil moisture deficit estimation at a catchment level and depicted the relation of soil moisture with the hydrological model based products.

Section V deals with the chapters related with future challenges in soil moisture retrieval and applications. Chapter 18 in this section by Munoz-Sabater et al. offered a detailed overview of soil moisture from microwave remote sensing and pointed out various obstacles in the production of soil moisture retrievals and its applications. At last, Chapter 19 written by Tebbs et al. supplied a discussion on the emerging and potential future applications of satellite-based soil moisture products and pointed out the current limitations of soil moisture applications and developments required in coming decades.

We believe that the book would be read by the people with a common interest in geo-spatial techniques, remote sensing, sustainable water resource development, applications and other diverse backgrounds within earth and environmental and hydrological sciences field. We do hope this book would be beneficial for the academician, scientists, environmentalists, meteorologists, environmental consultants, computing experts working in the area of water resources.

ACKNOWLEDGMENTS

Editors would like to thank all the contributing authors and anonymous reviewers for their time, talents, and energies and for adherence to a strict timeline and the staff at Academic Press, Elsevier for their patience and throughout support. Editors are also grateful to the publisher for the collaboration in accomplishing the preparation of this book.

ABOUT THE COVER

Sophisticated Soil Moisture and Ocean Salinity (or SMOS) satellite shown on the cover is provided by CESBIO/MIRA.

Prashant K. Srivastava
Maryland, United States

George P. Petropoulos
Wales, United Kingdom

Yann H. Kerr
Toulouse, France

Section I

Introduction

Chapter 1

Soil Moisture from Space: Techniques and Limitations

Y.H. Kerr*, J.-P. Wigneron†, A. Al Bitar*, A. Mialon* and P.K. Srivastava‡,§

*Centre d'Études Spatiales de la Biosphère, Toulouse, France, †INRA ISPA, Bordeaux, France, ‡NASA Goddard Space Flight Center, Greenbelt, MD, United States, §Banaras Hindu University, Banaras, India

1 INTRODUCTION

Soil moisture is certainly an important factor for everything linked to life on Earth. Without water life disappears, and with too much water (e.g., floods) non aquatic life may be severely affected. So the first use of "soil moisture" is to enable vegetation growth and impact weather through the heat and mass exchanges at the surface/atmosphere interface (Srivastava et al., 2013c). This "soil moisture" corresponds to water in both the surface and the root zone or vadose area. In a more "scientific" and hydrologic approach, soil moisture, in other words, water stored in the soil, plays many roles. Surface soil moisture controls the partitioning of rainfall into runoff and infiltration and thus impacts the amassing of water. Good infiltration usually means replenishing the water table, while runoff may mean both exportation of valuable water to other areas or even to the sea, and degradation of top soil through erosion. Also, when saturated, soil may transform heavy rainfalls into floods. Surface soil moisture is thus of great interest for improving the forecasting skills of run-off models which aim at flood risk prediction and/or water resources management (Entekhabi et al., 1999). Further consistent estimates of soil moisture from agricultural fields are required for efficient irrigation management and scheduling (Glenn et al., 2011; Lorite et al., 2012; Srivastava et al., 2013b).

Surface soil moisture is also important as it controls soil evaporation and vegetation transpiration and thus the heat and mass transfers between the earth and the atmosphere (van den Hurk et al., 2002). It is consequently very useful in weather forecast models through global circulation models (Entekhabi et al., 1996). It has been shown, for instance, that including actual soil moisture instead of simply using sea surface temperature may improve rain forecasts very significantly, especially in cases of extreme events (Entekhabi et al., 1996;

Satellite Soil Moisture Retrieval. http://dx.doi.org/10.1016/B978-0-12-803388-3.00001-2

3

van den Hurk et al., 2012). Finally, continuous monitoring of soil moisture on a large scale and over long periods of time gives a significant insight into climate changes (Srivastava et al., 2015).

But soil moisture is a very vague term and it is important to define it. The most common understanding of the term is "the total amount of water in the unsaturated zone." For practical reasons it is often separated into two components, surface soil moisture corresponding to the first centimeters (5 cm in general), and the root zone soil moisture, or second reservoir. Soil moisture is usually expressed in gravimetric units (g/cm^3) which is independent of soil characteristics (bulk density). It is, however, also commonly expressed in volumetric units: m^3/m^3. Sometimes it is expressed as a function of the wilting point and the field capacity. The two latter units are soil type dependent.

2 MEANS OF MEASURING SOIL MOISTURE

To achieve the goals mentioned previously, it is necessary to have access to soil moisture estimates. At a specific point in space and time this is relatively easy with gravimetric sampling. However, to have measurements representative of a larger area (such as a field), the procedure is already somewhat complex as it involves a dedicated sampling strategy. Moreover, because these measurements are time consuming, regional and, even more, global coverage is out of question. Provided one uses automatic probes (resistive, capacitive, time domain reflectometry, etc.) it is possible to achieve larger coverages but these approaches can only be confined to well-equipped and manned sites, as they require care and maintenance (Bircher et al., 2013). Finally these systems carry their own problems and inaccuracies (Chanzy et al., 1998; Escorihuela et al., 2006; Vaz et al., 2013). Also the in situ observations of soil moisture, such as those from probe or gravimetric measurements, do not represent the spatial distribution accurately as soil moisture is highly variable both spatially and temporally and therefore not suitable for regional or global applications. So, large scale monitoring of soil moisture can only rely on remote sensing from space approaches (Srivastava et al., 2013a, 2014).

2.1 Remotely Sensed Soil Moisture, The Main Approaches

Because of the high need for soil moisture measurements, a large number of approaches have been tested. For surface soil moisture, the first ones were based on short wave measurements and on the fact that soils get darker when wet. Obviously, due to atmospheric effects and potential cloud cover, as well as vegetation cover masking effects, this approach is bound to fail in most cases (see Kerr, 2007). A more promising feature is linked to latent heat effects. Wet soils have a higher thermal inertia and are "cooler" than dry soils. These properties led to various trials (thermal inertia monitoring, rate of heating in the morning, and surface temperature amplitude, etc.) (van den Hurk et al., 1997) to assess soil moisture indirectly. All those approaches proved to be somewhat disappointing due to factors inherent to optical remote sensing (atmospheric effects, cloud masking and forcing, and vegetation cover opacity) as well as to the fact that thermal infrared: (1) probes the very skin of soils and (2) this layer is dominated by the exchanges with the atmosphere (Kerr, 2007; Ochsner et al., 2013; Petropoulos et al., 2015). Consequently, to infer soil moisture from such measurements one needs to know exactly the forcings (wind, for instance, will change drastically the skin temperature of a wet soil). A brief review of soil moisture retrieval using optical/IR is provided by (Kerr, 2007; Petropoulos et al., 2015) and so not covered in this chapter.

As microwave systems measure the dielectric constant of soils which is directly related to the water content, research has quickly focused its efforts on assessing soil moisture with radars, scatterometers, or radiometers (Pellarin et al., 2003). These systems offer, when operated at low frequency, the added advantage of being all weather (measurements are not much affected by the atmosphere and clouds) and able to penetrate vegetation (Ulaby et al., 1986) and can operate during the night (Srivastava et al., 2015).

Finally, in an attempt to be exhaustive, a new approach, relying on measurements of the gravity field from space, is providing information on the changes in the total column of water with a spatial resolution of 500 km or more. This column includes the atmosphere, vegetation and surface water, possible snow or ice, soil moisture, and water storage. The first results certainly show a signal but its relationship with water storage has yet to be validated. Actually, many corrections have to be brought to the signal to separate water storage in the ground from surface water, atmospheric water, vegetation water content, or snow (Han et al., 2005).

2.2 Microwave as a Tool for Soil Moisture Monitoring: Current Status

Once it is established that the most promising way to access surface soil moisture is the use of microwave systems, it is necessary to assess the advantages and drawbacks of the different possible approaches. The most popular approach relies on the use of active systems such as Synthetic Aperture Radars (SARs). These systems, in use since 1977 with Sea Satellite (SEASAT), offer all weather measurements with a fine spatial resolution (tens of meters). Operationally, however, they suffer—as most high resolution systems—from a rather low temporal sampling (35 days for the European Remote Sensing satellite (ERS) for instance) which is not really compatible with hydrologic requirements or weather forecast models (Wagner et al., 2007). With the advent of the Copernicus program and the Sentinel 1 satellites, this limitation is bound to decrease as with two Sentinel 1 satellites and a revisit of 12 days, one may expect to achieve an overall 6 day revisit capacity. But the most complicated characteristic of SAR systems is the coherent nature of the signal itself and the interactions with the diffusing medium. SAR images are affected by speckle and by the scattering at the surface.

The scattering can be due to the vegetation cover (distribution of water in the canopy) or to the soil's surface (surface scattering when wet, and volume scattering when dry). The direct consequence of these perturbations is a signal at least as much sensitive to surface roughness as to moisture itself, not mentioning vegetation (see also Wigneron et al., 1999). Obviously these effects are frequency dependent.

All these inherent difficulties might explain that, although several SAR flew since 1977, they were not used in a standard and routine fashion, nor was any real soil absolute moisture mapping really done. Actually, to avoid the roughness and vegetation perturbations, an approach relying on change detection, hence relative, has been used with relative success (Gorrab et al., 2015; Kim et al., 2012; Moran et al., 2002; Tomer et al., 2015). This works as long as either vegetation or soil roughness do not change. Also, as long as Sentinel 1 and 2 are not operational, again, temporal coverage will often be an issue.

The use of scatterometers offers an interesting trade off. The spatial resolution is much coarser (tens of kilometers) but with a much wider swath allowing reasonably frequent coverage (every 2–6 days on average depending on the system). It also offers the advantage of being less subject to speckle (Magagi and Kerr, 1997). Consequently, several authors routinely produce soil wetness index maps of many areas of the world with scatterometers using change detection approaches. The effect of vegetation is, however, still significant and actually corresponds to most of the signal as the frequencies currently available are C-band on (MetOp) and higher. So the most interesting results were obtained over arid and semiarid regions where vegetation and soil moisture are very highly correlated anyway. It should be also noted that over dry areas (arid and semiarid) the retrievals are often erroneous; areas where soil moisture is prone to be of importance to users (Wagner et al., 2013; Al-Yaari et al., 2014b). The influence of surface roughness is also significant (Magagi and Kerr, 2001) and it is best dealt with by using, here again, change detection.

The advanced scatterometer (ASCAT) gives one of the freely available global soil moisture data sets derived from the backscatter measurement. The ASCAT onboard MetOp is a real-aperture and an active microwave remote sensing instrument with very good radiometric accuracy (about 0.3 dB) and stability. It is a C-band scatterometer operated at a frequency of 5.255 GHz using six sideways-looking, vertically polarized antennae. The main objective of installing an ASCAT is to monitor wind speed and direction over the oceans, though it is also used for studying soil moisture. Currently, with ASCAT on board MetOp, the signal is scaled between minimum and maximum measurements as long as the whole soil moisture range is explored.

The last possibility in the microwave domain is to use radiometers (Schmugge, 1998). The technique is old (Newton et al., 1982) and well mastered as many sensors and notably sounders rely on passive microwaves. To infer soil moisture, these systems are bound to offer the best compromise if used at low frequency, as demonstrated in the early 1970s with the very short SKY-LAB mission (Eagleman and Ulaby, 1975; Jackson et al., 2004). To be efficient, one needs to work in a protected frequency band to avoid unwanted man-made emissions and radio frequency interferences (RFIs) and needs an instrument sensitive to soil moisture while atmosphere is transparent and vegetation plays a limited role. At L-band the emissivity may vary from almost 0.5 for a very wet soil to almost 1 for a very dry one giving a range of 80–100 K for an instrument sensitivity usually of the order of 1 K. As the signal is not coherent, surface roughness and vegetation structure play a reduced role when compared to an active system (Ulaby and Long, 2014). So, one may wonder why L-band radiometry was not used extensively before when it had been proved to be most efficient during ground and airborne measurements (Jackson and Schmugge, 1991; Schmugge et al., 1988). This is due to an inherent limitation: the spatial resolution is proportional to the antenna diameter and inversely proportional to the wavelength. At 21 cm, to achieve a 40 km resolution from an altitude of 750 km requires and antenna of about 8 m in diameter which is a very significant technical challenge. So the research was performed with higher frequency systems as available on the scanning multichannel microwave radiometer (SMMR) (6.6 GHz) (Kerr and Njoku, 1990), the special sensor microwave/imager (SSM/I) (19 GHz), the Earth Observing Advanced Microwave Scanning Radiometer (AMSR-E) (6.8 GHz) (Njoku and Li, 1999) and now its follow on, the Advanced Microwave Scanning Radiometer 2 (AMSR-2) (Imaoka et al., 2010; Kim et al., 2015). Even though the frequency is not ideal, good results were obtained with the SMMR (in spite of a very poor resolution due to important side lobes) and now AMSR-E/2 (Al-Yaari et al., 2014b; Gruhier et al., 2008; Leroux et al., 2014; Njoku and Li, 1999). The limitations are mainly linked to the fact that the vegetation becomes rapidly opaque, the frequency is not protected and thus bound to be polluted by RFIs, and the single angular measurement makes it difficult—in several cases—to separate vegetation and soil contributions to the signal.

3 SATELLITE MISSIONS

Considering the necessity of making L-band measurements, several approaches have been tested to overcome the antenna size issue. The first was initiated in the early 1990s with the idea to apply radio astronomy techniques (very large arrays and very large baseline interferometers) to Earth remote sensing (Martin-Neira et al., 2014). The one-dimensional concept, electronically scanned thinned array radiometer (ESTAR) was implemented as an aircraft version and proved to fulfill the requirements (Le Vine et al., 1994). It is a system, deployable in space as a sort of large rake and offering—at the cost of a reduced sensitivity—an acceptable spatial resolution. In parallel, another approach using inflatable (or umbrella-like deploying technology) was studied at the Jet Propulsion Laboratory (JPL). The concepts appeared to be complex to deploy and to run, or to offer too limited measurements (single angle and frequency). By 1991, a small group started to work for the European Space Agency (ESA) on the development of a similar instrument but working in two dimensions (Goutoule, 1995;

Goutoule et al., 1996). The concept was named Microwave Imaging Radiometer with Aperture Synthesis (MIRAS) and an airborne prototype was made and operated (Bayle et al., 2002). From then on the concept evolved into a more tailored instrument which was proposed to the Centre National d'Etudes Spatiales (CNES) in 1997 and then to the ESA (Kerr, 1998) in the framework of the Earth Explorer Opportunity mission under the name of Soil Moisture and Ocean Salinity (SMOS) mission. SMOS was launched in Nov. 2009 (Kerr et al., 2001, 2010). Similarly, a mission was proposed and selected by NASA based on a deployable and rotating antenna related to both a radiometer and a radar and called Soil Moisture Active Passive (SMAP) (Entekhabi et al., 2010, 2014). The satellite was launched in Jan. 2015. Note that the SMAP radar broke down on Jul. 7, 2015 because of power supply failure on its high-power amplifier. Both SMOS and SMAP passive instruments have similar characteristics in spite of the very different concept (spatial resolution of more than 40 km, 3 day revisit with time of overpass at 6:00 am/6:00 pm local solar time) but with noticeable differences. SMOS measures the angular signatures of the brightness temperature, while SMAP measures at only one angle (40 degree) but with a better sensitivity (less than 1 K vs. more than 3 K for SMOS over land surfaces). SMAP also has a sophisticated RFI detection system (Spencer et al., 2013) whose necessity was identified during the first months of the SMOS mission (Daganzo-Eusebio et al., 2013; Khazaal et al. 2014; Oliva et al., in press; Soldo et al., 2014, 2015).

The SMOS mission is the first polar orbiting satellite launched that, operating at L-band, measures both sea surface salinity and soil moisture. It is an ESA-led Earth Explorer Mission with contributions from the CNES (France) and the Centro para el Desarrollo Teccnologico Industrial (CDTI) in Spain. It was launched on Nov. 2, 2009. It aims at collecting continuous global observation of near-surface soil moisture and vegetation opacity over land, sea surface salinity and extreme wind speed over the oceans, and thin sea ice at high latitudes. SMOS is on a 755 km polar sun-synchronous orbit with an equator crossing time of 6:00 am (ascending) and 6:00 pm (descending). The instrument on board SMOS is a 2D interferometer operating at L-band (1404–1423 MHz). It measures the emission from the surface in a fully polarimetric mode and with a large number of view angles (from 0 to 55 degree). The SMOS level 2 soil moisture products are defined on the ISEA 4H9 grid (i.e., Icosahedral Snyder Equal Area projection with aperture 4, resolution 9 and its shape of cells as hexagon) (Kerr et al., 2010). Each point (or node) of this grid is known as a Discrete Global Grid (DGG) that has fixed coordinates and is assigned with an identifier the "DGG Id." Level 3 and 4 products are defined on the Equal-Area Scalable Earth (EASE)-grid 2 projection. Data is available through the ESA portal for level 1 and 2 data and through the Centre Aval de Traitement des Données SMOS (CATDS) for level 3 and 4.

The SMAP was the first Earth observation satellite mission developed by NASA on the recommendation of the National Research Council's Earth Science Decadal Survey, Earth Science and Applications from Space: National

Imperatives for the Next Decade and Beyond in 2007. It was launched on Jan. 31, 2015. Likewise, SMOS, the SMAP mission designed to collect continuous global observation of near-surface soil moisture (0–5 cm depth) and freeze/ thaw state every 2–3 days at hydrometeorology (10 km) and hydroclimatology (40 km) scales. The SMAP is a for-synchronous satellite that runs from pole to pole and crosses Earth's terminator at the equator. The SMAP's orbit is 685 km above Earth's surface. Because Earth spins while SMAP revolves, swaths from each orbit are offset from each other and, after 8 days, the same swath is repeated. Over 2–3 days (2 days at the poles and 3 days at the equator), the gaps between swaths are compensated for and thus a global map of soil moisture is generated. The instrument onboard SMAP consists of a simultaneous active (in the form of radar) and passive (in the form of radiometer) microwave instruments at L-band frequency (1.2–1.4 GHz range). The radiometer instrument of the SMAP satellite is based on a single feed and deployable 6 m reflector antenna system that rotates around the nadir axis. This rotation antenna enables conical scanning at a constant incidence angle (40 degree). The radiometer (1.41 GHz) measures the intensity of microwave radiation emitted from the Earth's surface (i.e., brightness temperature) at a spatial resolution of approximately 40 km. The radiometer acquires measurements in four channels, vertical (V) polarization, horizontal (H) polarization, third, and fourth stokes. The radar (1.26 GHz) transmits microwave radiation in two linear polarizations and measures the scene backscatter at 1–3 km spatial resolution in multiple polarimetric channels. Each channel receives and transmits polarized radiation in different orientations: HH, VV and HV, or VH. Radiometric-based measurement algorithms provide better soil moisture estimation under vegetated conditions. The SMAP baseline data products are made available publicly through NASA's two Distributed Active Archive Centers (DAACs), the Alaska Satellite Facility, and the National Snow and Ice Data Center for all other products.

One should also mention the Aquarius instrument (Le Vine et al., 2014) which, even though designed to measure ocean salinity, was also used over land to infer soil moisture at a coarse spatiotemporal resolution (Bindlish et al., 2015). The three L-band mission characteristics are given in Table 1. Note that, on Jun. 7, 2015, the satellite carrying Aquarius suffered a power supply failure, ending the mission.

It is worth mentioning at this level that other passive microwave instruments, though not specifically designed for soil moisture retrievals (see AMSR specifications in Table 2 for instance) can show good performances (Al-Yaari et al., 2014b).

4 SOIL MOISTURE RETRIEVAL FROM SPACE USING PASSIVE MICROWAVES

Even though several techniques are available, we will focus here on the most promising, the approach using passive microwaves at low frequency and in particular at L-band.

TABLE 1 L-Band Radiometer Characteristics

	SMOS	Aquarius	SMAP
Platform	Proteus	SAC-D	SMAP
Altitude (km)	755	685	685
Equator crossing solar time (ascending orbit)	6:00 am	6:00 pm	6:00 pm
Antenna diameter (m)	Equivalent to 8	3	6
Swath (km)	960	300	1000
Dates of operation	Nov. 2, 2009	Jun. 11, 2011 to Nov. 7, 2015	Jan. 31, 2015
Radiometer	Interferometer	Push broom	Conical scan
Spatial resolution (3 dB)	27–55 km (mean ~43 km)	94×76 km 120×84 km 156×96 km	38×49 km
Polarization	Full	Three stokes	Full
Bandwidth (MHz)	1404–1423	1400–1427	
Incidence angle (degree)	0–60	29.36, 38.49, 46.29	40

4.1   Surface Soil Moisture

Soil moisture is a very easy target at L-band as the amplitude of the signal can reach 100 K for the whole soil moisture range. The problem is nevertheless complicated by the presence of vegetation, topography, freeze/defreeze cycles, snow cover, or water bodies. To cope with these, a very specific retrieval algorithm was developed (Kerr et al., 2012) and implemented from launch. It relies on the multiangular capacity of the SMOS instrument (Kerr et al., 2014).

The signal measured by a passive microwave sensor at L-band, over a non-frozen and snow free surface, is mainly a function of soil moisture, vegetation opacity, and effective surface temperature. Other surface characteristics like soil texture and roughness also play an important role. Atmospheric contribution (including clouds and rain) and galactic reflection can be easily neglected over land surfaces at L-band.

At a typical spatial resolution (40 km on average for SMOS or SMAP), the surface contribution to the L-band signal is rarely uniform. A surface area corresponding to a pixel often consists of some water bodies, low and possibly high vegetation fields, possibly frozen and snow covered surfaces, topography, and

TABLE 2 AMSR Series: Characteristics

	AMSR-J		AMSR-E		AMSR-2	
Platform	ADEOS-II		AQUA		GCOM-W1	
Altitude (km)	802.9		705		700	
Equator crossing solar time (ascending orbit)	10:30 pm		1:30 pm		1:30 pm	
Antenna diameter (m)	2		1.6		2	
Swath (km)	1600		1450		1450	
Dates of operation	Dec. 14, 2002 to Oct. 25, 2003		May 4, 2002 to Oct. 4, 2011		May 18, 2012	
Frequencies and resolutions All V and H except channel at 23.8 GHz (V only)	Central frequency (GHz)	Spatial resolution (3-dB) km × km	Central frequency (GHz)	Spatial resolution (3-dB) km × km	Central frequency (GHz)	Spatial resolution (3-dB) km × km
	6.93	75 × 43	6.93	75 × 43	6.93	62 × 35
					7.3	62 × 35
	10.65	51 × 29	10.65	51 × 29	10.65	42 × 24
	18.7	27 × 16	18.7	27 × 16	18.7	22 × 14
	23.8	19 × 11	23.8	19 × 11	23.8	19 × 11
	36.5	14 × 8	36.5	14 × 8	36.5	12 × 7
	89.0	6 × 4	89.0	6 × 4	89.0	5 × 3

others. Consequently, any physically based retrieval algorithm has to be able to account for a number of features in the observed area. Some surface characteristics, such as soil texture and land use, are obtained from static maps while others, such as temperature and snow, are obtained from forecasts. Therefore, it is a significant challenge to determine with an adequate accuracy and globally the surface characteristics, and use them in the retrieval process to infer the desired characteristics. The SMOS and SMAP missions differ slightly at this level. Both use a physically based approach using state of the art models but SMOS uses the multiangular information for assessing vegetation contribution while SMAP relies on the use of satellite derived vegetation indices.

The SMOS algorithm uses an iterative approach which aims at minimizing a cost function whose main component is the sum of the squared weighted differences between measured and modeled brightness temperature data, for a collection of incidence angles. This is achieved by finding the best-suited set of parameters which drive the direct TB model (e.g., soil moisture and vegetation characteristics using the L-MEB model) (Wigneron et al., 2007).

Despite the apparent simplicity of the soil moisture retrieval principle, the modeling of the radiometric signal is complex and requires close attention to many details. The SMOS "pixels" can correspond to rather large, inhomogeneous surface areas involving many land use types (crop fields, forests, water bodies, urban areas, etc.), each with their own characteristics. Moreover, the radiometric signal is impacted by the directional pattern of the SMOS interferometric radiometer. Therefore, modeling of the SMOS radiometric signal involves both the modeling of the ground target and the antenna for a variety of incidence angles. The modeling of the ground target involves estimation of various parameters (such as surface temperature and soil moisture) at various positions within the target. The antenna pattern is represented through a weighting function which depends on the incidence angle. It is important to note that, thanks to the "reconstruction" principle, the observations are computed for any given point of the surface (with an accuracy of 400 m) and it is always the same center. So, from one acquisition to the next, the instrument footprint is exactly the same.

As for any pixel, there might be a large variety of surface types, not all of which are characterized by the same set of parameters. It is, therefore, not realistic to carry out the same retrieval everywhere. It is understood that all surface types, regardless of whether they support the retrieval of a given set of parameters, do contribute to the SMOS signal according to a given model. However, estimation of soil moisture is only meaningful over certain surface types. For instance, while a lake contributes to the radiometric signal, it is not parameterized by soil moisture. In order to facilitate the retrieval process, a node is divided generally into two areas, one where the retrieval will take place and one where the contributions to the overall node signal need to be estimated but no retrievals will be performed. This latter part is then considered to have fixed contribution (*default contributions*) and the retrieval is made on the remaining—*dominant*—area. For instance, if there is an area of low vegetation with a dense forest and a lake, we will estimate the contribution of the lake and that of the forest using either external data or predetermined values of the surface characteristics: the *reference values*. This default contribution will be assumed constant in the modeled signal, and the retrieval is performed only on the remaining, dominant part.

Once the dominant area is ascertained, its content must be taken into account. Actually, a surface area contributing to the SMOS signal may include a large variety of surface types, such as cultivated agriculture, grasslands, and forests. These land-use classes are grouped together based on their L-band microwave emission properties. All surface types are aggregated into a small number (about 10) of generic classes having the same modeling characteristics and using similar parameters. A target area is subdivided into a number of units (4×4 km^2 by default) and each unit is defined by a collection of such aggregated fractions. The SMOS pixels correspond to various views of a node as seen from different angles. The spatial extents of two different views of the same target are generally different. The difference increases as the viewing angles are further apart.

The dominant part does not always cover nice rolling hills of green pastures (also known as "nominal"). The soil can be frozen or covered with snow or rocks. The target could be an island within a sea or have a large urban or mountainous component, not to mention marshes or rice fields. These nonnominal, or exotic, cases need to be modeled differently. The exotic surfaces could be either complementary (i.e., there is no overlap between two classes of this type) or supplementary (they necessarily overlap with complementary classes). For instance, surface characteristics can be supplementary when two "special" cases are present at the same place and same moment (i.e., topography and water body or forest). They can be supplementary when they exclude one another such as water body and forest.

When it is not possible or relevant to retrieve soil moisture, it may be possible to retrieve other parameters of interest. For instance, one can retrieve dielectric constant parameters (using the so-called Cardioid approach) (Kerr et al., 2014).

The algorithm has been undergoing continuous validation using ground sites and subsequent improvements. As already stated, RFI constitutes an issue, so results obtained in Europe and Asia are always difficult to interpret. Sites in the United States have proved to be much more useful (Albitar et al., 2012; Jackson et al., 2012). The results are available (SMOS special issue May 2012 and RSE special issue 2016). Currently version 620 is being run and results are very satisfying over normal land surfaces (flattish area with low vegetation and no RFI).

The SMAP mission is currently in its first year and four retrieval algorithms are being tested and evaluated. The most used are one (single channel algorithm) based on only one channel (the V polarization) as selected at the end of the commissioning phase and one using the two polarizations (dual channel algorithm). First intercomparison exercises show that both SMOS and SMAP are very coherent in their results and show similar quality. SMAP has the added advantage of an efficient RFI filtering, enabling a more exhaustive spatial coverage while SMOS seems to be more able to cope with very dense vegetation, at least for the time being (SMAP handbook).[1]

Globally, the SMOS retrieval algorithm performs well in terms of coverage with the big caveat of the RFI which affects mainly observations in Europe and Asia. The quality of the results is almost within expectations ($0.04 \text{ m}^3/\text{m}^3$), which is encouraging when one considers that the satellite was the first spaceborne interferometric system and has been operating for 6 years. Good results are obtained over RFI-free areas and very encouraging results are obtained over moderately dense forests (Rahmoune et al., 2014; Vittucci et al., in this issue).

Soil moisture retrievals are in the correct range with possible underestimation linked to RFI as could be expected. However, in some cases and after a heavy rainfall event, the soil moisture values obtained seem too high (sometimes exceeding $0.65 \text{ m}^3/\text{m}^3$). This may be due to ponding effects, saturation of the upper soil layer and can indicate flooding. In some cases, it can be linked to retrieval errors linked to erroneous auxiliary data.

Another point of concern is the way cold areas are processed. If freeze/thaw is easily detected (soil appears suddenly dry when it freezes and vegetation becomes transparent when frozen) (Matzler, 1994; Schwank et al., 2004), the main issue at this level is the fact that a high spatial resolution is required to monitor freeze/thaw, especially in the transition areas. Snow cover is more complex. When dry, snow is almost transparent, and L-band is sensitive to the relatively warm soil underneath. However, when the snow is wet it is rather opaque. All the intermediary cases (both in terms of snow state and spatial distribution) make retrieval in transition areas very difficult and the product is prone to being erroneous. Another source of concern is linked to water bodies. There are no available dynamic maps of water bodies at a fine enough resolution and water bodies do change with time, be it through seasonal variations, floods, or even tides. An error of 2% on the water body surface can lead to an error in soil moisture corresponding to $0.01 \text{ m}^3/\text{m}^3$ (Kerr et al., 2012).

Quantitative evaluation of the L2 algorithm has been described in several papers for both SMOS and SMAP. An example is given in Fig. 1 and described further in (Kerr et al., submitted for publication) and for two different cases; a semiarid environment (Walnut Gulch) and a temperate one (Little Washita). The charts show the temporal evolution of the soil moisture for SMOS, a scatterometer (ASCAT), model outputs (European Center for Medium-Range Weather Forecasts, ECMWF and Modern Era-Retrospective analysis for Research and Applications (MERRA)) and in situ data. Below is the Cumulative Distribution

---

1. http://smap.jpl.nasa.gov/system/internal_resources/details/original/178_SMAP_Handbook_FINAL_1_JULY_2014_Web.pdf.

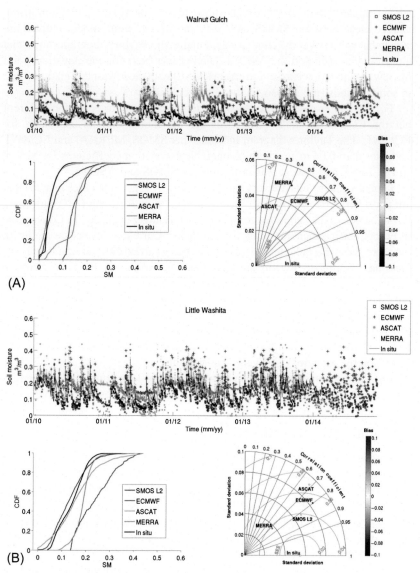

FIG. 1 Typical results of retrievals over a US Watershed (see Jackson et al., 2012): (A) for Walnut Gulch and (B) for Little Washita.

Function (CDF) plot and a Taylor's diagram for the same measurements. These types of plots allow for a good estimation of the retrievals quality even though not all aspects can be covered (in particular the spatial representation). The temporal behaviors over this example of arid environment shows that the ASCAT tends to have a strange behavior with large errors associated (Wagner et al., 2013), while models tend to largely overestimate soil moisture when compared to in situ. The CDF shows that the distribution of soil moisture is very close between SMOS, ASCAT, and in situ while model outputs are somewhat wetter over the entire range. Finally, the Taylor's diagram shows that almost in contradiction the ASCAT data behaves better on the statistical point of view for the root mean square error (RMSE) with the lowest correlation while in some periods (pre monsoon) the error is about 100%.

For the wetter case (Little Washita), temporal behavior is not too different but the CDF does show differences for the model outputs especially for ECMWF. The Taylor's diagram indicates that SMOS, MERRA, and ECMWF do better than 0.04 m^3/m^3 with comparable correlation coefficients for SMOS and ECMWF and lowest bias in the case of SMOS and ASCAT.

Globally, the results are very satisfying for low vegetation cases in areas that are not overly affected by RFI, such as in the US or Australia (Al-Yaari et al., 2014a,b; Kerr et al., submitted for publication; Rudiger et al., 2014). Conversely, in Europe for instance, the results are degraded. Currently, results are mediocre over forested areas, but intensive research is being carried out, and significant progress has been made. Results are very good over arid areas. From the range of results obtained, it can be said that, depending roughly on the area and RFI levels, the soil moisture estimate accuracy ranges between 0.02 and 0.06 m^3/m^3 with, in some cases, even higher values; while correlation coefficients between SMOS and SM retrievals and ground measurements can range from 0.5 to 0.85. Usually SMOS/SMAP fare very well as well, and all instruments (with the caveat of spatial resolution for Aquarius) give very similar results.

Fig. 2 shows an example of a global monthly soil moisture map. Large features, such as the monsoon in Bengal or Sahel, and dry deserts, are visible.

4.2 Root-Zone Soil Moisture

Root-zone soil moisture constitutes an important variable for agricultural, hydrological, and weather forecast models. Monitoring the availability of root-zone soil moisture is critical for predicting trends in agricultural markets and food availability in food-insecure regions. The most significant caveat of remotely sensed soil moisture is that direct measurement only concerns the surface layer. For instance, a few mm at X-band, and $\sim$2–4 cm at L-band are probed on average (depending on soil characteristics and condition). Microwave radiometers, like the L-band instrument on board the ESA's SMOS and NASA SMAP missions, are being designed to provide estimates of near-surface soil moisture (0–5 cm). However, it is necessary to know the available water in the unsaturated zone. Only one direct approach can currently be considered, i.e., to use even lower frequencies in wavelengths of several meters to reach the water table, as a very large penetration depth is required. This leads to large problems in terms of spatial resolution (a few 100 km) as well as ionospheric effects, etc. Therefore, this option is not feasible now. The indirect approach could be to rely on assimilation techniques, i.e., using models to infer

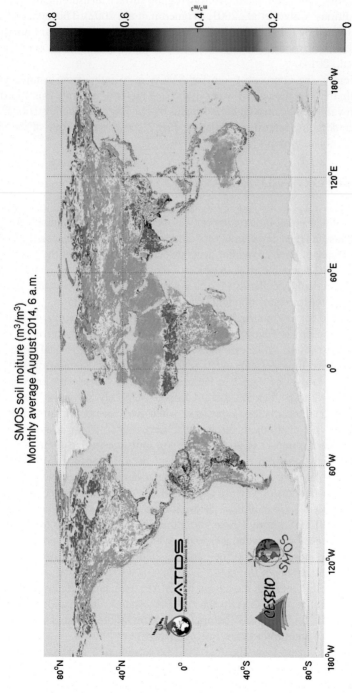

SMOS soil moiture (m³/m³)
Monthly average August 2014, 6 a.m.

FIG. 2 Global map of soil moisture derived from SMOS data for Aug. 2014.

what the root-zone soil moisture is from regular surface measurements and forcing conditions (Calvet et al., 2001; Wigneron et al., 2002). This approach has been validated by using both simulations and ground data. The real limitations are linked to the model's ability and input data quality (Dumedah et al., 2015; Ford et al., 2014).

Currently, the approach taken for SMOS is to infer, from a three layer approach, the total amount of water in the first meter of ground. This approach has been validated against model estimates by over several ground stations, and is now undergoing intensive validation. Fig. 3 shows an example of the root-zone soil moisture map.

5 WAY FORWARD

Over land, SMOS has also proven to open up a new set of possibilities. The first evidence of new possibilities to appear was the SMOS' ability to quickly monitor flood events, and their extent in spite of cloud cover. In many instances, SMOS has delivered maps of floods providing accurate details, such as during the Mississippi floods in 2011, where the actual position of broken levies could be spotted. It was also used successfully in Jan. 2011 to infer whether hurricane Yasi would, or would not, cause significant new flooding (linked to actual soil moisture).

It has also been demonstrated that SMOS (or SMAP) can follow rainfall by simply measuring soil moisture. Assimilation algorithms are currently being developed to provide actual estimates of rainfall events. Operational flood forecasting systems in large-scale catchments have also been demonstrated (Crow et al., 2011; Louvet et al., 2015; Pellarin et al., 2013). Research by Wanders et al. (2014) shows that the assimilation of remotely sensed soil moisture improves flood forecasting, especially when used in combination with the assimilation of distributed discharge observations, while Lievens et al. (2015) shows the impact of assimilating soil moisture for basin modeling.

Furthermore, the real time measurement of soil moisture from satellites makes possible quick and accurate determinations of the projected irrigation dates and the amounts of water to apply, and also eliminates the need for extensive computations required to determine crop water use (Campbell and Campbell, 1982). From root-zone soil moisture, the natural step forward is to provide information on potential droughts which can be assessed after the event by simply monitoring soil moisture and vegetation drying, and also deriving a drought index, as shown in Fig. 4.

Last but not least, with the advent of SMAP, work has started on the potential synergisms between the two sensors. For instance, inter calibration of the SMOS, SMAP, and Aquarius (before the mission failure) sensors are currently done routinely and products are intercompared.

Obviously, apart from the fact that the unsaturated zone is only partly probed, there are some requirements which are not fully fulfilled. The main requirement is spatial resolution. Some needs, notably in hydrology, can only

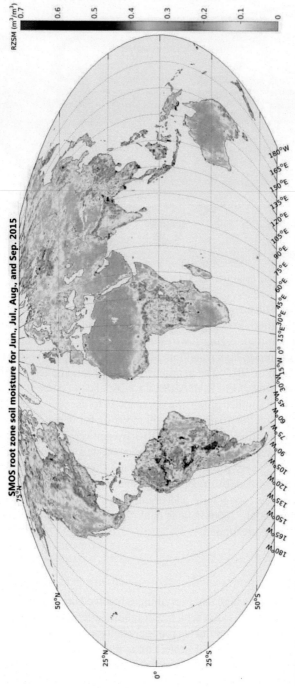

FIG. 3 Root-zone soil moisture (RZSM) for Jun., Jul., Aug., and Sep. 2015 as estimated from SMOS data.

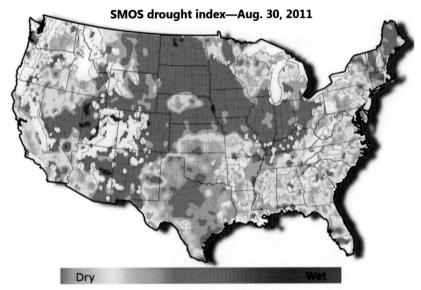

FIG. 4 SMOS drought index over the USA for Aug. 2011.

be resolved by having a better spatial resolution. However, a high temporal sampling is also required. From space, this is not straightforward. SMAP offered a potential solution by merging low-resolution absolute and accurate estimates (L-band radiometer) to high resolution SAR data. But this was before the SMAP radar broke down in 2015. The approach has also been applied to other synergetic associations (SMOS+RADARSAT for instance (Tomer et al., submitted for publication)). Another promising approach is to use an optical wavelength sensor at high resolution to distribute soil moisture for water resource management purposes (Merlin et al., 2012, 2013; Srivastava et al., 2013a; Petropoulos et al., 2013). For instance, the approach has been tested in Australia and in Spain (Catalonia), and is very promising as it is based on the evaporative demand. However, it is limited by cloud cover, and thus, data availability. Other approaches rely on topography and drainage, or rooting depth, to distribute moisture (Merlin et al., 2006; Pellenq et al., 2003) and now the real life validation has only to be performed.

6 CAVEATS

It is not intended to say here that everything is resolved and fine. There are still a number of outstanding issues, which will require attention before an accurate, and global, soil moisture product is routinely delivered. Some, such as RFI, can be a general issue, especially if protection is reduced in the future, which is a concern (Islam et al., 2015). SMAP has taken an aggressive approach, but if the data is well filtered, the quality is degraded or altogether absent by simple lack of measurements. The specific issues identified are currently being tackled and several references in the literature identify them. Currently, the following issues are well identified. The most stringent is pixel heterogeneity, with varying land use having very significant differences in behavior. The presence of free water within the pixel, for instance, has to be accurately known to reach the overall accuracy of $0.04 \text{ m}^3/\text{m}^3$ in soil moisture. And water bodies can be variable as a function of season or weather. Vegetation is not totally transparent at L-band, and when the integrated water content is above $4–5 \text{ kg/m}^2$, soil moisture retrievals are challenging. Litter on the ground may behave as a black body, strongly masking the soil's signal. During rain events, water interception by the canopy might artificially increase the apparent vegetation water content. Topography will induce an altered angular behavior (Mialon et al., 2008), snow and frozen soils will induce different signals which, if not accounted for, will produce wrong estimates (Schwank et al., 2004). Urban areas and rocks are not fully assessed in terms of emissivity. Finally, and generally speaking, good SM retrievals will require *a priori* knowledge of the surface cover and state; and the quality of the retrievals will be linked to the quality of the input data.

7 CONCLUSIONS AND PERSPECTIVES

L-band radiometry was a new approach to a new type of measurements. The challenge of building, before launch, retrieval algorithms with no experience of L-band measurements, synthetic aperture radiometer, or even soil moisture/sea surface salinity, proved to be successful.

After a few years in operation, the results have been outstanding. The soil moisture results are very good for the homogeneous, low vegetation, targets. Good results are also starting to emerge over forested areas.

The main issue encountered was the RFI pollution, but things are improving in this domain.

The next issue is to ensure data continuity, especially considering the various operational or preoperational applications using the L-band data. Nevertheless, SMOS and SMAP still does not fulfill all the current needs, and ways forward must be sought. The most important need is to improve the spatial resolution. Here, the SMOS concept is close to an optimum, as increasing the arm's length will improve the spatial resolution, but it significantly degrades the sensitivity, to the point that it would not be useful anymore. So other techniques, such as disaggregation, will have to be found. To be more efficient, an SMOS-like instrument might gain from either being multifrequency, or having a coupled active system for SMAP.

Based upon this success, the next challenge will be twofold.

- First, to keep on improving the accuracy and the validity domain of the retrieval algorithms (extending over more and more surface types).
- Secondly, to keep on developing new products and new research venues (cryopshere is one obvious target) and implementing them.

SMOS and SMAP also demonstrated the need for absolute estimates of surface variables. For soil moisture, for instance, change detection may have demonstrated its usefulness, but to really use the information, one needs absolute values which can currently only be obtained from low frequency radiometry.

CNES has initiated new research activities whose goals are to develop a new mission which would fulfill all of the SMOS requirements, but with a ten times better spatial resolution and improved sensitivity (factor of three for salinity applications), thus, paving the way toward more applications in water resources management, coastal area monitoring, basin hydrology, or even thin sea ice monitoring (Kaleschke et al., 2012). The concept, named SMOS next, is based on merging spatial and temporal 2D interferometry and is currently undergoing phase 0 at CNES with a proof of concept experiment funded by the R&D program.

REFERENCES

Albitar, A., Kerr, Y.H., Merlin, O., Richaume, P., Sahoo, A., Wood, E.F., 2012. Evaluation of SMOS soil moisture products over continental US using the SCAN/SNOTEL network. IEEE Geosci. Remote Sens. Lett. 50, 1572–1586.

Al-Yaari, A., Wigneron, J.P., Ducharne, A., Kerr, Y., de Rosnay, P., de Jeu, R., Govind, A., Al Bitar, A., Albergel, C., Munoz-Sabater, J., Richaume, P., Mialon, A., 2014a. Global scale evaluation of two satellite-based passive microwave soil moisture datasets (SMOS and AMSR-E) with respect to Land Data Assimilation System estimates. Remote Sens. Environ. 149, 181–195.

Al-Yaari, A., Wigneron, J.P., Ducharne, A., Kerr, Y.H., Wagner, W., De Lannoy, G., Reichle, R., Al Bitar, A., Dorigo, W., Richaume, P., Mialon, A., 2014b. Global-scale comparison of passive (SMOS) and active (ASCAT) satellite based microwave soil moisture retrievals with soil moisture simulations (MERRA-Land). Remote Sens. Environ. 152, 614–626.

Bayle, F., Wigneron, J.-P., Kerr, Y.H., Waldteufel, P., Anterrieu, E., Orlhac, J.-C., Chanzy, A., Marloie, O., Bernardini, M., Sobjaerg, S., Calvet, J.-C., Goutoule, J.-M., Skou, N., 2002. Two-dimensional synthetic aperture images over a land surface scene. IEEE Trans. Geosci. Remote Sens. 40, 710–714.

Bindlish, R., Jackson, T., Cosh, M., Zhao, T., O'Neill, P., 2015. Global soil moisture from the Aquarius/SAC-D satellite: description and initial assessment. IEEE Geosci. Remote Sens. Lett. 12, 923–927.

Bircher, S., Skou, N., Kerr, Y.H., 2013. Validation of SMOS L1C and L2 products and important parameters of the retrieval algorithm in the Skjern River catchment, Western Denmark. IEEE Trans. Geosci. Remote Sens. 51, 2969–2985.

Calvet, J.-C., Noilhan, J., Wigneron, J.-P., Kerr, Y., 2001. Root-zone soil moisture analysis using microwave radiometry. In: IGARSS2001, Sydney, Australia.

Campbell, G., Campbell, M., 1982. Irrigation scheduling using soil moisture measurements: theory and practice. Adv. Irrig. 1, 25–42.

Chanzy, A., Chadoeuf, J., Gaudu, J.C., Mohrath, D., Richard, G., Bruckler, L., 1998. Soil moisture monitoring at the field scale using automatic capacitance probes. Eur. J. Soil Sci. 49, 637–648.

Crow, W.T., van den Berg, M.J., Huffman, G.J., Pellarin, T., 2011. Correcting rainfall using satellite-based surface soil moisture retrievals: the Soil Moisture Analysis Rainfall Tool (SMART). Water Resour. Res. 47.

Daganzo-Eusebio, E., Oliva, R., Kerr, Y.H., Nieto, S., Richaume, P., Mecklenburg, S.M., 2013. SMOS radiometer in the 1400–1427-MHz passive band: impact of the RFI environment and approach to its mitigation and cancellation. IEEE Trans. Geosci. Remote Sens. 51, 4999–5007.

Dumedah, G., Walker, J.P., Merlin, O., 2015. Root-zone soil moisture estimation from assimilation of downscaled Soil Moisture and Ocean Salinity data. Adv. Water Resour. 84, 14–22.

Eagleman, J.R., Ulaby, F.T., 1975. Remote-sensing of soil-moisture by Skylab radiometer and scatterometer sensors. J. Astronaut. Sci. 23, 147–159.

Entekhabi, D., Rodriguez-Iturbe, I., Castelli, F., 1996. Mutual interaction of soil moisture state and atmospheric processes. J. Hydrol. 184, 3–17.

Entekhabi, D., Asrar, G.R., Betts, A.K., Beven, K.J., Bras, R.L., Duffy, C.J., Dunne, T., Koster, R.D., Lettenmaier, D.P., McLaughlin, D.B., Shuttleworth, W.J., van Genuchten, M.T., Wei, M.Y., Wood, E.F., 1999. An agenda for land surface hydrology research and a call for the second international hydrological decade. Bull. Am. Meteorol. Soc. 80, 2043–2058.

Entekhabi, D., Njoku, E.G., O'Neill, P.E., Kellogg, K.H., Crow, W.T., Edelstein, W.N., Entin, J.K., Goodman, S.D., Jackson, T.J., Johnson, J., Kimball, J., Piepmeier, J.R., Koster, R.D., Martin, N., McDonald, K.C., Moghaddam, M., Moran, S., Reichle, R., Shi, J.C., Spencer, M.W., Thurman, S.W., Tsang, L., Van Zyl, J., 2010. The Soil Moisture Active Passive (SMAP) mission. Proc. IEEE 98, 704–716.

Entekhabi, D., Yueh, S., ONeill, P., Kellogg, K., et al., 2014. SMAP Handbook. NASA CalTech, Pasadena, CA. JPL Publication JPL 400-1567.

Escorihuela, M.J., de Rosnay, P., Kerr, Y.H., 2006. Temperature Dependency of Bound Water Spectral Parameters and Its Influence in Soil Moisture Measurements. EGU, Wien. EGU General assembly.

Ford, T.W., Harris, E., Quiring, S.M., 2014. Estimating root zone soil moisture using near-surface observations from SMOS. Hydrol. Earth Syst. Sci. 18, 139–154.

Glenn, E.P., Neale, C.M.U., Hunsaker, D.J., Nagler, P.L., 2011. Vegetation index-based crop coefficients to estimate evapotranspiration by remote sensing in agricultural and natural ecosystems. Hydrol. Process. 25 (26), 4050–4062.

Gorrab, A., Zribi, M., Baghdadi, N., Mougenot, B., Fanise, P., Chabaane, Z.L., 2015. Retrieval of both soil moisture and texture using TerraSAR-X images. Remote Sens. 7, 10098–10116.

Goutoule, J.M., 1995. MIRAS spaceborne instrument and its airborne demonstrator. In: Proceedings of the Consultative Meeting on Soil Moisture and Ocean Salinity (SMOS), ESA WPP-87, April 20–21. ESTEC, Noordwijk, pp. 56–70.

Goutoule, J.M., Anterrieu, E., Kerr, Y.H., Lannes, A., Skou, N., 1996. MIRAS Microwave Radiometry Critical Technical Development. MMS, Toulouse.

Gruhier, C., de Rosnay, P., Kerr, Y., Mougin, E., Ceschia, E., Calvet, J.C., Richaume, P., 2008. Evaluation of AMSR-E soil moisture product based on ground measurements over temperate and semi-arid regions. Geophys. Res. Lett. 35.

Han, S.-C., Shum, C.K., Jekeli, C., Alsdorf, D., 2005. Improved estimation of terrestrial water storage changes from GRACE. Geophys. Res. Lett. 32L07302.

Imaoka, K., Kachi, M., Kasahara, M., Ito, N., Nakagawa, K., Oki, T., 2010. Instrument performance and calibration of AMSR-E and AMSR2. Int. Arch. Photogramm. Remote. Sens. Spat. Inf. Sci. 38.

Islam, T., Srivastava, P.K., Dai, Q., Gupta, M., Zhuo, L., 2015. An introduction to factor analysis for radio frequency interference detection on satellite observations. Meteorol. Appl. 22, 436–443.

Jackson, T.J., Schmugge, T.J., 1991. Vegetation effects on the microwave emission of soils. Remote Sens. Environ. 36, 203–212.

Jackson, T.J., Hsu, A.Y., Van de Griend, A., Eagleman, J.R., 2004. Skylab L-band microwave radiometer observations of soil moisture revisited. Int. J. Remote Sens. 25, 2585–2606. http://dx.doi.org/10.1080/01431160310001647723.

Jackson, T.J., Bindlish, R., Cosh, M., Zhaoa, T., Starks, P.J., Bosch, D.D., Seyfried, M., Moran, M.S., Goodrich, D., Kerr, Y.H., Leroux, D., 2012. Validation of Soil Moisture and Ocean Salinity (SMOS) soil moisture over watershed networks in the US. IEEE Geosci. Remote Sens. Lett. 50, 1530–1543.

Kaleschke, L., Tian-Kunze, X., Maaß, N., Mäkynen, M., Drusch, M., 2012. Sea ice thickness retrieval from SMOS brightness temperatures during the Arctic freeze-up period. Geophys. Res. Lett. 39.

Kerr, Y.H., 1998. The SMOS Mission: MIRAS on RAMSES. A Proposal to the Call for Earth Explorer Opportunity Mission. CESBIO, Toulouse.

Kerr, Y.H., 2007. Soil moisture from Space: Where are we? Hydrogeol. J. 15, 117–120.

Kerr, Y.H., Njoku, E.G., 1990. A semiempirical model for interpreting microwave emission from semiarid land surfaces as seen from space. IEEE Trans. Geosci. Remote Sens. 28, 384–393.

Kerr, Y.H., Waldteufel, P., Wigneron, J.-P., Martinuzzi, J.-M., Font, J., Berger, M., 2001. Soil moisture retrieval from space: the Soil Moisture and Ocean Salinity (SMOS) Mission. IEEE Trans. Geosci. Remote Sens. 39, 1729–1735.

Kerr, Y.H., Waldteufel, P., Wigneron, J.P., Delwart, S., Cabot, F., Boutin, J., Escorihuela, M.J., Font, J., Reul, N., Gruhier, C., Juglea, S.E., Drinkwater, M.R., Hahne, A., Martin-Neira, M., Mecklenburg, S., 2010. The SMOS mission: new tool for monitoring key elements of the global water cycle. Proc. IEEE 98, 666–687.

Kerr, Y.H., Waldteufel, P., Richaume, P., Wigneron, J.P., Ferrazzoli, P., Mahmoodi, A., Al Bitar, A., Cabot, F., Gruhier, C., Juglea, S., Leroux, D., Mialon, A., Delwart, S., 2012. The SMOS soil moisture retrieval algorithm. IEEE Geosci. Remote Sens. Lett. 50, 1384–1403.

Kerr, Y., Waldteufel, P., Richaume, P., Ferrazzoli, P., Wigneron, J.-P., 2014. SMOS level 2 processor soil moisture algorithm theoretical basis document (ATBD) V4. SM-ESL (CBSA), Toulouse, p. 142.

Kerr, Y.H., Al-Yaari, A., Rodriguez-Fernandez, N., Parrens, M., Molero, B., Leroux, D., Bircher, S., Mahmoodi, A., Mialon, A., Richaume, P., Delwart, S., Al Bitar, A., Pellarin, T., Bindlish, R., Jackson, T.J., Rudiger, C., Waldteufel, P., Mecklenburg, S., Wigneron, J.-P., 2016. Overview of SMOS performance in terms of global soil moisture monitoring after six years in operation. Remote Sens. Environ. (in press).

Khazaal, A., Cabot, F., Anterrieu, E., Soldo, Y., 2014. A kurtosis-based approach to detect RFI in SMOS image reconstruction data processor. IEEE Trans. Geosci. Remote Sens. 52, 7038–7047.

Kim, S.-B., Tsang, L., Johnson, J.T., Huang, S., van Zyl, J.J., Njoku, E.G., 2012. Soil moisture retrieval using time-series radar observations over bare surfaces. IEEE Trans. Geosci. Remote Sens. 50, 1853–1863.

Kim, S., Liu, Y.Y., Johnson, F.M., Parinussa, R.M., Sharma, A., 2015. A global comparison of alternate AMSR2 soil moisture products: why do they differ. Remote Sens. Environ. 161, 43–62.

Le Vine, D.M., Griffis, A.J., Swift, C.T., Jackson, T.J., 1994. ESTAR: a synthetic aperture microwave radiometer for remote sensing applications. Proc. IEEE 82.

Le Vine, D.M., Dinnat, E.P., Lagerloef, G.S.E., de Matthaeis, P., Abraham, S., Utku, C., Kao, H., 2014. Aquarius: status and recent results. Radio Sci. 49, 709–720.

Leroux, D.J., Kerr, Y.H., Al Bitar, A., Bindlish, R., Jackson, T.J., Berthelot, B., Portet, G., 2014. Comparison between SMOS, VUA, ASCAT, and ECMWF soil moisture products over four watersheds in US. IEEE Trans. Geosci. Remote Sens. 52, 1562–1571.

Lievens, H., Al Bitar, A., Verhoest, N.E.C., Cabot, F., De Lannoy, G.J.M., Drusch, M., Dumedah, G., Franssen, H.J.H., Kerr, Y., Tomer, S.K., Martens, B., Merlin, O., Pan, M., van den Berg, M.J., Vereecken, H., Walker, J.P., Wood, E.F., Pauwels, V.R.N., 2015. Optimization of a radiative transfer forward operator for simulating SMOS brightness temperatures over the Upper Mississippi Basin. J. Hydrometeorol. 16, 1109–1134.

Lorite, I.J., García-Vila, M., Carmona, M.A., Santos, C., Soriano, M.A., 2012. Assessment of the irrigation advisory services' recommendations and farmers' irrigation management: a case study in Southern Spain. Water Resour. Manag. 26 (8), 2397–2419.

Louvet, S., Pellarin, T., al Bitar, A., Cappelaere, B., Galle, S., Grippa, M., Gruhier, C., Kerr, Y., Lebel, T., Mialon, A., Mougin, E., Quantin, G., Richaume, P., de Rosnay, P., 2015. SMOS soil moisture product evaluation over West-Africa from local to regional scale. Remote Sens. Environ. 156, 383–394.

Magagi, R.D., Kerr, Y.H., 1997. Retrieval of soil moisture and vegetation characteristics by use of ERS1 Windscatterometer over arid and semi arid areas. J. Hydrol. 188–189, 361–384.

Magagi, R.D., Kerr, Y.H., 2001. Estimating surface soil moisture and soil roughness from ERS-1 winscatterometer data over semi-arid area: use of the co-polarisation ratio. Remote Sens. Environ. 75 (3), 432–445.

Martin-Neira, M., LeVine, D.M., Kerr, Y., Skou, N., Peichl, M., Camps, A., Corbella, I., Hallikainen, M., Font, J., Wu, J., Mecklenburg, S., Drusch, M., 2014. Microwave interferometric radiometry in remote sensing: an invited historical review. Radio Sci. 49, 415–449.

Matzler, C., 1994. Passive microwave signatures of landscapes in winter. Meteorog. Atmos. Phys. 54, 241–260.

Merlin, O., Chehbouni, A.G., Kerr, Y.H., Goodrich, D., 2006. A downscaling method for distributing surface soil moisture within a microwave pixel: application to Monsoon '90 data. Remote Sens. Environ. 101, 379–389.

Merlin, O., Rudiger, C., Al Bitar, A., Richaume, P., Walker, J.P., Kerr, Y.H., 2012. Disaggregation of SMOS soil moisture in Southeastern Australia. IEEE Trans. Geosci. Remote Sens. 50, 1556–1571.

Merlin, O., Escorihuela, M.J., Mayoral, M.A., Hagolle, O., Al Bitar, A., Kerr, Y., 2013. Self-calibrated evaporation-based disaggregation of SMOS soil moisture: an evaluation study at 3 km and 100 m resolution in Catalunya, Spain. Remote Sens. Environ. 130, 25–38.

Mialon, A., Coret, L., Kerr, Y.H., Secherre, F., Wigneron, J.P., 2008. Flagging the topographic impact on the SMOS signal. IEEE Trans. Geosci. Remote Sens. 46, 689–694.

Moran, M.S., Hymer, D.C., Qi, J., Kerr, Y., 2002. Comparison of ERS-2 SAR and Landsat TM imagery for monitoring agricultural crop and soil conditions. Remote Sens. Environ. 79, 243–252.

Newton, R.W., Black, Q.R., Makanvand, S., Blanchard, A.J., Jean, B.R., 1982. Soil-moisture information and thermal microwave emission. IEEE Trans. Geosci. Remote Sens. 20, 275–281.

Njoku, E.G., Li, L., 1999. Retrieval of land surface parameters using passive microwave measurements at 6–18 GHz. IEEE Trans. Geosci. Remote Sens. 37, 79–93.

Ochsner, T.E., Cosh, M.H., Cuenca, R.H., Dorigo, W.A., Draper, C.S., Hagimoto, Y., Kerr, Y.H., Larson, K.M., Njoku, E.G., Small, E.E., Zreda, M., 2013. State of the Art in Large-Scale Soil Moisture Monitoring. Soil Sci. Soc. Am. J. 77 (6), 1888–1919. http://dx.doi.org/10.2136/sssaj2013.03.0093.

Oliva, R., Daganzo, E., Richaume, P., Kerr, Y., Cabot, F., Soldo, Y., Anterrieu, E., Reul, N., Gutierrez, A., Barbosa, J., Lopes, G., 2016. Status of SMOS RFI in the 1400–1427 MHZ passive band after 5 years of mission. Remote Sens. Environ. (in press).

Pellarin, T., Wigneron, J.-P., Calvet, J.-C., Berger, M., Douville, H., Ferrazzoli, P., Kerr, Y.H., Lopez-Baeza, E., Pulliainen, J., Simmonds, L.P., Waldteufel, P., 2003. Two-year global simulation of L-band brightness temperatures over land. IEEE Trans. Geosci. Remote Sens. 41, 2135–2139.

Pellarin, T., Louvet, S., Gruhier, C., Quantin, G., Legout, C., 2013. A simple and effective method for correcting soil moisture and precipitation estimates using AMSR-E measurements. Remote Sens. Environ. 136, 28–36.

Pellenq, J., Kalma, J., Boulet, G., Saulnier, G.-M., Wooldridge, S., Kerr, Y., Chehbouni, A., 2003. A disaggregation scheme for soil moisture based on topography and soil depth. J. Hydrol. 276, 112–127.

Petropoulos, G.P., Griffiths, H., Dorigo, W., Xaver, A., Gruber, A., 2013. Surface soil moisture estimation: significance, controls and conventional measurement techniques chapter 2, pages 29–48. In: Petropoulos, G.P., (Ed.), Remote Sensing of Energy Fluxes and Soil Moisture Content. Taylor and Francis, ISBN: 978-1-4665-0578-0.

Petropoulos, G.P., Ireland, G., Barrett, B., 2015. Surface soil moisture retrievals from remote sensing: evolution, current status, products & future trends. Phys. Chem. Earth http://dx.doi.org/ 10.1016/j.pce.2015.02.009.

Rahmoune, R., Ferrazzoli, P., Singh, Y.K., Kerr, Y.H., Richaume, P., Al Bitar, A., 2014. SMOS retrieval results over forests: comparisons with independent measurements. IEEE J. Sel. Top. Appl. Earth Obs. Remote Sens. 7, 3858–3866.

Rudiger, C., et al., 2014. Toward vicarious calibration of microwave remote-sensing satellites in arid environments. IEEE Trans. Geosci. Remote Sens. 52, 1749–1760. http://dx.doi.org/10.1109/ tgrs.2013.2254121.

Schmugge, T.J., 1998. Applications of passive microwave observations of surface soil moisture. J. Hydrol. 212–213, 188–197.

Schmugge, T.J., Wang, J.R., Asrar, G., 1988. Results from the push broom microwave radiometer flights over the Konza Prairie in 1985. IEEE Trans. Geosci. Remote Sens. 26, 590–597.

Schwank, M., Stähli, M., Wydler, H., Leuenberger, J., Mätzler, C., Flühler, H., 2004. Microwave L-band emission of freezing soil. IEEE Trans. Geosci. Remote Sens. 42, 1252–1261.

Soldo, Y., Khazaal, A., Cabot, F., Richaume, P., Anterrieu, E., Kerr, Y.H., 2014. Mitigation of RFIS for SMOS: a distributed approach. IEEE Trans. Geosci. Remote Sens. 52, 7470–7479.

Soldo, Y., Cabot, F., Khazaal, A., Miernecki, M., Slominska, E., Fieuzal, R., Kerr, Y.H., 2015. Localization of RFI sources for the SMOS mission: a means for assessing SMOS pointing performances. IEEE J. Sel. Top. Appl. Earth Obs. Remote Sens. 8, 617–627.

Spencer, M.W., Chen, C.W., Ghaemi, H., Chan, S.F., Belz, J.E., 2013. RFI characterization and mitigation for the SMAP radar. IEEE Trans. Geosci. Remote Sens. 51, 4973–4982.

Srivastava, P.K., Han, D., Ramirez, M.A., Islam, T., 2013a. Machine learning techniques for downscaling SMOS satellite soil moisture using MODIS land surface temperature for hydrological application. Water Resour. Manag. 27, 3127–3144.

Srivastava, P.K., Han, D., Rico-Ramirez, M.A., Al-Shrafany, D., Islam, T., 2013b. Data fusion techniques for improving soil moisture deficit using SMOS satellite and WRF-NOAH land surface model. Water Resour. Manag. 27, 5069–5087.

Srivastava, P.K., Han, D., Ramirez, M.A., Islam, T., 2013c. Appraisal of SMOS soil moisture at a catchment scale in a temperate maritime climate. J. Hydrol. 498, 292–304.

Srivastava, P.K., O'Neill, P., Cosh, M., Kurum, M., Lang, R., Joseph, A., 2015. Evaluation of dielectric mixing models for passive microwave soil moisture retrieval using data from ComRAD ground-based SMAP simulator. IEEE J. Sel. Top. Appl. Earth Obs. Remote Sens. 8 (9), 4345–4354.

Srivastava, P.K., Han, D., Rico-Ramirez, M.A., O'Neill, P., Islam, T., Gupta, M., 2014. Assessment of SMOS soil moisture retrieval parameters using tau-omega algorithms for soil moisture deficit estimation. J. Hydrol. 519, 574–587.

Srivastava, P.K., Han, D., Rico-Ramirez, M.A., O'Neill, P., Islam, T., Gupta, M., Dai, Q., 2015. Performance evaluation of WRF-Noah Land surface model estimated soil moisture for hydrological application: synergistic evaluation using SMOS retrieved soil moisture. J. Hydrol. 529, 200–212.

Tomer, S.K., Al Bitar, A., Sekhar, M., Zribi, M., Bandyopadhyay, S., Sreelash, K., Sharma, A.K., Corgne, S., Kerr, Y., 2015. Retrieval and multi-scale validation of soil moisture from multi-temporal SAR data in a semi-arid tropical region. Remote Sens. 7, 8128–8153.

Tomer, S.K., Al Bitar, A., Sekhar, M., Merlin, O., Zribi, M., Bandyopadhyay, S., Sreelash, K., Sharma, A.K., Corgne, S., Kerr, Y.H., 2016. MAPSM: a conceptual spatio-temporal algorithm to merge active and passive soil moisture. Remote Sens. Environ. (submitted for publication).

Ulaby, F.T., Long, D.G., 2014. Microwave Radar and Radiometric Remote Sensing. The University of Michigan Press, Ann Arbor, MI.

Ulaby, F.T., Moore, R.K., Fung, A.K., 1986. Microwave Remote Sensing—Active and Passive. Artech House, Norwood, MA.

van den Hurk, B., Bastiaanssen, W.G.M., Pelgrum, H., van Meijgaard, E., 1997. A new methodology for assimilation of initial soil moisture fields in weather prediction models using Meteosat and NOAA data. J. Appl. Meteorol. 36, 1271–1283.

van den Hurk, B., Graham, L.P., Viterbo, P., 2002. Comparison of land surface hydrology in regional climate simulations of the Baltic Sea catchment. J. Hydrol. 255, 169–193.

van den Hurk, B., Doblas-Reyes, F., Balsamo, G., Koster, R.D., Seneviratne, S.I., Camargo, H., 2012. Soil moisture effects on seasonal temperature and precipitation forecast scores in Europe. Clim. Dyn. 38, 349–362.

Vaz, C.M.P., Jones, S., Meding, M., Tuller, M., 2013. Evaluation of standard calibration functions for eight electromagnetic soil moisture sensors. Vadose Zone J. 12.

Vittucci, C., Ferrazzoli, P., Kerr, Y., Richaume, P., Rahmoune, R., Guerriero, L., Vaglio Laurin, G., 2016. SMOS retrieval over forests: exploitation of optical depth and tests of soil moisture. Remote Sens. Environ.. (in this issue).

Wagner, W., Bloschl, G., Pampaloni, P., Calvet, J.C., Bizzarri, B., Wigneron, J.P., Kerr, Y., 2007. Operational readiness of microwave remote sensing of soil moisture for hydrologic applications. Nord. Hydrol. 38, 1–20.

Wagner, W., Hahn, S., Kidd, R., Melzer, T., Bartalis, Z., Hasenauer, S., Figa-Saldana, J., de Rosnay, P., Jann, A., Schneider, S., Komma, J., Kubu, G., Brugger, K., Aubrecht, C., Zueger, J., Gangkofner, U., Kienberger, S., Brocca, L., Wang, Y., Bloeschl, G., Eitzinger, J., Steinnocher, K., Zeil, P., Rubel, F., 2013. The ASCAT soil moisture product: a review of its specifications, validation results, and emerging applications. Meteorol. Z. 22, 5–33.

Wanders, N., Karssenberg, D., Roo, A., De Jong, S., Bierkens, M., 2014. The suitability of remotely sensed soil moisture for improving operational flood forecasting. Hydrol. Earth Syst. Sci. 18, 2343–2357.

Wigneron, J.P., Ferrazzoli, P., Calvet, J.C., Kerr, Y.H., Bertuzzi, P., 1999. A parametric study on passive and active microwave observations over a soybean crop. IEEE Trans. Geosci. Remote Sens. 37, 2728–2733.

Wigneron, J.-P., Chanzy, A., Calvet, J.-C., Olioso, A., Kerr, Y., 2002. Modeling approaches to assimilating L-band passive microwave observations over land surfaces. J. Geophys. Res. 107. http://dx.doi.org/10.1029/2001JD000958.

Wigneron, J.P., Kerr, Y., Waldteufel, P., Saleh, K., Escorihuela, M.J., Richaume, P., Ferrazzoli, P., de Rosnay, P., Gurney, R., Calvet, J.C., Grant, J.P., Guglielmetti, M., Hornbuckle, B., Matzler, C., Pellarin, T., Schwank, M., 2007. L-band Microwave Emission of the Biosphere (L-Meb) model: description and calibration against experimental data sets over crop fields. Remote Sens. Environ. 107, 639–655.

Chapter 2

Available Data Sets and Satellites for Terrestrial Soil Moisture Estimation

P.K. Srivastava[*,†], V. Pandey[†], S. Suman[†], M. Gupta[*] and T. Islam[‡,§]

[*]*NASA Goddard Space Flight Center, Greenbelt, MD, United States,* [†]*Banaras Hindu University, Banaras, India,* [‡]*NASA Jet Propulsion Laboratory, Pasadena, CA, United States,* [§]*California Institute of Technology, Pasadena, CA, United States*

1 INTRODUCTION

Soil moisture is an important component in the hydrological cycle, both on a small agricultural scale and in large-scale modeling of land/atmosphere interaction (Srivastava, 2013; Petropoulos et al., 2015a). Soil moisture is defined as the water contained in the unsaturated soil zone or vadose zone (Hillel, 1998). In other words, soil moisture is the water held within soil particles that plays an important role in a large number of applications such as numerical weather prediction, flood forecasting, agricultural drought assessment, water resources management, greenhouse gas accounting, civil protection, and epidemiological modeling of waterborne diseases, as well as several weather- and climate-related studies such as runoff potential estimation, water budgeting for irrigation planning, and scheduling of irrigation (Srivastava et al., 2013a, b; Petropoulos et al., 2014; Srivastava et al., 2015b). Soil moisture is divided into two zones, surface soil moisture and root zone soil moisture, on the basis of their heights from the surface. Surface soil moisture is the first soil moisture layer present in the upper 10 cm of layer of soil followed by the root water zone, which is readily available groundwater to plants and is generally considered to be present in the upper 200 cm of soil.

Soil moisture is a key variable of the climate system because of its significant impact on global water, energy, and biogeochemical cycles; therefore, it was recognized as an essential climate variable (ECV) in 2010 (http://www.esaCCISoilMoisturewebsite.htm) (Srivastava et al., 2013a). Soil moisture is the main source of water for major environmental phenomenon such as

Satellite Soil Moisture Retrieval. http://dx.doi.org/10.1016/B978-0-12-803388-3.00002-4

evapotranspiration. Evapotranspiration, which is the aggregate evaporation rate from vegetation and bare soil, is a major component of the atmospheric water cycle, and is responsible to return about 60% of the total land precipitation back to the atmosphere (Oki and Kanae, 2006; Petropoulos et al., 2015b). In practice, often only a fraction of soil moisture is relevant or measurable and thus soil moisture is always considered with a given soil volume. The expression for estimation of volumetric soil moisture θ (m^3 H_2O/m^3 soil) in the soil volume V is defined as follows:

$$\theta = \frac{\nu}{V}$$

where θ = volumetric soil moisture; ν = volume of water in soil volume; V = soil volume.

The equation is applicable on multiple scales, from a range of a few cubic centimeters to several cubic kilometers, depending on the measurement method or research application. In land surface and hydrological models, for which the soil is divided into discrete soil layers, θ is often expressed as (mm_{H_2O}/mm_{soil}) (Seneviratne et al., 2010).

Soil moisture interacts and controls the atmosphere through the proportion of rainfall that percolates, runs off, or evaporates from land. Therefore, the societal benefits of the various available data sets for soil moisture are expected to be large. The following section describes available data sets for soil moisture measurement.

2 IN SITU DATA SETS FOR SOIL MOISTURE

In situ soil moisture data sets are the mainstay of calibrating and validating land surface models and satellite-based soil moisture retrievals (Jackson et al., 2012; Srivastava et al., 2015a). Moreover, trends in climate or land cover change related with the hydrological cycle is clear from a collection of long-term series of in situ soil moisture measurements themselves (Dorigo et al., 2011a). Based on the importance of in situ soil moisture retrievals, the number of meteorological networks measuring soil moisture is continuously increasing, validating, and centralizing their data (Petropoulos et al., 2013). The details of few are provided below:

2.1 International Soil Moisture Network

The International Soil Moisture Network (ISMN; http://www.ipf.tuwien.ac.at/insitu) is an international operation that serves as a centralized global in situ soil moisture data hosting facility. The Global Energy and Water Cycle Experiment (GEWEX) coordinated this network in alliance with the Group on Earth Observations (GEO), the Committee on Earth Observation Satellites (CEOS), and the European Space Agency (ESA). In addition, scientists and associated networks from around the world have made the ISMN possible through their voluntary contributions. Furthermore, the ISMN is operated in cooperation with the Global Soil Moisture Databank of Rutgers University as the Global Soil Moisture Data Bank has closed and all files have been transferred to the ISMN (https://ismn.geo.tuwien.ac.at).

In ISMN, first globally available in situ soil moisture measurements are collected from operational networks and validation campaigns, and then harmonized, and finally made available to users through a web interface. Moreover, downloads are provided according to common standards for data and metadata (Dorigo et al., 2011a,b; 2013). Many scientific communities endorsed ISMN as a valuable resource for testing and improving satellite-derived soil moisture products and studying climate-related trends. The ISMN seems to evolve as the integrated distribution platform for in situ soil moisture measurements.

Presently there are approximately 48 (Department of Geodesy and Geoinformation, 2015) cooperative networks that deliver databases to ISMN and more than 1600 stations located primarily in the United States, Europe, Asia (only historical), and Australia. Some currently added networks of ISMN are DAHRA, LAB-net, ORACLE, PBO_H2O, SOILSCAPE, WEGENERNET, and WSMN (https://ismn.geo.tuwien.ac.at/news/newsletter/ismn-newsletter-10-april-2015/). Overall, ISMN is an open source network validated from 1952 to the present. The number of data sets is rapidly expanding; this means that both the number of stations and network as well as the time period covered by them are still growing (Dorigo et al., 2011a,b; 2013). Data collecting networks share their soil moisture data sets with the ISMN on a voluntary basis and without any tax. Then, these soil moisture data are automatically transformed into common volumetric soil moisture units and checked for outliers and implausible values and finally available for users on the web to download (Dorigo et al., 2011a,b).

The consciousness and willingness to realize such an integrated soil moisture measuring system is being made possible through the active participation of international organizations such as the World Climate Research Programme (WCRP), GEWEX, GEO, and others; the appreciable support of the space agencies, and the voluntary efforts of many individual scientists (Dorigo et al., n.d.). The main purpose of the ISMN is to provide calibration and validation of satellite observed soil moisture data sets. This task continuously increases and becomes important. Consequently, a large number of global soil moisture data sets have become available in recent years from microwave radiometers (passive) such as Soil Moisture and Ocean Salinity (SMOS), Advanced Microwave Scanning Radiometer for Earth Observing (AMSR-E), WindSat, Tropical Rainfall Measuring Mission (TRMM), and Scanning Multichannel Microwave Radiometer (SMMR), and from scatterometers (active) such as the European Remote Sensing Satellite Scatterometer (ERS SCAT) and the Advanced Scatterometer (ASCAT) (https://ismn.geo.tuwien.ac.at/satellites/). In addition to ISMN, various other field campaigns are conducted with the same and more advanced objectives.

2.2 Field Campaigns

The proliferation and implementation of a remote sensing system that will observe global soil moisture will require advancements in technology as well as in research. Many aspects of research require calibration and validation for application. These requirements can be accomplished through a controlled large-scale field experimentation campaign. The contributions of a large number of local and regional organizations and programs contribute to the success of this large-scale field campaign that requires significant resources.

Soil moisture field campaigns have their application in a broad range of scientific areas, focusing on technology development and demonstration, and providing educational experiences for students. Moreover, the data acquired by a field campaign has a primarily emphasis on developing map-based products. A number of field campaigns for soil moisture are held and are the basis of model estimates of flood and drought events and for studying the dynamics of soil moisture in space and time (Balsamo et al., 2009). In this chapter we put forth some valuable field campaigns, but not limited to:

2.2.1 Soil Moisture Experiments Series

The Soil Moisture Experiments (SMEX) series, organized at the Southern Great Plains (SGP) site, proved very successful in addressing a broad range of scientific and instrumentation applications. The data acquired by the SMEX series have been used in studies that went well beyond algorithm-based research, primarily due to an emphasis on developing map-based products.

The approach used in SMEX has been to collect ground-based samples of soil moisture in conjunction with aircraft flights at the same time as satellite overpasses. Whereas the purpose of SMEX is to validate microwave remote sensing, soil moisture products provide unique experimental data sets. Decades of scientific contributions are the basis of these data sets, which have significantly improved our knowledge of soil moisture variability within satellite pixel footprints. The objective and application of the SMEX series is briefly described here.

2.2.1.1 Soil Moisture Experiment 2002

Measurements for the Soil Moisture Experiment in 2002 (SMEX02) were held throughout the state of Iowa, USA, from mid-June to mid-July for about a 1-month period. After exploring the purpose of the experiment we tried to convey the objectives of SMEX02, which was to facilitate easy understanding of the land/atmosphere interaction, evaluation of better instrumental technology for soil moisture remote sensing, extension of algorithms for more complicated and challenging vegetation cover, and validation of AMSR-E satellite brightness temperature and soil moisture retrievals. This experiment supports the Aqua AMSR, NASA's GEWEX, and future satellite missions for terrestrial hydrology.

The Polarimetric Scanning Radiometer on C and X band microwave frequencies (PSR/CX) system was first demonstrated in Iowa during the Summer 2002 joint NASA-NOAA-USDA SMEX02. NASA found how to expand scattered measurements done from probing soil moisture with global measurements from the AMSR-E sensor. Extension of the AMSR-E observations and algorithms to more challenging vegetation conditions typical of agricultural regions, and integration of land surface and atmospheric boundary layer measurements to better understand moisture feedback mechanisms, are some additional scientific goals of SMEX02.

2.2.1.2   Soil Moisture Experiment 2003

Soil Moisture Experiment 2003 (SMEX03) is a continuation of the SMEX02 campaign in different geographic regions. The SMEX03 campaign was conducted at American sites in Oklahoma, Georgia, and Alabama between June 23 and July 18, 2003; and in Brazil during December 1–14, 2003. Like SMEX02, the SMEX03 campaign provided validation data for soil moisture levels, soil type diversity, and vegetation covers ranging from well-understood grass and wheat in Oklahoma to new observations of the Amazon rainforests. It also provided a test bed for other new launched satellite instruments such as the Envisat, Advanced Synthetic Aperture Radar (ASAR), and aircraft-based prototype satellite instruments.

SMEX03 uses as its principal airborne instrument the Environmental Technology Laboratory (ETL) PSR/CX. PSR/CX is designed to provide unique conically scanned airborne microwave imagery having full polarization capability and high spatial resolution. It is also used to detect and reject the anthropogenic radio interference that badly impacts many AMSR-E measurements. The small pixel size of PSR/CX ($\sim$0.5 km compared with $\sim$7 5 km for AMSR-E) provides an important bridge across the large range of sampling scales from that of the SMEX03 ground stations to the AMSR-E satellite imagery.

2.2.1.3   Soil Moisture Experiment 2004

The 2004 Soil Moisture Experiment (SMEX04) builds on the basis of preceding experiments (i.e., SMEX02 and SMEX03) by special focus on topography, vegetation, and strengthening the soil moisture components of the North American Monsoon Experiment (NAME). This experiment is also abbreviated as the SMEX04/NAME campaign. The most important purpose of NAME is using soil moisture and surface temperature data to improve prediction of summer precipitation, which is greatly dependent on convection. Whereas in situ soil moisture networks, aircraft mapping of soil moisture, intensive sampling concurrent with aircraft mission, and satellite products are four reciprocal elements of SMEX04.

Therefore, the combined perspective of SMEX04 and NAME (SMEX04/NAME) is to improve the terrestrial hydrological component of NAME by facilitating soil moisture development. Furthermore, it requires specific activities such as provision of soil moisture data sets from the in situ network over Arizona, development of an equivalent network within a study region in Mexico, and soil moisture data retrieval from available satellite sensors on Aqua and TRMM. From the Advanced Microwave Scanning Radiometer (AMSR) and future low frequency instruments, SMEX04/NAME will address important algorithm and validation issues for existing satellite-based soil moisture products.

2.2.1.4 Soil Moisture Experiment 2005

Soil Moisture Experiment 2005 (SMEX05) is the last series of SMEX. This experiment was part of a campaign known as Polarimetry Land Experiment (POLEX). The field data for POLEX measurements were collected from June 15, 2005, to July 3, 2005, using in situ manual sampling, pyrometers, temperature probes, and theta probes. The experiment was conducted specially for addressing algorithm development and validation related to all current and scheduled soil moisture satellite systems. There were many other purposes associated with the establishment of the SMEX05/POLEX campaign: exploration of unique polarimetric information from satellites such as WindSat and Conical Scanning Microwave Imager/Sounder (CMIS) for soil moisture, enhancement of AMSR-E soil moisture validation, and mitigation of radio frequency interference (RFI) for CMIS risk reduction are some main purposes.

Both ground and aircraft measurements of field experiments provide a strong connection between spatial scales, necessary for both algorithm development and validation. The parameters for POLEX measurement involve volumetric and gravimetric soil moisture, bulk density, and surface and soil temperature. These data are provided in ASCII text files and are available for users via FTP.

2.2.2 Canadian Experiment for Soil Moisture in 2010

The Canadian Experiment for Soil Moisture in 2010 (CanEx-SM10) was held from May 31, 2010, to June 17, 2010, over agricultural and forested sites located in Saskatchewan, Canada. The CanEx-SM10 was build specifically to support validation activities of the ESA's SMOS over land and to develop soil moisture retrieval algorithms in Canada. As part of Canada's involvement in NASA's Soil Moisture Active and Passive (SMAP) mission that was launched in January 2015, CanEx-SM10 is extended to include the validation of SMAP through collaboration with US researchers.

Spaceborne microwave measurements for CanEx-SM10 are done by SMOS, AMSR-E, ASAR-Envisat, RADARSAT-2, and Advanced Land Observing Satellite-Phased Array type L-band Synthetic Aperture Radar (ALOS-PALSAR) along with airborne measurements using passive and active instruments including an L-band radiometer mounted onboard Environment Canada's Twin

Otter aircraft and NASA's L-band Uninhabited Aerial Vehicle Synthetic Aperture Radar (UAVSAR) flown in a Gulfstream III piloted aircraft. In addition, the experiments have provided field measurements of soil moisture, surface temperature, and other surface characteristics such as vegetation, roughness, bulk density, Leaf Area Index (LAI), and so on, at a time close to satellite and airborne acquisitions to support the validation of SMOS and newly launched validation activities of SMAP.

Over 50 people participated to the field and aircraft component of CanEx-SM10, which took place over an agricultural site located in Kenaston (Saskatoon, Saskatchewan) and a forested site, which is the Boreal Ecosystem Research and Monitoring Sites (BERMS) also located in Saskatchewan. These sites were about 33 km × 71 km in measurement, which covered approximately two SMOS pixels, and were selected to test SMOS and UAVSAR data and soil moisture retrieval algorithms over very different soil and vegetation conditions. The field campaign soil moisture data collected was complemented by the observations from the permanent existing soil moisture measurement networks managed by European Country and the University of Guelph (U of G) located at the Kenaston site. Over the BERMS site, a temporary network of about 20 stations was installed by the U.S. Department of Agriculture (USDA) to collect hourly soil moisture data.

CanEx-SM10 is funded by the Natural Sciences and Energy Research Council (NSERC), Environment Canada (EC), the Canadian Space Agency (CSA), and Agriculture and Agri-Food Canada (AAFC) in Canada and by NASA in the United States. For the development of large-scale soil moisture retrieval algorithms in Canada, it was the first attempt to set up soil moisture observations simultaneously for satellite and aircraft microwave measurements.

The specific objectives of CanEx-SM10 were to develop soil moisture retrieval algorithms from microwave data of SMOS, RADARSAT-2, ALOS-PALSAR, NASA's UAVSAR, and airborne data from EC's radiometer; compare L-band active and passive microwave data with field measurements; improve environmental forecast models by assimilation of SMOS data in land surface systems; and develop scaling methodologies for SMOS coarse resolution data.

2.2.3 Soil Moisture Active Passive Validation Experiment

2.2.3.1 SMAPVEX08

The first Soil Moisture Active Passive Validation Experiment (SMAPVEX) was conducted during 2 weeks in early October 2008 at the Eastern Shore of Maryland and Delaware. The primary sensor for this campaign was the Passive Active L-band System (PALS). However, after PALS several other new instruments will become part of the experiment. SMAP-VEX08 is part of the overall SMAP mission Phase A activities designed to test soil moisture instrument retrieval algorithms for L-band radiometers.

The SMAP-VEX08 mission validated and demonstrated two new NASA L-band radiometers and a re-fly of a Global Positioning System (GPS) reflectometer. After the instrument validation and demonstration phase, the Wallops Flight Facility (WFF) P-3B aircraft participated in a multiplatform, mixed altitude soil moisture mapping experiment coordinated by the USDA. The National Environmental Policy Act (NEPA), through the NEPA's Environmental Checklist for Research and Development Projects, evaluated WFF's support of the SMAP-VEX08 mission. The impacts from the SMAP-VEX08 mission are not considered very robust and vast; rather, it was likely to be transient and negligible.

2.2.3.2 SMAPVEX12

The main objective behind designing Soil Moisture Active Passive Validation Experiment 2012 (SMAPVEX12) was to provide extended-duration measurements that exceed those of any past field experiments. When compared with previous airborne experiments of field campaigns held for soil moisture retrieval, SMAPVEX12 emerges as unique and valuable. Although the primary focus of SMAPVEX12 was directed toward development, enhancement, and assessment of SMAP soil moisture retrieval algorithms, SMAP will also attempt to estimate moisture under forest cover. Data obtained during SMAP-VEX12 has proved very useful to enhance and develop radar forest scattering models and to test the current radar-based soil moisture retrieval algorithm, both of which have proved essential to successful SMAP postlaunch validation.

The SMAPVEX12 experiment was conducted at a site located in southern Manitoba centered on the town of Elm Creek, Canada. The SMAPVEX12 site was suitable to facilitate efficient aircraft flights over forest and agricultural landscapes. This experiment commenced on June 7, 2012, and ended approximately 6 weeks later on July 19, 2012. So, SMAPVEX12 lasted 43 days during which soil moisture and vegetation conditions significantly varied. The campaign started at the period of early crop development and finished at the point where crops had reached maximum biomass.

About 75 researchers and scholars from Canada and the United States were involved in the collection of field and aircraft data during SMAPVEX12. During the course of the 6-week campaign, the PALS acquired 17 days of data. Among these 17 days of data the previous 13 days the UAVSAR was also flown. Sampling days were dedicated to either soil moisture or vegetation/surface roughness depending on whether aircraft flights took place. The SMAPVEX12 campaign resulted in 700 soil cores from which the volumetric water content was determined.

Data acquired from the SMAPVEX12 experiment provided high-quality soil moisture and vegetation data, captured over a wide range of conditions coincident with airborne active and passive microwave acquisitions. Preliminary results of SMAPVEX12 for soil moisture retrieval are to encourage promising expectations that SMAP will be a key source of data applicable for fulfilling the needs of vegetation communities (agriculture and forest). This is a valuable data set and will be valuable in developing robust algorithm retrieval methods for SMAP and other future satellite missions.

2.2.4 Soil Moisture Active Passive Experiments

The Soil Moisture Active Passive Experiments (SMAPEx) was a series of airborne field experiments done in Australia. These experiments were designed to provide prototype SMAP data for development and validation of soil moisture retrieval algorithms applicable for the SMAP mission. The fieldwork of SMAPEx was done in the Yanco intensive study area, a semiarid agricultural and grazing area located in the Murrumbidgee catchment in southeastern Australia. Equivalent to a SMAP radiometer, the main flight of SMAPEx covers a 36 km × 38 km study area, with an altitude of approximately 3000 m (AGL). It has 2–3 days revisit time to simulate SMAP observations both in spatial and temporal resolution.

The SMAPEx network involves 24 surface monitoring stations and profile monitoring stations. Along with soil moisture, other parameters such as soil temperature, vegetation biomass and structure, and surface roughness are also measured in the SMAPEx campaign to support airborne observation. Instrument used for SMAPEx data retrieval are the Polarimetric L-band Multibeam Radiometer (PLMR; 1.41 GHz) and the Polarimetric L-band Imaging Scatterometer (PLIS; 1.26 GHz), with additional thermal infrared radiometer and multispectral sensors for monitoring of vegetation properties and surface temperature.

The series of SMAP experiments were timed to cover various climatic conditions as well as various stages of the crop-growing season. The first experiment (i.e., SMAPEx-1) was conducted in austral winter from July 5–10, 2010. Weather conditions allowed observations of a moderately wet winter in the range 0.15–0.25 m^3/m^3 soil moisture.

Likewise, the second campaign (SMAPEx-2) was conducted in the austral summer from December 4–8, 2010. In these 4 days of experiment, a total of 25 h of scientific flights were conducted with concurrent ground sampling. Extreme rainfall occurred in the study area prior to the campaign leading to an unusually wet soil moisture condition (0.25–0.33 v/v) for that time of year. During the experimental phase there is no rainfall in the study area, yielding a dynamic range of 0.05–0.10 m^3/m^3 soil moisture.

The third campaign (SMAPEx-3) was 3-week long and occurred from September 5–23, 2011, in the austral springtime. Moderate rainfall (~35 mm) was recorded in the study area prior to 1 week of starting the experiment, while soft showers (~4 mm) were experienced during the first week of the field experiment. These fluctuations in climatic condition lead to soil moisture values in the range of 0.05–0.4 m^3/m^3. A particular objective of SMAPEx-3 was to restore a long-term series of active and passive microwave data for change detection of algorithms to estimate soil moisture from future SMAP missions.

The fourth campaign (SMAPEx-4) was also a 3 week program, which started around the end of April and finished at the end of May 2015. For comparing the observations from satellite and airborne, SMAPEx involves two types of flights, one flight for SMAP and the other flight for Aquarius. These observations include intensive spatial soil moisture sampling as well as regional soil moisture sampling in the SMAPEx field area.

2.3 FLUXNET sites

FLUXNET sites estimate soil moisture as a core parameter. CarboEurope is one of the parts of FLUXNET, which currently forms the largest global network for measuring micrometeorological flux and soil moisture. CarboAfrica was established in 2006 with the purpose of providing in situ data sets of meteorological parameters and soil moisture that increases Africa's role in the global carbon cycle. The main objective of this project is to form flux data from existing eddy covariance sites in Africa, therefore providing an excellent opportunity for validation of land surface models. CarboAfrica has about twelve eddy covariance measurement sites that also provide soil moisture. These dataset are available through the CarboAfrica network website.

3 SATELLITE DATA SETS FOR SOIL MOISTURE

Soil moisture is a parameter that is highly variable both spatially and temporally. Although in situ measurement gives invaluable and accurate knowledge regarding both near surface and subsurface soil moisture content, they are limited to specific locations and hence incapable of representing large-scale spatial distribution (Petropoulos et al., 2015b). For estimation of any events global data of soil moisture is required, so that we are forced to use satellite-based large-scale spatial and temporal data sets of soil moisture with the limitation of near surface measurement. That is the reason two dedicated space missions were proposed for enhanced global soil moisture measurement. The first was the ESA, which launched the SMOS mission in November 2009 and the second was NASA, which launched the SMAP mission in January 2015. In addition to these two, in this chapter we explore some other earlier satellites that retrieve soil moisture data but not limited to:

3.1 The Scanning Multichannel Microwave Radiometer (SMMR)

The SMMR onboard NASA's Nimbus-7 Pathfinder satellite provided service from October 25, 1978 to August 20, 1987. As per Rüdiger et al. (2009), it was the first microwave instrument operated for long duration and within adequate wavelengths. The SMMR provides data every other day. It was a 10-channel instrument that receives electromagnetic waves on both horizontally and vertically polarized radiation. The spatial resolution is 25 km for all channels; that is, 6.6, 10.7, 18, 21, and 37 GHz. A parabolic antenna of SMMR is 79 cm in diameter and maintains a constant nadir angle of 42 degrees, resulting in an incidence angle of 50.3 degrees at Earth's surface. This provides a 780 km swath of the Earth's surface.

R. Laughlin and K. Richter at Goddard Space Flight Center came up with the idea of developing SMMR. This idea come from the experience gained by a wide variety of experiments carried out in the laboratory, in the field, and onboard air and spacecraft using microwave radiometers over a wide wavelength range. A team led by J. Johnston shaped the design and accomplished the fabrication of SMMR at the Jet Propulsion Laboratory. The purpose of SMMR is to serve ocean circulation parameters such as sea surface temperatures, low altitude winds, water vapor and cloud liquid water content, sea ice extent, sea ice concentration, snow cover, snow moisture, soil moisture, rainfall rates, and differentiation of ice types. The processed data stream of SMMR was divided into three categories. First, the data was received by the meteorological operations control center (MetOCC); then, the user-formatted output tape from MetOCC is transferred to and processed by the science and applications computer center (SaCC); and finally, SaCC derives the required geophysical parameters from the radiometric data.

3.2 Advanced Microwave Scanning Radiometer for Earth Observing System (AMSR-E/2)

AMSR-E is a modified version of AMSR that flew on ADEOS-II. AMSR-E was launched in May 2002 and provided by Japan Aerospace Exploration Agency (JAXA) with the full cooperation of US and Japanese scientists, for flight onboard NASA's Earth Observing System (EOS) Aqua platform. AMSR-E is essential for Aqua's mission, which is dedicated to the observation of climate and hydrology. It has 12 channels and 6 frequencies ranging from 6.9 to 89.0 GHz and a passive dual-polarized microwave radiometer system that sensed faint microwave emissions from the Earth's surface and atmosphere.

The AMSR-E instrument provides measurement of geophysical atmospheric variables such as oceanic, land, and atmospheric variables for Earth's energy and water cycle. This involves precipitation rate, cloud liquid water, water vapor, sea surface winds, sea surface temperature, sea ice concentration, snow water equivalent, surface wetness, wind speed, and soil moisture. For weather, ecosystem, and climate modeling as well as knowledge of land surface hydrology, soil moisture is a key state variable.

Horizontally and vertically polarized radiations are measured by AMSR-E separately at 6.9, 10.7, 18.7, 23.8, 36.5, and 89.0 GHz. In other words, spatial resolution of the individual measurement varies from 5.4 km at 89 GHz to

56 km at 6.9 GHz. The AMSR-E satellite rotates continuously around an axis parallel to the local spacecraft vertical at 40 resolutions per minute (rpm). It has swath width of 1445 km. The spacecraft subsatellite point travels 10 km with an interval of 1.5 s. Even though the instantaneous field-of-view for each channel is different, active scene measurements are recorded at equal intervals of 10 km (5 km for the 89 GHz channels) along the scan. The spatial resolution of AMSR-E data doubles that of SMMR and Special Sensor Microwave/Imager (SSM/I) data. Also, AMSR-E combines into one sensor all the channels that SMMR and SSM/I had individually. So, in this way AMSR-E improves upon past microwave radiometers like SMMR and SSM/I.

Global and continuous observation is performed with fine spatial resolution by using one of the largest ever microwave radiometer antennas. The long-term geophysical record will play an important role in climate change monitoring and will provide valuable information for understanding the Earth's climate system, including water and energy circulation. Near real-time products will be used to investigate satellite data assimilation into weather forecasting models and to contribute to improved forecasting accuracy.

The AMSR-E instrument on board the Aqua satellite has failed to produce data since October 2011, because of some problem with the rotation of its antenna. So, for continuation the Advanced Microwave Scanning Radiometer 2 (AMSR2) which is loaded onto the Global Change Observation Mission-Water (GCOM-W) was launched by JAXA in May 2012 and JAXA started providing the AMSR data to the National Oceanic and Atmospheric Administration (NOAA) in February 2013.

3.3 Advanced Scatterometer (ASCAT)

The ASCAT gives one of the freely available global soil moisture data sets derived from the backscatter measurement. ASCAT was launched sequentially twice from 2006 to the present. The first ASCAT was launched in October 2006, onboard ESA's European Organisation for the Exploitation of Meteorological Satellites (EUMETSAT) meteorological operational satellite (MetOp-A). It started to work properly from May 2007 and continues to work today, and its data are now available operationally. Subsequently, an identical ASCAT instrument was launched in September 2012 on the MetOp-B satellite, which is in operational phase. Further, a third satellite carrying an ASCAT instrument is scheduled for launch in 2018. Therefore, the three series of ASCAT's MetOp satellites can be expected to provide uninterrupted backscatter observations of ASCAT from 2006 to the 2020s as well.

The ASCAT is a real-aperture and an active microwave remote sensing instrument with very good radiometric accuracy (about 0.3 dB) and stability. It is a C-band scatterometer operated at a frequency of 5.255 GHz using six sideways looking, three vertical polarized antennae. That means the antenna transmits and receives electromagnetic pulses in vertical polarization only (VV). The measurements through antennae are made at 45, 90, and 135 degree azimuth angles with respect to the satellite track. It scans the globe in a push-broom mode over two 550 km wide swaths separated by a gap of about 360 km, observing the surface of the Earth with 30–50 km spatial resolution. As ASCAT operates continuously, it achieved global coverage in ~1.5 days. In Western Europe, measurements are generally obtained twice a day, one in the morning (descending orbit) and one in the evening (ascending orbit).

The main objective of installing ASCAT is to monitor wind speed and direction over the oceans, though it is also used for studying polar ice, soil moisture, and vegetation. But initially it was not foreseen for monitoring soil moisture over land. The ASCAT soil moisture measurement was developed by cooperation of both EUMETSAT and the Vienna University of Technology (TU Wien). Consequently, the operational near real-time (NRT) dissemination of ASCAT soil moisture data sets was started in December 2008. As ASCAT soil moisture data have been available since 2008 by instruments launched on both MetOp-A/B, it improves the spatial and temporal coverage of soil moisture over the globe.

3.4 Soil Moisture and Ocean Salinity (SMOS)

The SMOS is the major satellite remote sensing mission for soil moisture measurements. The SMOS mission was proposed by Centre d'Etudes Spatiales de la BIOsphère (CESBIO) headed by Yann H. Kerr and was launched on November 2, 2009, as a second Earth Explorer Opportunity after Gravity Field and Steady-State Ocean Explorer (GOCE). This mission is a collaborative program of the ESA, Centre National d'Etudes Spatiales (CNES) in France, and the Centro Para el Desarrollo Tecnologico Industrial (CDTI) Earth Observation in Spain. With the objective of providing soil moisture (SM) and ocean salinity (OS) maps, the SMOS mission will have almost exact forecasting of weather, disastrous event, and seasonal climate.

This mission has a novel measuring technique that basically depends on the microwave radiation around the low frequency emitted by soil and ocean parameters. For estimation of surface soil moisture and sea surface salinity (ocean salinity), microwave radiometry at 1.4 GHz (21 cm) and L-band is an ongoing technique with a suitable sensitivity. SMOS will achieve global coverage every 3 days.

The SMOS satellite has a single payload instrument, the Microwave Imaging Radiometer with Aperture Synthesis (MIRAS) coupled to a Proteus platform. The MIRAS instrument is a dual polarized L-band 2-D passive interferometer radiometer, spaceborne, polar orbiting, and operating in the 1.4 GHz protected band. It was designed to provide global information on surface soil moisture and sea surface salinity with 4% accuracy. This aspect is managed by the Hydros Project. Basically this instrument receives the radiations that are emitted from Earth's surface. The received radiation can be then related to the moisture content in the first few centimeters (about 0–5 cm depth) of soil over land and to salinity in the surface waters over the oceans. The ground resolution of MIRAS is 30 to 50 km at the swath edges, providing multiangular acquisitions.

The SMOS mission center providing data sets of level 1 and level 2 is located at ESAC Centre (ESA) in Villafranca. The specific data processing center providing data sets of level 3 and level 4 is developed by CNES and is operated by Ifremer in Plouzané. Further, the CATDS (Centre Aval de Traitement des Données) released these daily gridded surface soil moisture data sets at a global level. In the future the SMOS team wants to do work with farmers around the world, including the USDA to use as ground-based calibration for models determining soil moisture, as it hopefully will help to much better understand crop yields over a wide region.

3.5 Soil Moisture Active and Passive (SMAP) Mission

The SMAP mission was one of four missions recommended by the U.S. National Research Council Committee on Earth Science and Applications from Space in 2007. Delayed 2 years from its scheduled launch date, it was launched on January 31, 2015. It is one of the first Earth observation satellites being developed and managed by NASA, with participation by the Goddard Space Flight Center. SMAP mission built on the heritage and risk-reduction activities of NASA's Earth System Science Pathfinder (ESSP) Hydros mission. The SMAP mission will provide measurements of near-surface soil moisture (0–5 cm depth) data products at hydrometeorology (10 km) and hydroclimatology (40 km) scales with 3-day global revisit coverage using combined information from both the radiometer and the radar measurements.

To overcome the impact of vegetation on the soil parameter measurements, the SMAP satellite is using simultaneous active (in the form of radar) and passive (in the form of radiometer) microwave instruments at the L-band frequency (1.2–1.4 GHz range). Radiometric-based measurement algorithms provide better soil moisture estimation under vegetated condition; on the other hand, the radar-based measurements provide higher spatial resolution, subpixel roughness, and vegetation information. In this way, the strategy of combining radar and radiometric data can enhance the resolution capability and accuracy of the soil moisture estimates. Consequently, the SMAP measurements are expected to be very significant for societal purposes by contributing in the forecasting of weather and climate, especially the prediction of extreme hydrological events such as droughts and floods. Therefore this has ultimate impact on human health and crop yield.

The principle behind the SMAP mission is that the radar sends pulses of radio waves to the selected earth surface, which are scattered back a few microseconds later. The interpretation of pulses that are received gives an idea of the moisture level of the soil, even through moderate levels of vegetation. Because the radar sends and receives radio waves actively, active comes in the SMAP satellite. In addition, the radiometer senses self-emission of the Earth for the same area. The strength of emitted radio waves indicates temperature of the ground in that location. Because a radiometer detects temperature of the ground passively, so that's why passive comes in the SMAP satellite.

SMAP is sun-synchronous satellite that runs from pole to pole and crosses Earth's terminator at the equator. The SMAP's orbit is 685 km above the Earth's surface. Because Earth spins while SMAP revolves, swaths from each orbit are offset from each other and, after 8 days, the same swath is repeated. Over 2 to 3 days (2 days at the poles and 3 days at the equator), the gaps between swaths are compensated and a global map of moisture is generated. Measurements are obtained by SMAP satellite across a wide swath (i.e., 1000 km). Both radar and radiometer instruments of the SMAP satellite share a single feed and a deployable 6 m reflector antenna system that rotates around the nadir axis. This rotation antenna makes conical scanning at a constant incidence angle (40 degrees). The radiometer resolution is 40 km and the radar resolution varies from 1 to 3 km across the entire swath with 3-day global revisit.

SMAP has wide application in the fields of science, agriculture, and environmental management. These are the basis of Earth's health and sustainability. Moreover, for understanding the processes that link the water, carbon, and energy cycles to improving weather and for climate prediction models, SMAP gives vital environmental knowledge. Data products from the SMAP mission will be made available through a NASA-designated data center.

4 CONCLUSION

Soil moisture estimation has paramount importance for hydrological land surface processes and their management. Taking into consideration the importance of soil moisture and its application in monitoring of extreme hydrologic events such as drought and flood, runoff modeling, numerical weather prediction, landslide monitoring, vegetation monitoring, agricultural monitoring, epidemiological prediction, climate studies, groundwater modeling etc. This chapter provides a comprehensive study of mostly all the possible established and emerging facilities that retrieve soil moisture data from in situ (field method) to remote sensing methods. These includes but not limited to in situ network (ISMN, FLUXNET etc), field campaigns (SMEX series, CanEx, SMAPVEX etc) and satellites (ASCAT, Aquarius, AMSR E/2, SMOS, SMAP etc). The mentioned field campaigns, in situ networks and satellite in this chapter could be useful for researchers working in the field of soil moisture retrieval, validation and applications.

REFERENCES

Balsamo, G., Beljaars, A., Scipal, K., Viterbo, P., van den Hurk, B., Hirschi, M., Betts, A.K., 2009. A revised hydrology for the ECMWF model: verification from field site to terrestrial water storage and impact in the integrated forecast system. J. Hydrometeorol. 10, 623–643.

Department of Geodesy and Geoinformation, 2015. www.ismn.geo.tuwien.ac.at/news/newsletter/ismn (accessed on 31.12.15)

Dorigo, W., Gruber, A., Van Oevelen, P., Wagner, W., Drusch, M., Mecklenburg, S., Robock, A., Jackson, T., 2011a. The International Soil Moisture Network – an observational network for soil moisture product validations. In: 34th International Symposium on Remote Sensing of Environment. The GEOSS Era: Towards Operational Environmental Monitoring. Sydney, Australia, 10-15 April 2011. http://www.isprs.org/proceedings/2011/ISRSE-34/211104015Final00819.pdf.

Dorigo, W., Wagner, W., Hohensinn, R., Hahn, S., Paulik, C., Xaver, A., Gruber, A., Drusch, M., Mecklenburg, S., Oevelen, P.v., 2011b. The International Soil Moisture Network: a data hosting facility for global in situ soil moisture measurements. Hydrol. Earth Syst. Sci. 15, 1675–1698.

Dorigo, W., Xaver, A., Vreugdenhil, M., Gruber, A., Hegyiová, A., Sanchis-Dufau, A., Zamojski, D., Cordes, C., Wagner, W., Drusch, M., 2013. Global automated quality control of in situ soil moisture data from the International Soil Moisture Network. Vadose Zone J. 12. http://dx.doi.org/10.2136/vzj2012.0097.

Hillel, D., 1998. Environmental Soil Physics: Fundamentals, Applications, and Environmental Considerations. Academic Press, San Diego, CA, USA.

Jackson, T.J., Bindlish, R., Cosh, M.H., Tianjie, Z., Starks, P.J., Bosch, D.D., Seyfried, M., Moran, M.S., Goodrich, D.C., Kerr, Y.H., Leroux, D., 2012. Validation of soil moisture and ocean salinity (SMOS) soil moisture over watershed networks in the U.S. IEEE Trans. Geosci. Remote Sens. 50, 1530–1543.

Oki, T., Kanae, S., 2006. Global hydrological cycles and world water resources. Science 313, 1068–1072.

Petropoulos, G.P., Griffiths, H., Dorigo, W., Xaver, A., Gruber, A., 2013. Surface soil moisture estimation: significance, controls and conventional measurement techniques chapter 2, pages 29–48. In: Petropoulos, G.P., (Ed.), Remote Sensing of Energy Fluxes and Soil Moisture Content. Taylor and Francis, ISBN: 978-1-4665-0578-0.

Petropoulos, G.P., Ireland, G., Barrett, B., 2015a. Surface soil moisture retrievals from remote sensing: evolution, current status, products & future trends. Phys. Chem. Earth. http://dx.doi.org/10.1016/j.pce.2015.02.009.

Petropoulos, G., Ireland, G., Cass, A., Srivastava, P.K., 2015b. Performance assessment of the SEVIRI evapotranspiration operational product: results over diverse Mediterranean ecosystems. IEEE Sens. J. 15 (6), 3412–3423. http://dx.doi.org/10.1109/jsen.2015.2390031.

Petropoulos, G.P., Ireland, G., Srivastava, P.K., Ioannou-Katidis, P., 2014. An appraisal of the accuracy of operational soil moisture estimates from SMOS MIRAS using validated in situ observations acquired in a Mediterranean environment. Int. J. Remote Sens. 35, 5239–5250.

Rüdiger, C., et al., 2009. An intercomparison of ERS-Scat and AMSR-E soil moisture observations with model simulations over France. J. Hydrometeorol. 10 (2), 431–447.

Seneviratne, S.I., Corti, T., Davin, E.L., Hirschi, M., Jaeger, E.B., Lehner, I., Orlowsky, B., Teuling, A.J., 2010. Investigating soil moisture–climate interactions in a changing climate: a review. Earth Sci. Rev. 99, 125–161.

Srivastava, P.K., O'Neill, P., Cosh, M., Lang, R., Joseph, A., 2015a. Evaluation of radar vegetation indices for vegetation water content estimation using data from a ground-based SMAP simulator. In: TU2.Y1: Soil Moisture Algorithms and Downscaling, IGARSS, Milan, Italy.

Srivastava, P.K., 2013. Soil Moisture Estimation from SMOS Satellite and Mesoscale Model for Hydrological Applications. PhD Thesis, University of Bristol, Bristol, UK.

Srivastava, P.K., Han, D., Ramirez, M.A., Islam, T., 2013a. Appraisal of SMOS soil moisture at a catchment scale in a temperate maritime climate. J. Hydrol. 498, 292–304.

Srivastava, P.K., Han, D., Ramirez, M.A., Islam, T., 2013b. Machine learning techniques for downscaling SMOS Satellite soil moisture using MODIS land surface temperature for hydrological application. Water Resour. Manag. 27, 3127–3144.

Srivastava, P.K., Han, D., Rico-Ramirez, M.A., O'Neill, P., Islam, T., Gupta, M., Dai, Q., 2015b. Performance evaluation of WRF-Noah land surface model estimated soil moisture for hydrological application: synergistic evaluation using SMOS retrieved soil moisture. J. Hydrol. 529 (Part 1), 200–212.

Optical and Infrared Techniques & Synergies Between them

Chapter 3

Soil Moisture Retrievals Using Optical/TIR Methods

P. Rahimzadeh-Bajgiran[*] and A. Berg[†]

[*]*University of Maine, Orono, ME, United States,* [†]*University of Guelph, Guelph, ON, Canada*

1 INTRODUCTION

Current technological advances in satellite remote sensing (RS) have offered an alternative to field measurements of soil moisture (SM) and enabled us to monitor it at higher temporal and spatial resolutions at considerably lower cost and time. Since the 1970s a wide array of RS methods have been developed to investigate SM using different regions of electromagnetic spectrum from the optical to microwave region. RS methods have been relatively successful for surface SM estimation at a depth of 0–5 cm and the techniques have been shown to be more accurate over bare soil and less-vegetated soils (Sandholt et al., 2002; Carlson, 2007). Satellite-based SM estimations over dense vegetation cover and at the root zone depth are still a challenge.

There are three types of RS sensors that have the capabilities to be used for SM estimation: (a) passive optical, thermal, and optical/thermal sensors; (b) passive microwave sensors; and (c) active microwave sensors. SM estimation methods based on all these three types of sensors are powerful but have some limitations as described in the literature. Comprehensive reviews on the application of remotely sensed methodologies for the estimation of surface SM including the principles, advantages, and constraints as well as currently available sensors can be found in Carlson (2007), Moran et al. (2004), Owe et al. (2008), Verstraeten et al. (2008), and Wang and Qu (2009). This chapter will only focus on passive optical/thermal infrared (TIR) approaches to estimate SM.

SM estimation using passive optical/thermal (multispectral) sensors takes advantage of the information available from visible, near-infrared (NIR) and shortwave infrared wavebands (Muller and Décamps, 2000; Liu et al., 2002; Chen et al., 2014), only TIR wavebands (Schmugge, 1978), or combination of visible, NIR, and TIR bands known as the surface temperature/vegetation index method (Nemani et al., 1993; Gillies and Carlson, 1995; Carlson et al., 1995). Regardless of the shortcomings of each approach here, the

Satellite Soil Moisture Retrieval. http://dx.doi.org/10.1016/B978-0-12-803388-3.00003-6

common sensor-related limitations for group "a" are cloud contamination, lower accuracy over dense vegetation cover, and estimating only top surface SM due to minimal surface penetration (Moran et al., 2004). On the other hand, fine spatial resolution, sound temporal coverage, long-standing free of charge archival imagery, and long-term acquisition plan for many sensors of this category such as LANDSAT, Moderate Resolution Imaging Spectroradiometer (MODIS), and Advanced Very High Resolution Radiometer (AVHRR) are the most important advantages of these sensors, making them very useful for SM estimation.

Although both passive and active microwave sensors were originally developed to provide improved SM estimation from RS data compared to passive optical, TIR, or optical/TIR approaches and to overcome shortcomings related to atmospheric sensitivity, SM estimation methods based on using only passive optical or optical/thermal RS are still being increasingly used. Combination of optical/thermal methods with passive or active microwave has also been a current approach over the past years. This trend can be attributed to the following reasons:

(1) Approaches based on optical, TIR, and optical/TIR SM estimation perform adequately well over bare soil and low vegetation cover in arid and semi-arid regions; finding several cloud free days especially during water stress times of the year is an easy task in these regions.

(2) Optical/TIR data products are being improved and sensors such as LAND-SAT and AVHRR have long-standing free of charge archival imagery that provide an invaluable pool of data. New sensors having high-quality better spatial and spectral resolutions such as MODIS have been developed. This is being followed by the development of newer sensors with much finer resolutions (Sentinel satellites). These new opportunities enable us to improve and refine current methodologies derived from currently available sensors through the application of new more specifically designed sensors.

(3) Optical and thermal data alone or together can provide useful information for both active and passive microwave SM models leading to the development of a number of methods based on using the combination of passive and active microwave with optical/TIR sensors (Mattar et al., 2012; Jahan and Gan, 2015; Merlin et al., 2005; Temimi et al., 2010). Also, as passive microwave sensors have coarse spatial resolutions, their products have significant mixed pixel problems over heterogeneous land covers. Several efforts are being made to develop appropriate approaches for downscaling SM data retrieved from low-resolution microwave sensors. For example, MODIS and AVHRR optical/TIR approaches have been used to downscale passive microwave SM estimations (Chauhan et al., 2003; Merlin et al., 2008, 2010, 2012; Piles et al., 2011; Chakrabarti et al., 2015; Piles et al., in press) but downscaling methodologies still need to be improved.

Combined application of optical and TIR bands known as the surface temperature (T_s)/vegetation index (VI) method is a promising approach to estimate SM as T_s and vegetation have been found to have a complicated dependence on SM. Due to the importance and increasing application of optical/TIR RS approaches alone or in combination with other sensors for SM estimation and in particular for downscaling of SM data from passive microwave sensors, it seems helpful to provide a comprehensive and informative description of the T_s/VI concept and describe suggested methodologies, their advantages and limitations for future application.

2 OPTICAL/TIR MODEL HISTORY AND CONCEPT

2.1 History

Since the 1980s a number of satellite RS indices have been developed to investigate vegetation condition and amount for SM estimation and drought monitoring. AVHRR data, supplied by the National Oceanic and Atmospheric Administration (NOAA), were the first remotely sensed data used for vegetation condition and change monitoring and provided the ability to generate the normalized difference vegetation index (NDVI) and T_s products. The NDVI is a common VI that reflects vegetation amount and chlorophyll content and is one of the first remotely sensed indices successfully used for monitoring vegetation condition and drought detection at larger scales (Anyamba and Tucker, 2005; Kogan, 1990; Liu et al., 1994; Malo and Nicholson, 1990; Tucker et al., 1981; Vicente-Serrano et al., 2006). However, the NDVI has two main limitations when used for drought monitoring and soil/vegetation water stress detection; first there is an apparent time lag between precipitation and the NDVI response and second, the influence of significant precipitation events later in the growing season on the NDVI is small (Rahimzadeh-Bajgiran et al., 2009, 2012; Rundquist and Harrington, 2000; Wang et al., 2001).

Because the NDVI represents the vegetation amount and chlorophyll content rather than land water status, there has been a need for a more sensitive indicator than the NDVI for water stress monitoring. To overcome these problems, approaches combining vegetation indices and the T_s were proposed for determining land water status. The T_s is a good indicator of energy partitioning at the surface and can be used as a proxy for SM and vegetation water stress estimation for the reason that in dry conditions due to the lack of SM, both leaf temperature (as a result of stomata closure) and surface temperature will increase.

The earliest applications of the T_s for detecting land water stress and crop evapotranspiration (ET) rate were reported in the 1970s when aerial thermal scanners were used (Bartholic et al., 1972; Heilman et al., 1976). Following that, studies based on the application of thermal scanning techniques and energy balance analysis for plant physiology and stomata response and transpiration activities at the leaf level were conducted in the early 1980s (Omasa et al., 1981a,b; Omasa and Croxdale, 1991). Later, using remote thermal sensing devices, indices such as the crop water stress index (CWSI) were developed and used for irrigation scheduling (Idso et al., 1981; Jackson et al., 1981). Using concepts such as the CWSI at larger scales (using remotely sensed data) was problematic due to interference from soil background and mixed pixels. One of the approaches to solve this problem was the one used by Nemani et al. (1993) who made use of the fact that the NDVI is an indicator of green vegetation cover within the pixels, and therefore the slope of T_s versus NDVI plot gives a measure of stomata conductance and ET.

Over the past 40 years, the T_s/VI concept has been applied for various applications such as the estimation of ET, evaporative fraction (EF), and SM using different sensors such as the AVHRR, LANDSAT, MODIS, and so forth. A number of studies have documented the T_s/VI (mainly NDVI or fractional vegetation cover (FVC)) relationship and described the triangular/trapezoidal shape of the data falling between the T_s and the NDVI axes. Comprehensive reviews on the application of remotely sensed T_s/VIs for different applications such as ET, EF, and SM estimation can also be found in Petropoulos et al. (2009) and Li et al. (2009).

The application of the T_s/VI concept for SM estimation began with the work of Nemani et al. (1993) who found a strong negative relationship between T_s and the NDVI for all biome types studied with a clear change in the gradient between dry and wet days. The idea was further developed by Gillies and Carlson (1995) and Carlson et al. (1995) and led to the presentation of the universal triangular method to explore relationships between SM, T_s, and NDVI. Over the past years the T_s/VI concept has been applied and tested over different climatic conditions and land cover types from rangeland to agricultural lands at different scales.

2.2   The T_s/VI Concept

The T_s/VI relationship is related to the ET rate of the surface and is suited for monitoring vegetation and soil water status. The spatial T_s/VI space relationship was initially proposed by Goward et al. (1985). The slope of the T_s/VI relationship was studied by several researchers (Carlson et al., 1990; Friedl and Davis, 1994; Goward et al., 2002; Nemani and Running, 1989; Smith and Choudhury, 1991). Goetz (1997) reported that there is a negative correlation between the T_s and the NDVI and indicted that T_s can rise rapidly as a result of water stress. Other researchers developed the concept of the T_s/VI relationship and described the triangular/trapezoidal shape of the data falling between the T_s and the VI axes (Price, 1990; Moran et al., 1994; Gillies and Carlson, 1995; Carlson et al., 1995; Gillies et al., 1997). The triangle/trapezoid space of T_s/VI is a simplified representation of the energy balance model, which estimates actual ET from potential ET (Moran et al., 1994). Fig. 1 exhibits the theoretical basis of the T_s/VI concept.

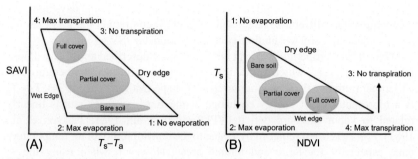

FIG. 1 Trapezoidal space of T_s/VI as presented by Moran et al., 1994 (A) and the triangular space of T_s/VI as presented by Sandholt et al., 2002 (B). SAVI is the soil-adjusted vegetation index and T_a is air temperature.

The triangular/trapezoidal shape of the data falling between the T_s and the VI, especially the NDVI, has been documented by a number of studies. The lower edge of this space (wet edge) is characterized by wet land surfaces with maximum ET whereas the upper edge (dry edge) of the scatter plot represents dry land with limited ET. Moran et al. (1994) described the data in the T_s/VI (here soil-adjusted vegetation index (SAVI)) space as a trapezoid and developed the water deficit index (WDI) based on the CWSI concept for soil and vegetation water deficit. On the basis of the triangular space, several RS models were developed for SM and ET estimation (Gillies and Carlson, 1995; Sandholt et al., 2002; Jiang and Islam, 2001).

To estimate SM, ET, or EF from T_s/VI space, dry and wet limiting edges in the triangular shape (Fig. 1B) or the four vertices (1–4) in Fig. 1A must be first defined. Here we will primarily focus on the dry edge/wet edge determination as many SM estimation models developed so far are based on the triangular model.

Researchers have applied two general approaches to determine dry and wet edges. The first approach is based on mainly RS data and use of linear least squares (Jiang and Islam, 1999; Sandholt et al., 2002) or nonlinear methods (Gillies and Carlson, 1995; Carlson et al., 1995) to fit the extreme points in the T_s/VI space. In the second approach, limiting edges will be determined theoretically based on the energy balance model using meteorological data such as vapor pressure deficit, wind speed, and aerodynamic resistance (Moran et al., 1994). One advantage of this approach is that the upper and lower limits of the T_s/VI space are determined close to the real land surface conditions not the observed conditions, but dependence of the method on ancillary data makes it less applicable for larger scales.

Using the T_s/VI model, different methods were suggested to form the two-dimensional scatter plot of T_s and vegetation. Petropoulos et al. (2009) summarized these approaches into five categories: (1) T_s and simple VI; (2) T_s and albedo; (3) T_s and T_a (air temperature) difference and VI; (4) the day-night surface temperature difference (ΔT_s) and VI; and (5) combining the T_s/VI feature space data with a soil-vegetation-atmosphere transfer (SVAT) model.

The surface temperature axis in the T_s/VI space described above is a simplified term of $T_s - T_a$, which is the difference between surface temperature and air temperature mostly seen as a trapezoidal shape (Moran et al., 1994). $T_s - T_a$ is a representative of energy exchange at the Earth surfaces and is linearly related to vapor pressure deficit (VPD) (Idso et al., 1981; Jackson et al., 1981). $T_s - T_a$ is often replaced by T_s due to the unavailability of air temperature data, especially for large and remote areas. However, when using the T_s instead of $T_s - T_a$ to estimate SM status, heterogeneity of the earth surfaces increases the uncertainty of the method to estimate SM (Rahimzadeh-Bajgiran et al., 2012). Therefore, the T_s/NDVI method should ideally be applied over smaller regions and those with negligible topographic variation. Stisen et al. (2008) and Wang et al. (2006) also considered temporal variations of T_s (day-night T_s difference $= \Delta T_s$) instead of only using T_s in the temperature axis of the scatter plot to better estimate EF. The diurnal T_s difference has a similar concept to that of thermal inertia; as soil thermal inertia is related to SM, it can be applicable to SM estimation as well (Cai et al., 2007; Verstraeten et al., 2006).

3 OPTICAL/TIR MODELS USED FOR SM ESTIMATION

Over the past decades, several models have been suggested for SM estimation from the T_s/VI model. Approaches to estimate SM based on T_s/VI were mainly developed during the 1990s. There are two main approaches to estimate SM. The first approach that is well-documented and studied by several researchers is based on direct estimation of SM from the T_s/VI space. One of the first models developed using this approach was the universal triangular method developed by Gillies and Carlson (1995) and Carlson et al. (1995), later modified by other researchers (Mallick et al., 2009; Sandholt et al., 2002; Vicente-Serrano et al., 2004; Zhang et al., 2015).

The second, although not as well documented in the RS literature, approach to estimate SM is through the relationship between SM and EF. To do this, EF needs to be estimated and then transformed into SM values using empirical equations. A number of models have been developed to estimate EF using the T_s/VI concept but they have been mainly used for ET estimation (Jiang and Islam, 2001; Nishida et al., 2003) and not for SM estimation. Also, a number of empirical models (Komatsu, 2003; Lee and Pielke, 1992) have been suggested to correlate EF to SM at different soil depths. More recently, remotely sensed EF was used for SM estimation using the SM/EF empirical relationship (Merlin et al., 2008, 2010, 2012; Rahimzadeh-Bajgiran et al., 2013). Below we will present these two approaches.

3.1 Direct Estimation of SM From T_s/VI Space

Using the concept suggested by Carlson et al. (1990, 1994), Gillies and Carlson (1995) introduced the triangle method for the retrieval of soil surface moisture availability (M_0), which is defined as the ratio of soil surface evaporation to the potential evaporation or as the ratio of soil water content to that at field capacity, and land-atmosphere energy fluxes using AVHRR satellite data. They proposed a nonlinear relationship between T_s and FVC estimated from the NDVI to estimate SM using a SVAT model simulation.

Fig. 2 represents a schematic description of the universal triangle method. The SVAT model (Carlson, 1986; Taconet et al., 1986) determines the relative shape of the isopleths of SM availability (M_0) within the triangle from zero M_0 (warm edge) on the right of the triangle to 1.0 M_0 (cold edge) of the triangle. To simplify the triangle concept and make the scatter plots from different images comparable, both the NDVI and the T_s are scaled using the following equations:

$$NDVI^* = \frac{NDVI - NDVI_0}{NDVI_s - NDVI_0} \tag{1}$$

$$T^* = \frac{(T - T_0)}{(T_s - T_0)} \tag{2}$$

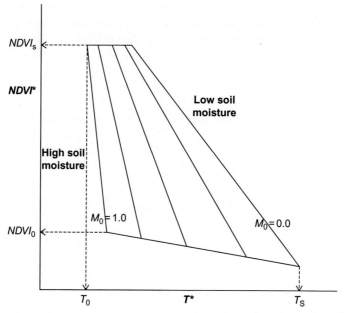

FIG. 2 The universal triangular model presented by Gillies and Carlson (1995) and Carlson et al. (1995). *(Adopted from Chauhan, N.S., Miller, S., Ardanuy, P., 2003. Spaceborne soil moisture estimation at high resolution: a microwave-optical/IR synergistic approach. Int. J. Remote Sens. 24, 4599–4622.)*

where $NDVI*$ (or FVC) is the scaled version of observed NDVI between bare soil ($NDVI_0$) to full vegetation cover ($NDVI_s$) and $T*$ is the scaled version of observed T_s between warmest (T_s, dry bare soil) to coldest corners (T_0, well-watered vegetation) of the triangle. The mathematical representation of the triangle model and the relationship between M_0, $NDVI*$ and $T*$ as presented by Gillies and Carlson (1995) is

$$M_0 = \sum_{i=0}^{i=n} \sum_{j=0}^{j=n} a_{ij} NDVI^{*(i)} T^{*(j)} \tag{3}$$

where i and j are the coefficients of the polynomial relationship between M_0, $NDVI*$, and $T*$. Carlson (2007) suggested a third-order polynomial relationship for M_0, $NDVI*$, and $T*$ as it gives a more reasonable representation of the data compared to a single polynomial one. Identification of the triangular shape in the pixel distribution requires a region with little topography and a large number of pixels over an area with a wide range of SM and FVC. Also, determination of the warm edge and the vegetation limits of bare soil and full cover assumes some subjectivity (Carlson, 2007). Triangle-derived SM values are called surface SM estimates and are related to the top 1–5 mm of the soil; however, remotely sensed derived SM values can be highly correlated with in situ measured SM values at a few to larger centimeters depth of soil. In many conditions, however, due to the rapid soil surface drying, the recharge of water from depth to surface will be reduced as a result of lower hydraulic conductivity in the dry surface layer. For this reason, remotely sensed derived SM values may correlate weakly to ground SM data at deeper layers (Carlson et al., 1994, 1995; Idso et al., 1975). In general, T_s is insensitive to SM content in the root zone. However, there are reports correlating RS-derived SM estimates to field SM data at different depths from a few millimeters to the root zone.

The triangular model was verified by Carlson et al. (1995) using NS001 multispectral scanner onboard a NASA C-130 aircraft for Mahantango Watershed in Pennsylvania. The derived remotely sensed SM values were compared with those obtained from the Push Broom Microwave Radiometer (PBMR) and in situ SM data at two depths (0.15 and 0.35 m). But low correlations were found between the three measurements. The authors attributed these discrepancies to the different soil depths evaluated for SM estimation using the microwave and triangle methods and that SM values of each sensor may be uncorrelated during conditions of rapid soil drying. Capehart and Carlson (1997) also verified the performance of this model using six satellite AVHRR (1 km resolution) images between April and July 1990 for the same study area as Carlson et al. (1995). AVHRR-derived SM estimates were compared with simulated SM data from a hydrological model at the depth of 0.5 cm and a poor relationship was observed. The relationship was improved by averaging over all studied regions (R^2 values ranging from 0.27 to 0.44 between retrieved SM values and field data). However, remotely sensed SM values were still reported to be underestimated using the triangle model. The poor correlation was attributed to the difference between both horizontal and vertical (soil depth) scale of the SM values obtained by the two methods. The process of rapid drying of the shallow surface soil layers after wetting by precipitation or irrigation also makes it difficult to relate satellite derived and field SM estimates. Gillies et al. (1997) suggested two major modifications to the triangle model to estimate SM. The transformation model suggested by them to determine the cold edge of the triangle demonstrated some improvement over the original triangle model with standard errors of 16% for SM availability. The retrieved SM estimation from airborne NS001 data were verified by SM field data at the depth of 0–5 cm from the First ISLSCP Field Experiment (FIFE) and MONSOON '90 field programs. Wang et al. (2007) applied the model suggested by Gillies and Carlson (1995) and Carlson et al. (1995) using 3 years of 1 km MODIS NDVI and T_s data to estimate SM in China using second-order polynomial equations. The MODIS-derived SM results were validated using the top 10 cm ground SM data. The result showed that MODIS NDVI and T_s were strongly correlated with the ground measured SM (R^2 values ranging from 0.42 to 0.95) and regression relationships were land cover and soil type dependent. A simple nonlinear interpolation function with fewer coefficients than those proposed by Carlson et al. (1994, 1995), Gillies and Carlson (1995) and Gillies et al. (1997) was suggested by Zhang et al. (2015) over arable, forest, and grassland land covers in North China Plain using MODIS NDVI and T_s products. The study used the energy balance equation to derive the theoretical limiting edges to reduce the uncertainties induced by the regressed maximum and minimum T_s determinations. Using the Noah land surface model (Noah LSM) simulations (Niu et al., 2011) under different types of atmospheric conditions and vegetation cover, the authors revealed nonlinear SM variations in the T_s/VI space, verifying the findings of Carlson et al. (1994, 1995). High estimation accuracy was observed between simulated SM data and the SM estimations from the established model with an R^2 higher than 0.90 and a root mean square error (RMSE) of approximately 0.02 $m^3\ m^{-3}$. The comparison between the retrieved SM content and the in situ measurements demonstrated that the R^2 ranged from 0.65 to 0.82 for 20 and 10 cm soil depths, respectively. The RMSE for 10-cm depth was found to be less than that at a 20-cm depth. However, the presented model suffers from some shortcomings as described by the authors.

Later, Chauhan et al. (2003) took a first step toward disaggregating passive microwave-derived SM to obtain high-resolution SM maps and suggested the application of the triangle model for SM estimation at higher spatial resolution. Using AVHRR and special sensor microwave imager (SSM/I) data for the Southern Great Plains (SGP) experiment region, they found good relationship between SM estimated from the triangle model and SM from SSM/I with RMSE of 5%. However, the authors excluded all highly vegetated pixels (NDVI above 0.4), which are those regions in the triangle model that the uncertainty in SM estimation would be higher. Piles et al. (2011) applied the triangle model and the method suggested by Chauhan et al. (2003) to downscale Soil Moisture and Ocean Salinity (SMOS) SM estimates using MODIS data.

Many researchers suggested several models based on a linear relationship between SM and T_s/VI or a simpler nonlinear relationship (Sandholt et al., 2002; Mallick et al., 2009; Wang et al., 2011; Zhang et al., 2015). One of the common approaches based on a linear relationship between SM and T_s/VI was the model suggested by Sandholt et al. (2002). The authors proposed the temperature vegetation dryness index (TVDI) to link T_s/VI space to SM using AVHRR data (Fig. 3). The TVDI is estimated using the following equation:

$$TVDI = \frac{T_s - T_{s_{min}}}{a + bNDVI - T_{s_{min}}} \tag{4}$$

where T_s is the observed surface temperature at a given pixel, $T_{s_{min}}$ is the minimum surface temperature in the T_s/NDVI space, defining the wet edge, NDVI is the observed NDVI, and a and b are parameters defining the dry edge, modeled as a linear fit to data $\left(T_{s_{max}} = a + bNDVI\right)$, where $T_{s_{max}}$ is the maximum surface temperature observation for a given NDVI. The TVDI for a given pixel in T_s/NDVI space is estimated from the ratio of line A to line B as presented in Fig. 3. Parameters a and b are estimated from an area large enough to represent the entire range of surface moisture content. The TVDI is lower for wet and

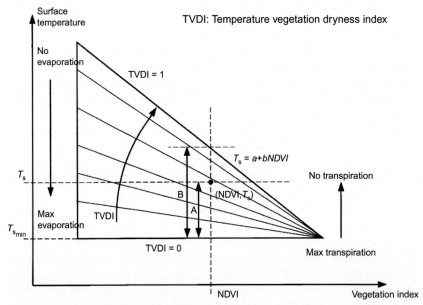

FIG. 3 Definition of TVDI. *(Reproduced by permission from Sandholt, I., Rasmussen, K., Andersen, J., 2002. A simple interpretation of the surface temperature/vegetation index space for assessment of surface moisture status. Remote Sens. Environ. 79, 213–224.)*

higher for dry conditions and varies between 0 and 1. TVDI represents relative SM rather its volumetric content. The wet edge is modeled as a horizontal line as shown in Fig. 3, which may cause an overestimation of TVDI at low NDVI values. Sandholt et al. (2002) applied NDVI and land surface temperature (LST) estimated from 24 cloud-free AVHRR 1 km imagery from 1990 (day of the year from 138 to 321) in northern Senegal to estimate TVDI (Eq. (4)). They present a good and clear demonstration of NDVI/T_s scatter plots and their changes over time for dry and rainy seasons. The TVDI results were compared with simulated SM values at root zone from the MIKE SHE hydrological model. The result showed that TVDI values were closely related to simulated SM ($R^2 = 0.70$). Also, similar spatial patterns were found in the TVDI and simulated SM maps. TVDI values closely correlated with simulated SM at a much higher depth (root zone) in this research. In conditions of extreme soil water deficiency remotely sensed SM value can be related to SM content at the root zone (Carlson et al., 1994). Several researchers also applied TVDI for SM estimation at deeper soil layers and at the root zone depth (Chen et al., 2011, 2015; Patel et al., 2009; Sun et al., 2012). A similar model to TVDI was suggested by Mallick et al. (2009) who introduced soil wetness index (SWI) to estimate volumetric SM from the T_s/NDVI space in cropped areas in four Indian states. The performance of the SWI was evaluated at field and landscape scales using Advanced Spaceborne Thermal Emission and Reflection Radiometer (ASTER) and MODIS data, respectively. The SM estimates from both satellites were verified by in situ SM data and the results were promising. The RMSEs of field-scale (ASTER) SM estimates were higher (0.039 m^3 m^{-3}) than that of the MODIS scale (0.033 m^3 m^{-3}). At both scales, the lowest error of SM estimates belonged to the NDVI range of 0.35–0.65. A comparison between SM from MODIS Aqua and the Advanced Microwave Scanning Radiometer for Earth Observing (AMSR-E) system was also conducted in the research. The results indicated good correlations between SM estimates for the two sensors with a 0.027 m^3 m^{-3} RMSE for FVCs less than 0.5. The model suffered from uncertainties similar to what was described by Sandholt et al. (2002) in addition to limitations caused by using only T_s in the model.

As the TVDI is based on only RS data making the index suitable for larger-scale applications. However, as it is assumed that the main source of variation in the TVDI is SM, air temperature is not considered in the model, which may increase the uncertainty of the TVDI for larger areas and higher vegetation covers. An inherent assumption when applying the TVDI is that Ta is constant for the subset or window over which the index is estimated. On the other hand, when using the TVDI to estimate SM status, heterogeneity of the Earth surfaces increases the uncertainty of the TVDI to estimate SM. Therefore, the TVDI is best applied in regions with little topography. Several researchers suggested various methods to improve the TVDI performance. For example, to correct for the effect of topography, Ran et al. (2005) used an approach to correct T_s with a digital elevation model (DEM) before constructing the AVHRR T_s/NDVI space. Hassan et al. (2007) proposed a correction method to use a DEM to infer local pressure from altitude and then transform surface temperature to potential temperature. A modified approach toward the TVDI concept, incorporating air temperature and a DEM to develop the improved TVDI (iTVDI) was also introduced by Rahimzadeh-Bajgiran et al. (2012) using MODIS imagery. Compared with the TVDI, results indicated that there were more statistically significant relationships between the iTVDI and recent precipitation and SM in the four studied land cover types. Stisen et al. (2008) and Wang et al. (2006) also suggested using the temporal variations of T_s (ΔT_s) instead of using only T_s in the temperature axis of the triangle model. Other limitations of the TVDI are well documented in Sandholt et al. (2002).

Wang et al. (2011) evaluated the capability of the WDI trapezoidal model developed by Moran et al. (1994) to estimate SM using SM and precipitation observations in the Walnut Gulch Experimental Watershed (WGEW) in Arizona. Using MODIS imagery the authors applied $T_s - T_a$ for the temperature axis and EVI (enhanced VI) for VI/FVC axis of the scatter plot to estimate WDI. In the WDI model, the four vertices of the trapezoid in Fig. 1A should be determined using meteorological data and the energy balance equation to define the wet and dry edges. The results showed that the WDI can capture temporal variations in surface SM well ($R = -0.92$ for the average of all sites in 10 days), but the capability of detecting spatial variation of SM was found to be poor for WGEW.

3.2 Models Based on T_s/VI and Empirical Equations

To estimate SM using these models, EF needs to be first estimated and then transformed into SM values using empirical equations. Various models were suggested to estimate EF and ET using remotely sensed T_s/VI space. Jiang and Islam (1999, 2001) estimated EF by using the T_s/VI concept to calculate the Priestley-Taylor parameter (Priestley and Taylor, 1972) and eventually ET using AVHRR data for the SGP. EF is defined as the ratio of ET to available energy and can be directly estimated from the last part of Eq. (5) (Jiang and Islam, 2001):

$$EF = \frac{LE}{R_n - G} = \phi \frac{\Delta}{\Delta + \gamma} \tag{5}$$

where LE is a representative of ET (Wm^{-2}), R_n is the net radiation (Wm^{-2}), G is the soil heat flux, and ϕ is the so-called Priestley-Taylor parameter, which is slightly different from the original Priestley-Taylor parameter (α) as α is generally applicable for wet surfaces whereas ϕ can be applied for a wide range of surface evaporative conditions (Jiang and Islam, 2001). Δ is the slope of saturated vapor pressure and air temperature (T_a) ($hPa\ K^{-1}$) and γ is the psychometric constant having the same dimension as Δ. The term $\Delta/(\Delta + \gamma)$, also called the air temperature control parameter, varying between 0.55 and 0.85 for air temperatures between 10 and 40°C is used to normalize the EF so that the seasonal variation of air temperature can be partly removed and better relationships between SM and EF can be established (Wang et al., 2006). Assuming that there are physical relationships between ϕ, SM, T_s, and VI or fraction, ϕ can be estimated through the T_s/VI space from RS data. In the Jiang and Islam (2001) model, ϕ can be estimated through the T_s/VI space from RS data using Fig. 4. The lower edge of the space (Wet edge = CD) is representative of wet land covers with maximum ET whereas the upper edge (Dry edge = AB) of the scatter plot represents dry land covers with reduced ET. The ϕ value ranges between zero (ϕ_{min}) for zero vegetation fraction with maximum temperature and 1.26 (ϕ_{max}) for maximum vegetation fraction and minimum temperature. The ϕ value for each pixel i is calculated by connecting point A to point E and extending it to point G. The ϕ value at point A is equal to ϕ_{min} and the wet edge has the maximum ϕ_{max}. Therefore, the length of line AG will be equal to $\phi_{max} - \phi_{min}$ whereas AE will be $\phi_i - \phi_{min}$. Using the similarity of triangles EFG and ACG, ϕ_i at any point in the scatter plot can be calculated as

$$\phi_i = \frac{T_{s_{max}} - T_{s_i}}{T_{s_{max}} - T_{s_{min}}}(\phi_{max} - \phi_{min}) + \phi_{min} \tag{6}$$

A two-step linear interpolation scheme will be used to estimate ϕ value for each of the pixels in an image (Jiang and Islam, 2001; Tang et al., 2010; Venturini et al., 2004; Wang et al., 2006). Finally, Eq. (5) is used to calculate EF values. Jiang and Islam (2001) obtained satisfactory estimation accuracies with a smaller number of input variables than the original Priestley-Taylor model.

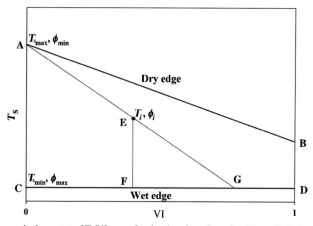

FIG. 4   Theoretical concept of T_s/VI space for ϕ estimation. *(Based on Wang, K.C., Li, Z.Q., Cribb, M., 2006. Estimation of evaporative fraction from a combination of day and night land surface temperatures and NDVI: a new method to determine the Priestley-Taylor parameter. Remote Sens. Environ. 102, 293–305.)*

The value of ϕ can be estimated from different combinations of temperatures and NDVI/FVC. Stisen et al. (2008) and Wang et al. (2006) used the difference between day and night surface temperature (ΔT_s) in the temperature axis of T_s/VI space in the Jiang and Islam (2001) model to estimate EF. Wang et al. (2006) suggested ΔT_s/NDVI spatial variations from the Aqua and Terra MODIS global daily products to estimate EF. They used ground-based measurements taken at the SGP of the United States to validate the MODIS EF retrievals. The EFs retrieved from the spatial variations of ΔT_s/NDVI show a distinct improvement over that retrieved from the T_s/NDVI model. The EF was retrieved with a mean relative accuracy of about 17% using the proposed ΔT_s/NDVI model. Using geostationary Meteosat Second Generation-Spinning Enhanced Visible Infrared Imager (MSG-SEVIRI) sensor data and nonlinear parameterization of ϕ values, Stisen et al. (2008) found RMSE and R^2 of 0.13 and 0.63 for remotely sensed EF and 41.45 W m^{-2} and 0.66 for remotely sensed ET, respectively.

A different approach was taken by Nishida et al. (2003) to estimate EF from RS data; their method used a two-source model considering a landscape to consist of bare soil and vegetation to estimate EF and the T_s/VI model was applied for the estimation of EF of bare soil. The algorithm is primarily driven by RS data to meet the authors' ultimate goal, which was the production of EF globally. However as mentioned by the authors the suggested algorithm is flexible to include ground ancillary data where these are available. The EF retrieved from AVHRR data were verified with actual observations of EF at AmeriFlux stations (standard error of 0.17 and R^2 of 0.71). Venturini et al. (2008) presented a new formula to derive EF and ET maps from remotely sensed data and tested and validated the proposed formula over the SGP using MODIS imagery. Estimates of ET showed an overall RMSE and bias of 33.89 and −10.96 Wm, respectively.

To estimate SM from EF the first step is to choose a sound model to estimate EF. All the above models presented a sound estimation of EF from remotely sensed data. Merlin et al. (2008, 2010, 2012) estimated SM from MODIS data using the EF/SM relationship. They applied the EF model suggested by Nishida et al. (2003) to estimate EF from MODIS data. Rahimzadeh-Bajgiran et al. (2013) applied the Jiang and Islam (2001) model and the improvements suggested by Wang et al. (2006) to retrieve EF from MODIS data.

The second step in SM estimation is to select a sound model that can best correlate remotely sensed EF to SM. A number of empirical models ranging from simple linear to more sophisticated exponential and cosine models (Crago, 1996; Deardorff, 1978; Jacquemin and Noilhan, 1990; Komatsu, 2003; Kondo et al., 1990; Lee and Pielke, 1992; Noilhan and Planton, 1989) have been suggested to correlate EF to SM at different soil depths. Merlin et al. (2010, 2012) applied the Komatsu (2003), Lee and Pielke (1992) model and the Noilhan and Planton (1989) models to downscale SM estimates from passive microwave sensors using MODIS data. The case study presented below (Rahimzadeh-Bajgiran et al., 2013) will summarize how SM can be estimated through EF using MODIS data.

4 CASE STUDY: ESTIMATION OF SM USING OPTICAL/TIR RS IN THE CANADIAN PRAIRIES

4.1 Introduction

Canadian Prairies are located in the northern region of the North American Great Plains and are characterized by semi-arid to sub-humid climate. The three Prairie Provinces (Alberta, Saskatchewan, and Manitoba) account for approximately 80% of Canada's cropland area, making them agriculturally, socioeconomically, and environmentally important. However, the area is prone to drought conditions due to its location on the leeward side of the Rocky Mountains, and its distance from the moderating influence of large water bodies. The southwestern Canadian prairies are semiarid, receiving on average around 350–400 mm of precipitation annually with the majority falling between April and June and are highly prone to frequent and severe droughts. In this case study SM was estimated using MODIS satellite imagery through evaluating the relationship between SM and EF. The EF was first estimated based on Jiang and Islam (2001) with modifications applicable to the study region. Three different T_s/VI combinations were evaluated for obtaining the Priestley-Taylor parameter, which was used for EF estimation. Two different empirical models to obtain SM from EF were also tested and discussed and correlation analysis was performed between estimated and field measured SM data to find the most accurate SM estimation approach.

4.2 Materials and Methods

4.2.1 Study Area and Data

The study area covers parts of Canadian Prairies over Saskatchewan and Alberta having a semiarid continental climate with cold winters and warm summers. Most of the precipitation occurs in May to July and averages around 350 mm annually. Satellite data used in this study were MODIS Terra (10:30 am) and Aqua (1:30 am, 1:30 pm) daily surface temperature having 1 km resolution (MOD11A1 and MYD11A1, respectively) and MODIS Terra 7-day NDVI composite product supplied by Agriculture and Agri-Food Canada (AAFC) retrieved from daily 250 m surface reflectance data (MOD09GQ). For this research, MODIS Terra-day and Aqua-day surface temperatures for 4 days during growing season (day of year (DOY) 137, 148, 232, and 260) and MODIS Aqua-night for 3 days (DOY 148, 232, and 260) were selected. This was done to obtain three cloud-free images per day for the days with available SM data and to cover a wide range of vegetation conditions and enable us to have a comparison between the performances of the three different approaches to estimate EF and SM on the same days.

Air temperature data were obtained from the National Center for Atmospheric Environmental Prediction (NCEP)/North America Regional Reanalysis (NARR) having 32 km spatial resolution and 3 h temporal resolution. Air temperature data at the height of 2 m were used in this research. NARR air temperature data revalidated with available climatic data in Saskatchewan (10 meteorological stations) in DOY 148 in 2008 and DOY 260 during MODIS Aqua and Terra satellite overpasses and found accuracies around $\pm 0.6°C$ ($R^2 = 0.97$, p-value < 0.0001). Wind speed data were also needed for one of the SM models Komatsu (2003) evaluated in this study. However, the NARR data have been shown to underestimate wind speed values (Rasmussen et al., 2011). Therefore, station-based meteorological data of wind speed were used for this research. All imagery used in this research was resampled to 500 m resolution before performing the analyses. A network of SM probes installed by the University of Guelph were used for the field data. The network consists of 16 stations with SM monitoring probes installed horizontally at depths of 5, 20, and 50 cm. For this analysis, surface SM from the 5 cm probe was used to match those estimated by the techniques described below. Further description of the region and instrumentation can be found in Champagne et al. (2010) and Magagi et al. (2013). SM data covered the period between May 1, 2008, and October 31, 2008, with no data available from the end of June to mid August. On average, 11 stations were operating for each day.

4.2.2 Methodology

To estimate SM using these models, EF needs to be first estimated and then transformed into SM values using empirical equations. The Jiang and Islam (1999, 2001) model was used in this research to estimate EF by using the schematic presented in Fig. 4 and Eqs. (5) and (6). Three different variables were examined to be used as the y axis (T_s axis); (1) $T_s - T_a$ where T_s is derived from MODIS Terra-day acquisition and T_a is NARR 3-h average air temperature around satellite overpass times; (2) $T_s - T_a$ where T_s is derived from MODIS Aqua-day acquisition; and (3) ΔT_s where the difference between day and night temperatures of MODIS Aqua was used. Terra-day, Aqua-day, and Aqua-night T_s values correspond to 10:30 am, 1:30 pm, and 1:30 am measurement times, respectively. The ΔT_s approach (Approach 3) was previously developed by Wang et al. (2006) and had provided improved relationships with field measured EF compared to the original model (Jiang and Islam, 2001) and therefore has been used here as a reference to compare with our approaches. The x axis of the scatter plot is vegetation fraction as calculated from NDVI values according to Gillies and Carlson (1995) where $NDVI_{min}$ and $NDVI_{max}$ were 0.11 and 0.87, respectively.

For SM estimation using satellite data it is crucial to select a model that can best correlate remotely sensed EF to SM. Here the Komatsu (2003) and Lee and

Pielke (1992) models were used for the estimation of SM from EF data. Komatsu (2003) suggested a relationship to describe EF based on SM:

$$EF = 1 - \exp(-\theta/\theta_c) \tag{7}$$

where EF is the evaporative fraction, θ is the volumetric SM, and θ_c is the characteristic volumetric water content depending on the soil type and wind speed calculated using Eq. (8).

$$\theta_c = \theta_{c0}(1 + \gamma/r_a) \tag{8}$$

where θ_{c0} and γ are two soil dependent parameters estimated from Komatsu (2003) and r_a is the aerodynamic resistance over bare soil determined using wind speed and soil roughness data and was considered to be 0.005 m for bare soil (Nishida et al., 2003).

The second model is suggested by Lee and Pielke (1992) as presented in Eq. (9):

$$EF = \begin{cases} \dfrac{1}{4}\left[1 - \cos\left(\dfrac{\theta}{\theta_{fc}}\pi\right)\right]^2 & \theta < \theta_{fc} \\ 1 & \theta \geq \theta_{fc} \end{cases} \tag{9}$$

where θ and θ_{fc} are the volumetric SM and the volumetric SM at field capacity, respectively. Assuming that all different soils should behave the same at some fixed soil-water characteristics, a reference point (soil field capacity) is used in this model. Based on the soil texture of the study area varying from silt loam and clay loam, θ_{fc} was assumed to be 0.3 and 0.35 vol/vol, respectively (Saxton and Rawls, 2006).

4.3 Results

4.3.1 Comparing EF Estimations Retrieved From Three Different Approaches

EF maps calculated using Eq. (6) for 2 days (DOY 148 in 2008 and DOY 260 in 2008) for the 60 km by 60 km covering the SM networks are presented in Fig. 5. For all days studied in this research, the Aqua-day approach resulted in the highest estimated EF values whereas those derived from the Aqua-DayNight approach exhibited the lowest values. The Terra-day approach gave intermediate values. All three approaches exhibited similar trends as is seen in Fig. 5 for both DOY 148 in 2008 and DOY 260 in 2008.

4.3.2 SM Estimation From Evaporative Fraction

A simulated representation of the Komatsu (2003) and Lee and Pielke (1992) models along with data points for Terra-day and Aqua-day estimated EF and field SM values for 2 days (DOY 137 in 2008 and DOY 148 in 2008) is presented in Fig. 6. It was observed that these data were better fitted to the cosine model (Eq. (9)) presented by Lee and Pielke (1992) whereas the Komatsu (2003)

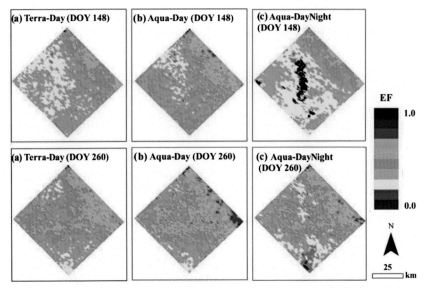

FIG. 5 Comparison of EF maps retrieved through three different approaches for DOY 148 in 2008 and DOY 260 in 2008. *(Reproduced from Rahimzadeh-Bajgiran, P., Berg, A., Champagne, C., Omasa, K., 2013. Estimation of soil moisture using optical/thermal infrared remote sensing in the Canadian Prairies. ISPRS J. Photogramm. Remote Sens. 83, 94–103.)*

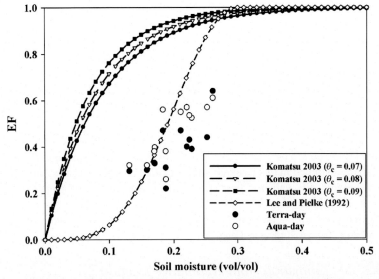

FIG. 6 Theoretical representation of EF/SM relationships used in the present study. θ_c in Komatsu (2003) for agricultural soil ranges between 0.07 and 0.09 based on varying wind speeds of 2–6 m/s. θ_{fc} is 0.3 vol/vol for Lee and Pielke (1992). *(Reproduced from Rahimzadeh-Bajgiran, P., Berg, A., Champagne, C., Omasa, K., 2013. Estimation of soil moisture using optical/thermal infrared remote sensing in the Canadian Prairies. ISPRS J. Photogramm. Remote Sens. 83, 94–103.)*

exponential model (Eq. (7)) resulted in very low SM estimations for all EF values. As explained by Komatsu (2003), this model is more appropriate for wet soils in thin layers and the shape of the curve approaches that of Lee and Pielke (1992) at higher soil depths. Therefore, Eq. (9) was used in this case study to estimate SM from EF.

4.3.3 Correlations Between Estimated SM and Field Data

Results of correlation analyses performed between estimated SM data obtained from Eq. (9) and field measurements at depths of 5 cm for the three different approaches in days DOY 137, 148, 232, and 260 in 2008 are tabulated in Table 1. Aqua-DayNight data for DOY 137 in 2008 were not available due to cloud contamination. All three approaches resulted in statistically significant correlations between estimated and observed SM for all days. For each day, estimated SM obtained from $T_s - T_a$ Aqua-day generally had better correlations with field data confirmed by higher coefficients of determination. SM values estimated from ΔT_s Aqua-DayNight resulted in the lowest R^2 values compared with the other two approaches. Coefficients of determination were found to be higher for DOY 137 and DOY 148 as compared with the other 2 days. This can be

TABLE 1 Coefficients of Determination (R^2), Statistical Significance (p Value), and Average Root Mean Square Errors ($RMSE_{ave}$) for Each Day and Satellite Data Combination

Approach	Satellite Data	DOY	R^2	p Value	$RMSE_{ave}$ (vol/vol%)
$T_s - T_a$	Terra-day	137	0.66	0.0025	6.9
	Aqua-day	137	0.67	0.0019	6.4
ΔT_s	Aqua-DayNight	137	–	–	–
$T_s - T_a$	Terra-day	148	0.77	0.0004	6.5
	Aqua-day	148	0.77	0.0004	6.2
ΔT_s	Aqua-DayNight	148	0.71	0.0012	7.2
$T_s - T_a$	Terra-day	232	0.42	0.0425	4.6
	Aqua-day	232	0.48	0.0259	4.7
ΔT_s	Aqua-DayNight	232	0.42	0.0417	3.9
$T_s - T_a$	Terra-day	260	0.57	0.0046	4.4
	Aqua-day	260	0.61	0.0025	5.0
ΔT_s	Aqua-DayNight	260	0.50	0.0101	4.6

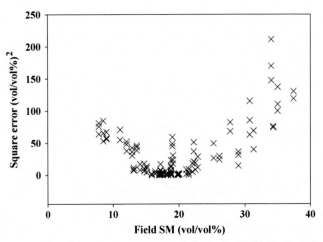

FIG. 7 Square errors of the estimation of SM as a function of observed SM values. *(Reproduced from Rahimzadeh-Bajgiran, P., Berg, A., Champagne, C., Omasa, K., 2013. Estimation of soil moisture using optical/thermal infrared remote sensing in the Canadian Prairies. ISPRS J. Photogramm. Remote Sens. 83, 94–103.)*

attributed to the amount of vegetation fraction that was the lowest for those days and the highest for DOY 232 in 2008, where the lowest R^2 values were obtained.

Average RMSE values for the three approaches of estimations and various days ranged between 3.9 and 6.9 vol/vol%. However, as presented in Fig. 7 where square errors for all days and approaches are plotted versus field SM data, the highest errors of estimation correspond to high and low SM values where the number of samples were limited. Intermediate SM values in the range of 10–20 vol/vol% resulted in low RMSE (around 3.0 vol/vol%) and consequently more accurate estimations. Further work to improve the accuracy of existing empirical models to convert EF to SM or development of remote sensing based empirical models can enhance the overall accuracy of SM estimation over the entire range of SM.

5 SUMMARY AND FUTURE OUTLOOK

Regardless of differences in complexity and assumptions, all T_s/VI methods presented above for the estimation of SM have the following characteristics in common:

(1) The models can be affected by atmospheric conditions and cloud contamination.
(2) The performance and accuracy of the models are lower over dense vegetation cover.
(3) The models are only able to estimate top surface SM due to the passive optical/thermal sensors minimal surface penetration.

Other common sources of uncertainty to estimate SM from T_s/VI models as summarized by Sandholt et al. (2002) are (1) The effect of satellite viewing angle on T_s and VI retrieval; (2) Potential error in T_s estimation due to emissivity effect; (3) Influence of SM from deeper layers on surface SM; and (4) Effect of T_s and VI dependence on land cover types.

While methods such as Carlson et al. (1995), Gillies and Carlson (1995), and Gillies et al. (1997) suggest nonlinear relationships between SM, T_s, and VI, many others assume SM to be changing linearly in the T_s/VI space. However, it seems that as also studied and confirmed by several researchers, the SM is nonlinearly related to T_s and linear assumption of the relationship will be too much of a simplification of the triangle model, leading to SM estimation inaccuracy. This simplification might, however, be inevitable should a more practically feasible model be desired. Also, $T_s - T_a$ or ΔT_s (e.g., T_s from MODIS Aqua and MODIS Terra) seems to be a better alternative than using only T_s in the temperature axis of the T_s/VI space. Again, using only the T_s in the temperature axis is another simplification to make modeling more practical. Finally, although NDVI is a common VI to be used for the triangle model, many researches recommend using FVC for the vegetation axis of the model. Other vegetation indices such as SAVI and EVI can be potentially used as well.

Passive optical/thermal sensors are increasingly being enhanced with the introduction of newer manufacturing technologies and sensor designs. Comparing the quality of imagery obtained from MODIS and AVHRR, for instance, confirms the advantages of MODIS data over AVHRR data. These include higher spatial resolutions; narrower spectral bandwidths, which can help avoid atmospheric water vapor influences; more consistent viewing angles; and onboard calibration for the optical channels making them more stable over time (for AVHRR data there exists different sensor calibrations and drift between NOAA satellites). More recent SM estimations using newer satellite images tend to provide more reliable estimations promising a better outlook for the future of optical/thermal SM estimation methodologies. However, it seems to be imperative to evaluate the performance of various T_s/VI approaches over a range of geographical settings with different ecological and environmental characteristics to establish a more comprehensive method that can estimate SM with reasonable accuracy.

REFERENCES

Anyamba, A., Tucker, C.J., 2005. Analysis of Sahelian vegetation dynamics using NOAA-AVHRR NDVI data from 1981–2003. J. Arid Environ. 63 (3), 596–614.

Bartholic, J.F., Wiegand, C.L., Namken, L.N., 1972. Aerial thermal scanner to determine temperatures of soils and of crop canopies differing in water stress. Agron. J. 64, 603–608.

Cai, G., Xue, Y., Hu, Y., Wang, Y., Guo, J., Luo, Y., Wu, C., Zhong, S., Qi, S., 2007. Soil moisture retrieval from MODIS data in Northern China Plain using thermal inertia model. Int. J. Remote Sens. 28, 3567–3581.

Capehart, W.J., Carlson, T.N., 1997. Decoupling of surface and near-surface soil water content: a remote sensing perspective. Water Resour. Res. 33, 1383–1395.

Carlson, T.N., 1986. Regional-scale estimates of surface moisture availability and thermal inertia using remote thermal measurements. Remote Sens. Rev. 1 (2), 197–247.

Carlson, T.N., 2007. An overview of the "triangle method" for estimating surface evapotranspiration and soil moisture from satellite imagery. Sensors 7, 1612–1629.

Carlson, T.N., Perry, E.M., Schmugge, T.J., 1990. Remote estimation of soil moisture availability and fractional vegetation cover for agricultural fields. Agr. Forest. Meteorol. 52, 45–69.

Carlson, T.N., Gillies, R.R., Perry, E.M., 1994. A method to make use of thermal infrared temperature and NDVI measurements to infer surface soil water content and fractional vegetation cover. Remote Sens. Rev. 9, 161–173.

Carlson, T.N., Gillies, R.R., Schmugge, T.J., 1995. An interpretation of methodologies for indirect measurement of soil-water content. Agr. Forest. Meteorol. 77, 191–205.

Chakrabarti, S., Bongiovanni, T., Judge, J., Nagarajan, K., Principe, J.C., 2015. Downscaling satellite-based soil moisture in heterogeneous regions using high-resolution remote sensing products and information theory: a synthetic study. IEEE Trans. Geosci. Remote Sens. 53 (1), 85–101.

Champagne, C., Berg, A.A., Belanger, J., McNairn, H., de Jeu, R., 2010. Evaluation of soil moisture derived from passive microwave remote sensing over agricultural sites in Canada using ground-based soil moisture monitoring networks. Int. J. Remote Sens. 31, 3669–3690.

Chauhan, N.S., Miller, S., Ardanuy, P., 2003. Spaceborne soil moisture estimation at high resolution: a microwave-optical/IR synergistic approach. Int. J. Remote Sens. 24, 4599–4622.

Chen, J., Wang, C.Z., Jiang, H., Mao, L.X., Yu, Z.R., 2011. Estimating soil moisture using temperature-vegetation dryness index (TVDI) in the Huang-huai-hai (HHH) plain. Int. J. Remote Sens. 32, 1165–1177.

Chen, C., Valdez, M.C., Chang, N., Chang, L., Yuan, P., 2014. Monitoring spatiotemporal surface soil moisture variations during dry seasons in central America with multisensor cascade data fusion. IEEE J. Sel. Topics Appl. Earth Observ. Remote Sens. 7 (11), 4340–4355.

Chen, S., Wen, Z., Jiang, H., Zhao, Q., Zhang, X., Chen, Y., 2015. Temperature vegetation dryness index estimation of soil moisture under different tree species. Sustainability 7, 11401–11417.

Crago, R.D., 1996. Comparison of the evaporative fraction and the Priestley-Taylor alpha for parameterizing daytime evaporation. Water Resour. Res. 32, 1403–1409.

Deardorff, J.W., 1978. Efficient prediction of ground surface-temperature and moisture, with inclusion of a layer of vegetation. J. Geophys. Res. Oceans Atmos. 83, 1889–1903.

Friedl, M.A., Davis, F.W., 1994. Sources of variation in radiometric surface temperature over a tall-grass prairie. Remote Sens. Environ. 48, 1–17.

Gillies, R.R., Carlson, T.N., 1995. Thermal remote-sensing of surface soil-water content with partial vegetation cover for incorporation into climate-models. J. Appl. Meteorol. 34, 745–756.

Gillies, R.R., Carlson, T.N., Cui, J., Kustas, W.P., Humes, K.S., 1997. A verification of the 'triangle' method for obtaining surface soil water content and energy fluxes from remote measurements of the normalized difference vegetation index (NDVI) and surface radiant temperature. Int. J. Remote Sens. 18, 3145–3166.

Goetz, S.J., 1997. Multi-sensor analysis of NDVI, surface temperature and biophysical variables at a mixed grassland site. Int. J. Remote Sens. 18 (1), 71–94.

Goward, S.N., Cruickshanks, G.D., Hope, A.S., 1985. Observed relation between thermal emission and reflected spectral radiance of a complex vegetated landscape. Remote Sens. Environ. 18 (2), 137–146.

Goward, S.N., Xue, Y., Czajkowski, K.P., 2002. Evaluating land surface moisture conditions from the remotely sensed temperature/vegetation index measurements: an exploration with the simplified simple biosphere model. Remote Sens. Environ. 79 (2–3), 225–242.

Hassan, Q.K., Bourque, C.P.A., Meng, F.-R., Cox, R.M., 2007. A wetness index using terrain-corrected surface temperature and normalized difference vegetation index derived from standard MODIS products: an evaluation of its use in a humid forest-dominated region of eastern Canada. Sensors 7, 2028–2048.

Heilman, J.L., Kanemasu, E.T., Rosenberg, N.J., Blad, B.L., 1976. Thermal scanner measurement of canopy temperatures to estimate evapotranspiration. Remote Sens. Environ. 5, 137–145.

Idso, S.B., Schmugge, T.I., Jackson, R., Reginato, R.J., 1975. The utility of surface temperature measurements for remote sensing of soil water studies. J. Geophys. Res. 80 (21), 3044–3049.

Idso, S.B., Jackson, R.D., Pinter, P.J., Reginato, R.J., Hatfield, J.L., 1981. Normalizing the stress-degree-day parameter for environmental variability. Agric. Meteorol. 24, 45–55.

Jackson, R.D., Idso, S.B., Reginato, R.J., Pinter, P.J., 1981. Canopy temperature as a crop water-stress indicator. Water Resour. Res. 17, 1133–1138.

Jacquemin, B., Noilhan, J., 1990. Sensitivity study and validation of a land surface parameterization using the Hapex-Mobilhy data set. Bound.-Layer Meteorol. 52, 93–134.

Jahan, N., Gan, T.Y., 2015. Lakshmi, V. (Ed.), Soil moisture retrieval from microwave (RADARSAT-2) and optical remote sensing (MODIS) data using artificial intelligence techniques. Remote Sensing of the Terrestrial Water Cycle, Geophysical Monograph 206. first ed. American Geophysical Union, John Wiley & Sons, Inc., pp. 255–275.

Jiang, L., Islam, S., 1999. A methodology for estimation of surface evapotranspiration over large areas using remote sensing observations. Geophys. Res. 26 (17), 2773–2776.

Jiang, L., Islam, S., 2001. Estimation of surface evaporation map over southern Great Plains using remote sensing data. Water Resour. Res. 37, 329–340.

Kogan, F.N., 1990. Remote-sensing of weather impacts on vegetation in nonhomogeneous areas. Int. J. Remote Sens. 11 (8), 1405–1419.

Komatsu, T.S., 2003. Toward a robust phenomenological expression of evaporation efficiency for unsaturated soil surfaces. J. Appl. Meteorol. 42, 1330–1334.

Kondo, J., Saigusa, N., Sato, T., 1990. A parameterization of evaporation from bare soil surfaces. J. Appl. Meteorol. 29, 385–389.

Lee, T.J., Pielke, R.A., 1992. Estimating the soil surface specific-humidity. J. Appl. Meteorol. 31, 480–484.

Li, Z.L., Tang, R., Wan, Z., Bi, Y., Zhou, C., Tang, B., Yan, G., Zhang, X., 2009. A review of current methodologies for regional evapotranspiration estimation from remotely sensed data. Sensors 9 (5), 3801–3853.

Liu, W.T.H., Massambani, O., Nobre, C.A., 1994. Satellite recorded vegetation response to drought in Brazil. Int. J. Climatol. 14 (3), 343–354.

Liu, W., Baret, F., Gu, X., Tong, Q., Zheng, L., Zhang, B., 2002. Relating soil surface moisture to reflectance. Remote Sens. Environ. 81, 238–246.

Magagi, R., Berg, A.A., Goita, K., Belair, S., Jackson, T.J., Toth, B., Walker, A., McNairn, H., O'Neill, P.E., Moghaddam, M., Gherboudj, I., Colliander, A., Cosh, M.H., Burgin, M., Fisher, J.B., Kim, S.-B., Mladenova, I., Djamai, N., Rousseau, L.-P.B., Belanger, J., Shang, J., Merzouki, A., 2013. Canadian experiment for soil moisture in 2010 (CanEx-SM10): overview and preliminary results. IEEE Trans. Geosci. Remote Sens. 51, 347–363.

Mallick, K., Bhattacharya, B.K., Patel, N.K., 2009. Estimating volumetric surface moisture content for cropped soils using a soil wetness index based on surface temperature and NDVI. Agr. Forest. Meteorol. 149, 1327–1342.

Malo, A.R., Nicholson, S.E., 1990. A study of rainfall and vegetation dynamics in the African Sahel using normalized difference vegetation index. J. Arid Environ. 19 (1), 1–24.

Mattar, C., Wigneron, J.P., Sobrino, J.A., Novello, N., Calvet, J.C., Albergel, C., Richaume, P., Mialon, A., Guyon, D., Jiménez Muñoz, J.C., Kerr, Y., 2012. A combined optical-microwave method to retrieve soil moisture over vegetated areas. IEEE Trans. Geosci. Remote Sens. 50 (5), 1404–1413.

Merlin, O., Chehbouni, G., Kerr, Y., Njoku, E.G., Entekhabi, D., 2005. A combined modeling and multi-spectral/multi-resolution remote sensing approach for disaggregation of surface soil moisture: application to SMOS configuration. IEEE Trans. Geosci. Remote Sens. 43, 2036–2050.

Merlin, O., Walker, J.P., Chehbouni, A., Kerr, Y., 2008. Towards deterministic downscaling of SMOS soil moisture using MODIS derived soil evaporative efficiency. Remote Sens. Environ. 112, 3935–3946.

Merlin, O., Al Bitar, A., Walker, J.P., Kerr, Y., 2010. An improved algorithm for disaggregating microwave-derived soil moisture based on red, near-infrared and thermal-infrared data. Remote Sens. Environ. 114, 2305–2316.

Merlin, O., Rudiger, C., Al, Bitar A., Richaume, P., Walker, J., Kerr, Yann H., 2012. Disaggregation of SMOS soil moisture in southeastern Australia. IEEE Trans. Geosci. Remote Sens. 50 (5), 1556–1571.

Moran, M.S., Clarke, T.R., Inoue, Y., Vidal, A., 1994. Estimating crop water-deficit using the relation between surface-air temperature and spectral vegetation index. Remote Sens. Environ. 49, 246–263.

Moran, M.S., Peters-Lidard, C.D., Watts, J.M., McElroy, S., 2004. Estimating soil moisture at the watershed scale with satellite-based radar and land surface models. Can. J. Remote. Sens. 30, 805 826.

Muller, E., Décamps, H., 2000. Modeling soil moisture-reflectance. Remote Sens. Environ. 76, 173–180.

Nemani, R.R., Running, S.W., 1989. Estimation of regional surface resistance to evapotranspiration from NDVI and thermal IR AVHRR data. J. Appl. Meteorol. 28, 276–284.

Nemani, R., Pierce, L., Running, S., Goward, S., 1993. Developing satellite-derived estimates of surface moisture status. J. Appl. Meteorol. 32, 548–557.

Nishida, K., Nemani, R.R., Running, S.W., Glassy, J.M., 2003. An operational remote sensing algorithm of land surface evaporation. J. Geophys. Res.-Atmos. 108. D94270.

Niu, G.-Y., et al., 2011. The community Noah land surface model with multiparameterization options (Noah-MP): 1. Model description and evaluation with local-scale measurements. J. Geophys. Res. 116. http://dx.doi.org/10.1029/2010JD015139. D12109.

Noilhan, J., Planton, S., 1989. A simple parameterization of land surface processes for meteorological models. Mon. Weather Rev. 117, 536–549.

Omasa, K., Croxdale, J.G., 1991. Image analysis of stomatal movements and gas exchange. In: Hader, D.P. (Ed.), Image Analysis in Biology. CRC Press, Boca Raton, pp. 171–193.

Omasa, K., Hashimoto, Y., Aiga, I., 1981a. A quantitative analysis of the relationship between O_3 sorption and its acute effect on plant leaves using image instrumentation. Environ. Control. Biol. 19 (3), 85–92.

Omasa, K., Hashimoto, Y., Aiga, I., 1981b. A quantitative analysis of the relationship between SO_2 or NO_2 sorption and their acute effect on plant leaves using image instrumentation. Environ. Control. Biol. 19 (2), 59–67.

Owe, M., de Jeu, R., Holmes, T., 2008. Multisensor historical climatology of satellite-derived global land surface moisture. J. Geophys. Res. Earth Surface. 113. F01002.

Patel, N.R., Anapashsha, R., Kumar, S., Saha, S.K., Dadhwal, V.K., 2009. Assessing potential of MODIS derived temperature/vegetation condition index (TVDI) to infer soil moisture status. Int. J. Remote Sens. 30 (1), 23–39.

Petropoulos, G., Carlson, T.N., Wooster, M.J., Islam, S., 2009. A review of T_s/VI remote sensing based methods for the retrieval of land surface energy fluxes and soil surface moisture. Prog. Phys. Geogr. 33, 224–250.

Piles, M., Camps, A., Vall-Llossera, M., Corbella, I., Panciera, R., Ruediger, C., Kerr, Y.H., Walker, J., 2011. Downscaling SMOS-derived soil moisture using MODIS visible/infrared data. IEEE Trans. Geosci. Remote Sens. 49, 3156–3166.

Piles, M., Petropoulos, G.P., Ireland, G., Sanchez, N., in press. A novel method to retrieve soil moisture at high spatio-temporal resolution based on the synergy of SMOS and MSG SEVIRI observations. Remote Sens. Environ.

Price, J.C., 1990. Using spatial context in satellite data to infer regional scale evapotranspiration. IEEE Trans. Geosci. Remote Sens. 28, 940–948.

Priestley, C.H.B., Taylor, R.J., 1972. Assessment of surface heat-flux and evaporation using large-scale parameters. Mon. Weather Rev. 100, 81–92.

Rahimzadeh-Bajgiran, P., Shimizu, Y., Hosoi, F., Omasa, K., 2009. MODIS vegetation and water indices for drought assessment in semi-arid ecosystems of Iran. J. Agric. Meteor. 65 (4), 349–355.

Rahimzadeh-Bajgiran, P., Omasa, K., Shimizu, Y., 2012. Comparative evaluation of the vegetation dryness index (VDI), the temperature vegetation dryness index (TVDI) and the improved TVDI (iTVDI) for water stress detection in semi-arid regions of Iran. ISPRS J. Photogramm. Remote Sens. 68, 1–12.

Rahimzadeh-Bajgiran, P., Berg, A., Champagne, C., Omasa, K., 2013. Estimation of soil moisture using optical/thermal infrared remote sensing in the Canadian Prairies. ISPRS J. Photogramm. Remote Sens. 83, 94–103.

Ran, Q., Zhang, Z.X., Zhou, Q.B., Wang, Q., 2005. Soil moisture derivation in China using AVHRR data and analysis of its affecting factors. In: IGARSS 2005: IEEE International Geoscience and Remote Sensing Symposium, Vols 1–8, Proceedings, pp. 4497–4500.

Rasmussen, D.J., Holloway, T., Nemet, G.F., 2011. Opportunities and challenges in assessing climate change impacts on wind energy—a critical comparison of wind speed projections in California. Environ. Res. Lett. 6.

Rundquist Jr., B.C., Harrington, J.A., 2000. The effects of climatic factors on vegetation dynamics of tallgrass and shortgrass cover. Geocarto Intern. 15 (3), 31–36.

Sandholt, I., Rasmussen, K., Andersen, J., 2002. A simple interpretation of the surface temperature/vegetation index space for assessment of surface moisture status. Remote Sens. Environ. 79, 213–224.

Saxton, K.E., Rawls, W.J., 2006. Soil water characteristic estimates by texture and organic matter for hydrologic solutions. Soil Sci. Soc. Am. J. 70, 1569–1578.

Schmugge, T., 1978. Remote-sensing of surface soil-moisture. J. Appl. Meteorol. 17, 1549–1557.

Smith, R.C.G., Choudhury, B.J., 1991. Analysis of normalized difference and surface temperature observations over southeastern Australia. Int. J. Remote Sens. 12 (10), 2021–2044.

Stisen, S., Sandholt, I., Norgaard, A., Fensholt, R., Jensen, K.H., 2008. Combining the triangle method with thermal inertia to estimate regional evapotranspiration – applied to MSG-SEVIRI data in the Senegal River basin. Remote Sens. Environ. 112, 1242–1255.

Sun, L., Sun, R., Li, X., Liang, S., Zhang, R., 2012. Monitoring surface soil moisture status based on remotely sensed surface temperature and vegetation index information. Agr. Forest. Meteorol. 166, 175–187.

Taconet, O., Carlson, T.N., Bernard, R., Vidal-Madjar, D., 1986. Evaluation of a surface vegetation parameterization using satellite measurements of surface temperature. J. Clim. Appl. Meteorol. 25 (11), 1752–1767.

Tang, R., Li, Z.-L., Tang, B., 2010. An application of the Ts-VI triangle method with enhanced edges determination for evapotranspiration estimation from MODIS data in arid and semi-arid regions: implementation and validation. Remote Sens. Environ. 114, 540–551.

Temimi, M., Leconte, R., Chaouch, N., Sukumal, P., Khanbilvardi, R., Brissette, F., 2010. A combination of remote sensing data and topographic attributes for the spatial and temporal monitoring of soil wetness. J. Hydrol. 388 (1–2), 28–40.

Tucker, C.J., Holben, B.N., Elgin, J.H., Mcmurtrey, J.E., 1981. Remote-sensing of total dry-matter accumulation in winter-wheat. Remote Sens. Environ. 11 (3), 171–189.

Venturini, V., Bisht, G., Islam, S., Jiang, L., 2004. Comparison of evaporative fractions estimated from AVHRR and MODIS sensors over south Florida. Remote Sens. Environ. 93, 77–86.

Venturini, V., Islam, S., Rodriguez, L., 2008. Estimation of evaporative fraction and evapotranspiration from MODIS products using a complementary based model. Remote Sens. Environ. 112 (1), 132–141.

Verstraeten, W.W., Veroustraete, F., van der Sande, C.J., Grootaers, I., Feyen, J., 2006. Soil moisture retrieval using thermal inertia, determined with visible and thermal spaceborne data, validated for European forests. Remote Sens. Environ. 101, 299–314.

Verstraeten, W.W., Veroustraete, F., Feyen, J., 2008. Assessment of evapotranspiration and soil moisture content across different scales of observation. Sensors 8, 70–117.

Vicente-Serrano, S.M., Pons-Fernandez, X., Cuadrat-Prats, J.M., 2004. Mapping soil moisture in the central Ebro river valley with Landsat and NOAA satellite imagery: a comparison with meteorological data. Int. J. Remote Sens. 25 (20), 4325–4350.

Vicente-Serrano, S.M., Cuadrat-Prats, J.M., Romo, A., 2006. Early prediction of crop production using drought indices at different time-scales and remote sensing data: application in the Ebro Valley (north-east Spain). Int. J. Remote Sens. 27 (3), 511–518.

Wang, L., Qu, J.J., 2009. Satellite remote sensing applications for surface soil moisture monitoring: a review. Front. Earth Sci. China 3, 237–247.

Wang, J., Price, K.P., Rich, P.M., 2001. Spatial patterns of NDVI in response to precipitation and temperature in the central Great Plains. Int. J. Remote Sens. 22 (18), 3827–3844.

Wang, K.C., Li, Z.Q., Cribb, M., 2006. Estimation of evaporative fraction from a combination of day and night land surface temperatures and NDVI: a new method to determine the Priestley-Taylor parameter. Remote Sens. Environ. 102, 293–305.

Wang, L., Qu, J.J., Zhang, S., Hao, X., Dasgupta, S., 2007. Soil moisture estimation using MODIS and ground measurements in eastern China. Int. J. Remote Sens. 28 (6), 1413–1418.

Wang, W., Huang, D., Wang, X.G., Liu, Y.R., Zhou, F., 2011. Estimation of soil moisture using trapezoidal relationship between remotely sensed land surface temperature and vegetation index. Hydrol. Earth Syst. Sci. 15, 1699–1712.

Zhang, D., Tang, R., Tang, B.-H., Wu, H., Li, Z.-L., 2015. A simple method for soil moisture determination from LST–VI feature space using nonlinear interpolation based on thermal infrared remotely sensed data. IEEE J. Sel. Topics Appl. Earth Observ. Remote Sens. 8 (2), 638–648.

Chapter 4

Optical/Thermal-Based Techniques for Subsurface Soil Moisture Estimation

M. Holzman* and R. Rivas[†]

*Consejo Nacional de Investigaciones Científicas y Técnicas – Instituto de Hidrología de Llanuras "Dr. Eduardo J. Usunoff", Rep. Italia 780, Azul, Buenos Aires, Argentina, [†]Comisión de Investigaciones Científicas de la provincia de Buenos Aires – Instituto de Hidrología de Llanuras "Dr. Eduardo J. Usunoff", Campus Universitario, Tandil, Buenos Aires, Argentina

1 INTRODUCTION

Soil moisture (SM) is a crucial variable of the soil-vegetation system that controls vegetation productivity, land surface and atmospheric interaction, and ecosystem water cycling. The assessment of such a variable through traditional methods (e.g., lysimeters, SM probes, in situ water balance) and the subsequent integration at regional scale is complex given the high variability due to topography, water table depth, rainfall, and vegetation cover (Petropoulos et al., 2014). Diverse remote sensing techniques of SM assessment have been developed over the last few decades and the results have been integrated to hydrological models and vegetation condition estimation (Bhattacharya et al., 2011; Chakrabarti et al., 2014; Crow et al., 2008; Crow et al., 2005; Holzman et al., 2014a). Between these techniques, microwave data have a direct relation with surface SM (2–5 cm) and can retrieve information under cloudy conditions (Li et al., 2010). Passive microwave provides reliable estimates, particularly over bare soil surfaces with high temporal resolution. The main disadvantages are the influence of vegetation cover and surface roughness on the sensed signal and the coarse spatial resolution (20–40 km), which makes them useful to monitor great basins. Active microwave provides fine spatial resolution, but the temporal resolution is coarse and the sensed signal is influenced by vegetation amount and surface roughness (Petropoulos et al., 2015). Finally, subsurface SM plays an important role in vegetation water cycling through the extraction from deep soil

Satellite Soil Moisture Retrieval. http://dx.doi.org/10.1016/B978-0-12-803388-3.00004-8

layer and generally microwave-based SM estimation is considered shallow for agricultural applications (Holzman et al., 2014a).

A group of methods that can provide a more detailed (250–1000 m) estimation of water status in the soil-vegetation system is based on the energy balance equation

$$R_n = LE + H + G \tag{1}$$

where R_n is the net radiation, LE is the latent heat flux (evapotranspiration (ET)), H is the sensible heat flux, and G is the soil heat flux. The lower LE, the higher H according to the more energy available for sensible heating of the surface or land surface temperature (LST). Depending on root zone SM availability, vegetation exerts physiological control over water loss through the stomatal resistance to transpiration (r_c) (Fig. 1). In this way, over vegetation cover the partitioning of the available incident energy ($R_n - G$) into sensible and latent heat fluxes is mostly governed by vegetation regulation (Mallick et al., 2009). Under low SM availability, most of the energy incident results in high H (LST) increasing the air and canopy temperature gradient (Fig. 1).

The LSTs have been widely used to evaluate the upward transport of sensible heat and then derive the energy involved in the ET process (Boulet et al., 2007; Friedl, 2002; Moran et al., 1994; Nutini et al., 2014; Rivas and

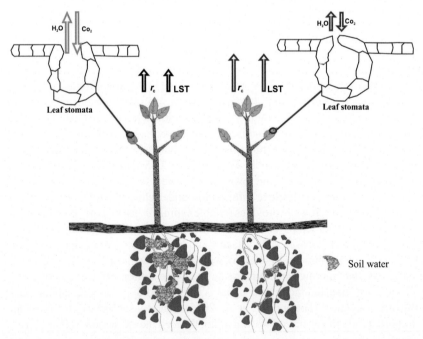

FIG. 1 Simplified scheme of the relationship between stomatal resistance (r_c) and LST under high (left) and low (right) root zone soil moisture availability.

Caselles, 2004). Thus, SM availability in root zone can be indirectly estimated over vegetated areas through the thermal response of the vegetation canopy to limitations in soil water without data of antecedent precipitation or SM storage capacity. While LST is an indicator of sensible heat flux, vegetation indices (VI) indicate the proportion of vegetation and its photosynthetic capacity. Hence, optical/thermal-based techniques allow the evaluation of subsurface water status from full to partially vegetated surfaces (see Section 2.2 about limitations of methods).

Several works have shown the high correlation between optical/thermal data and surface SM (10–20 cm depth) (Han et al., 2010; Holzman et al., 2014a; Mallick et al., 2009; Sandholt et al., 2002). In China, Han et al. (2010) found a coefficient of determination of 0.76 between surface SM and the temperature vegetation dryness index (TVDI) obtained from a moderate resolution imaging spectroradiometer (MODIS)/Terra normalized difference vegetation index (NDVI) and LST. Sandholt et al. (2002) reported consistent spatial agreement between TVDI and simulated SM in semiarid areas of Senegal. In Holzman et al. (2014b) we found $R^2 > 0.6$ between LST/VI monthly composite and SM at 10–20 cm depth in humid areas of the Argentine Pampas. These works, among others, show the consistency of LST/VI methods to estimate SM. However, depending on soil and vegetation characteristics, a decoupling between surface and subsurface SM could occur and few works have been done to analyze such process (Carlson et al., 1995; Holzman et al., 2014b). In this sense, the mismatch between sampling depth of conventional SM measurements and the depth contributing to satellite signal is a frequent explanation for deviations between estimated and observed data in methods such as microwave (≈ 5–10 cm depth) or optical data (≈ 1 cm penetration power). However, LST and optical data combination (low penetration power) has the ability to indirectly reflect deep SM through the relationship between canopy condition and root zone SM.

Different authors have used optical and thermal data to evaluate root zone SM (Akther and Hassan, 2011; Carlson et al., 1994; Chang et al., 2012; Crow et al., 2008; Holzman et al., 2014b; Li et al., 2010). Akther and Hassan (2011) analyzed the temporal dynamics of deep SM (up to 100 cm depth) with 8 days of composite data from MODIS in an agricultural and forest region of Canada. In Holzman et al. (2014b) we evaluate deep SM through LST and VI from MODIS/Aqua showing that the ability of the method depends on factors like soil and vegetation type, the heterogeneity of land wetness, and fractional vegetation cover. It should be noted that this approach needs to be calibrated to obtain direct values of SM content. However, it represents an attractive and simple procedure to evaluate spatial and temporal fluctuations of water status in root zone. The aim of this chapter is to analyze the aptitude of a simple method based on the relationship between remote sensed data of LST and VI from MODIS to retrieve subsurface SM in two different environmental conditions.

2 METHODOLOGY

2.1 Study Area

The Argentine Pampas is one of the most productive rainfed regions of agriculture and livestock in the world. The interannual and spatial variability of water availability results in important fluctuations of grain and cattle production at a regional scale. In general, the agriculture is based on a few crops that enable us to apply methods to evaluate vegetation condition using medium resolution images with an acceptable level of accuracy. Three agroclimatic zones were analyzed to assess the relationship between vegetation water stress and SM (Fig. 2): Sandy Pampas, the northern zone of Northern Hills, and Flooding Pampas. The climate of Sandy Pampas is temperate; the mean annual precipitation is frequently equal to potential ET (800 mm) and the mean annual temperature is 18°C. The soil type is Entic Hapludoll with sandy loam texture. The low water-retention capacity produces frequent water deficit, particularly in summer. The dominant summer vegetation is maize and soybean covering up to 30% of the total area.

In Northern Hills, the mean annual precipitation is about 1000 mm, significantly higher than potential ET (700 mm) with frequent water excess, especially during autumn and winter. The mean annual temperature is 16°C. The soil is Typic Argiudoll (silty loam to silty clay loam in subsuperficial horizons), with high water-retention capacity and a dense horizon (Bt) that frequently represents a limitation for vertical movement of water. The dominant vegetation in the analyzed zone is grassland (*Poaceae*, *Lolium*, *Festuca*), with shallow root system mostly concentrated in the first 30 cm of soil. The Flooding Pampas is characterized by very low slopes that favor the vertical movement of water (infiltration and ET). The mean annual precipitation (≈ 1100 mm) is higher than potential ET (≈ 800 mm) and the mean annual temperature is 17°C. Water excess is frequent and shallow water levels represent a common physical limitation for cropping systems. Thus, the main vegetation type is grassland and the dominant production activity is livestock. The most widespread soil in the western zone is Argiudoll. Natraquolls and Natraqualfs are dominant eastward in poorly drained areas.

2.2 Satellite Data, Estimation of TVDI, and Measurements

MODIS data were used because of the high temporal resolution useful to evaluate the day-to-day variations in SM. In addition, Aqua images from NASA and EOSDIS (2013) were used because it is approximately 2:00 pm overpass time allows us to analyze the period of maximum atmospheric evaporative demand during the day and thus the effect of SM availability on vegetation water status. TVDI was calculated with MODIS/Aqua product of daily LST level 3, version 5, at 1-km spatial resolution (MYD11A1). On the other hand, the Enhanced Vegetation Index (EVI) was obtained from bands 1 (red), 2 (near infrared), and 3 (blue) from MODIS/Aqua surface reflectance product level 2G, version 5,

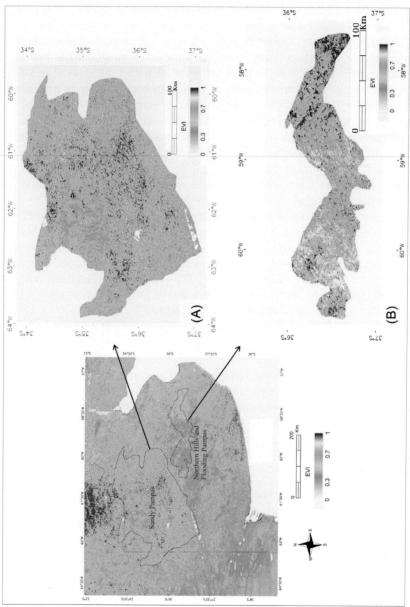

FIG. 2 Location of study area and the analyzed agroclimatic zones: (A) Sandy Pampas and (B) Northern Hills and Flooding Pampas.

at 500-m resolution (MYD09GA). The EVI was selected because it takes into account the effect of canopy background and has improved sensitivity into high biomass areas (Liu and Huete, 1995). TVDI was achieved at 500-m resolution resampling the LST images through the method proposed in Bayala and Rivas (2014), which is based on the difference between LST of dry and wet surfaces in the LST/VI relationship. Data products were used to reduce the data processing for atmospheric correction. Images were reprojected to geographic latitude/longitude coordinates, Datum WGS 84 using the MODIS Conversion Toolkit of ENVI software. Then, the two areas of interest in both zones were defined based on the MODIS/Aqua VIs, 16-day composite, version 5, at 250 m spatial resolution (MYD13Q1). Considering that vegetation cover in Tandil station is representative of a livestock region, the area of grassland in the northern part of Northern Hills and southern zone of Flooding Pampas was delimited to analyze the TVDI-SM relationship (Fig. 2). On the other hand, in Sandy Pampas cultivated areas were selected for the analysis, given that La Ydalina station is located in the corn region of the Argentine Pampas (Fig. 2).

Numerous works (Carlson et al., 1995; Holzman et al., 2014a; Moran et al., 1994; Nutini et al., 2014) have shown a consistent negative relationship between VIs and LST, which varies according to SM availability and therefore, to the partitioning of incoming energy in latent and sensible heat fluxes. Based on the VI/LST scatterplot, Sandholt et al. (2002) proposed a normalization of water status through the TVDI, having values of 0 (maximum soil wetness and potential ET) and 1 (under limited water availability):

$$TVDI = \frac{LST - LST_{min}}{LST_{max} - LST_{min}} \tag{2}$$

where LST is the observed surface temperature at a given pixel, LST_{min} is the minimum temperature and represents maximum SM availability and ET. $LST_{max} = aVI + b$ is the maximum temperature (minimum ET) for a given VI value, modeled as a linear fit to VI. The "a" and "b" parameters are the intercept and slope of the linear regression of LST_{max}.

A number of possible error sources in soil wetness estimation from TVDI should be noted:

(1) The parameters of TVDI should be estimated on the basis of pixels from a region with uniform atmospheric forcing.
(2) A wide range of water status must be present in the study area for an accurate definition of TVDI parameters. Otherwise, if wet/dry conditions are dominant the LST_{max}/LST_{min} can be underestimated/overestimated. This limitation of the method is usually overcome with the use of medium resolution sensors (>250 m spatial resolution), whose large swath width easily offers the required heterogeneity of water availability (saturated to dry surfaces) and vegetation cover (bare soil to full vegetation). With images of higher resolution, other models should be applied to estimate water status

(e.g., Moran et al., 1994) because the concept that LST decreases with an increase of VI is often not linear.

(3) Cloud cover and shadow restrict the calculation of TVDI because of the lack of data or variations in net radiation. In this sense, the use of high-quality images is required.

(4) In bare soils only the near surface SM could be estimated and the influence of deep SM on top surface soil layer can occur.

(5) Different vegetation type should have different sensitivity to deep-water availability, so this effect should be taken into account for estimation of short-time root zone SM variations.

Given that the definition of TVDI parameters have a degree of uncertainty depending of water condition in the study area, for each daily LST/EVI images the LST_{max} and LST_{min} were calculated considering the scatter plots of the semiarid and humid zones of the Argentine Pampas, respectively. In this way, the true minimum (complete stomatal closure) and maximum ET rate were defined for the study area (Holzman et al., 2014a; Stisen et al., 2008). LST_{max} were obtained using the least squares method, with a significance level of 5%. LST_{min} values were obtained based on the LST/EVI scatter plots of humid area by averaging a group of points with minimum LST for each EVI value. As shown in previous works (Holzman et al., 2014a,b), daily TVDI parameters were compared to define extreme LST_{max} (maximum slope and intercept) and LST_{min} (minimum LST_{min}). Thus, TVDI was computed based on these extreme values and, hence, TVDI series were normalized for the study period.

The normalized series of TVDI were compared with daily ground-based SM. The measurements were made between January 1 to February 27, 2012, in Campus Tandil station (37°19′S, 59°05′W, Northern Hills) and La Ydalina station (35°09′S, 61°07′W, Sandy Pampas). Based on cloud masks accompanying the MYD11A1 product, cloudy observations were excluded from the analysis. In this way, 19 and 25 cloud-free values for Campus Tandil and La Ydalina were used for the analysis, respectively. On Campus Tandil station, calibrated EC-H$_2$O (Decagon Devices, Inc.) SM probes were used to measure relative soil water content at 10, 40, and 60 cm depths. These sensors measure the integrated dielectric constant (mV) at 10 cm depth, which is directly related to the volumetric water content. These sensors were previously intercalibrated with lysimeters and moisture probes showing an accuracy of 97% (Carmona et al., 2011). Such depths were selected to analyze the water dynamics in different horizons. On La Ydalina station, SM was measured at 5, 60, and 120 cm depths through a tensiometer incorporated to a Davis Vantage Pro2 (Davis Instruments Corporation). Relative soil water content was obtained using a previous moisture characteristic curve for this zone. Then, the TVDI-subsurface SM relationship was evaluated comparing 3 × 3 pixels TVDI values with relative SM measurements (Fig. 3). Finally, maps of subsurface SM were achieved for the agroclimatic zones using the obtained adjustments of TVDI-water content (see equations in Table 1).

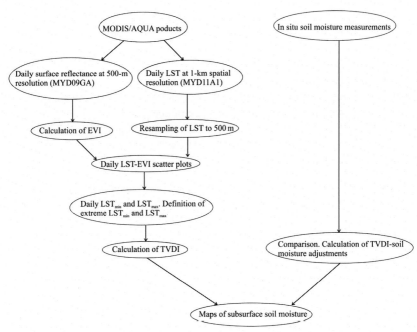

FIG. 3 Flowchart of the methodology.

TABLE 1 Linear Adjustments Between Daily TVDI and In Situ Soil Moisture (SM)

La Ydalina Station	Campus Tandil Station
5 cm depth: no correlation[a]	10 cm depth: $SM = -25.2TVDI + 21.7$ $(R^2 = 0.69)$
60 cm depth: $SM = -12.69TVDI + 19.6$ $(R^2 = 0.70)$	40 cm depth: $SM = -20.9TVDI + 28.3$ $(R^2 = 0.70)$
120 cm depth: $SM = -7.23TVDI + 19.5$ $(R^2 = 0.76)$	60 cm depth: no correlation

[a]Although a high R^2 value was observed, it has no physical sense under full vegetation and sandy soil with quick drying process.

3 RESULTS AND DISCUSSION

3.1 TVDI Parameters

Fig. 4 shows the temporal fluctuation of daily LST_{min}, slope and intercept of LST_{max} in relation to rainfall data of Santa Rosa aerodrome (36°56'S, 64°26'W) (Fig. 4A), and Junín aerodrome (34°55'S, 60°50'W). A high temporal

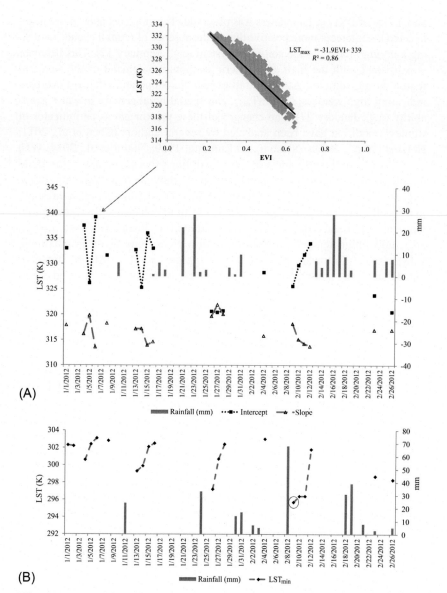

FIG. 4 Temporal fluctuation of (A) LST_{max} and (B) LST_{min} parameters in relation to rainfall events in Santa Rosa aerodrome and Junín aerodrome meteorological stations, respectively. The extreme LST_{max} (January 6) and LST_{min} (February 9) are indicated by a circle and the scatterplot of LST/EVI of extreme LST_{max} is included.

variability of LST_{max} parameters was observed. The slope becomes more negative and the intercept more positive during the absence of rainfall events, indicating a drying process of the soil-plant system (e.g., February 12). This behavior is consistent with the high atmospheric evaporative demand of the semiarid region during summer. After rainfall events, the slope and intercept decrease indicating high wetness, flat LST_{max}, and spatial homogeneity in water availability (e.g., January 27). The changes of these parameters as an indicator of wetness condition have been analyzed by several authors (Chen et al., 2011; Holzman et al., 2014a; Nemani and Running, 1989; Nutini et al., 2014; Wan et al., 2004) and the usefulness to follow a drying process and define threshold of vegetation damage in different region worldwide (e.g., Mekong Delta, Argentine Pampas, and Africa) has been shown. The definition of that threshold in a specific productive region should be useful to enhance early warning systems and determine drought management strategies. On the other hand, LST_{min} also shows high temporal variability with a clear trend to reach maximum values (≈ 303 K) toward the end of a drying process (Fig. 4B). It should be noted that during a rainfall event there is a lack of data because of cloud cover, which represents a limitation of the method. An alternative may be to analyze the compatibility of water stress indices and microwave data to derive wetness condition under cloudy periods. In this sense, Cho et al. (2014) tested the relationship between microwave sensor-derived SM and the TVDI in the Sahel showing high correlation between them.

Because the LST_{max} and LST_{min} are different for each image, extreme values of these parameters were defined to standardize TVDI values. Extreme LST_{max} indicating the driest condition during the study period was defined as the maximum slope and intercept. The linear fit was $LST_{max} = -31.9\text{EVI}$ (± 2) $+ 339(\pm 3)$ (January 6). The minimum LST_{min} was 296 (± 2) K (February 9). These values were consistent with previous works carried out in the study area at daily and monthly levels (Holzman et al., 2014a,b). The TVDI series were calculated using these extreme parameters. It should be noted that this kind of standardization is useful for a multitemporal study. Otherwise, the stress index values are not comparable between different images.

3.2 Comparison of TVDI and Subsurface Soil Moisture Measurements

According to numerous works (Chen et al., 2011; Cho et al., 2014; Han et al., 2010; Holzman et al., 2014a; Mallick et al., 2009; Wang et al., 2004), the relationship between LST/VI stress indices and surface SM (≈ 5 cm depth) could be assumed as negative linear. However, few studies have analyzed the aptitude of optical/thermal data to estimate root zone water content and the decoupling between surface and subsurface soil layers. Once the TVDI series were standardized, these values were compared with SM measurements obtained from the two stations of the agroclimatic zones (Fig. 5). The water stress index showed high sensitivity to subsurface SM, increasing as soil water availability

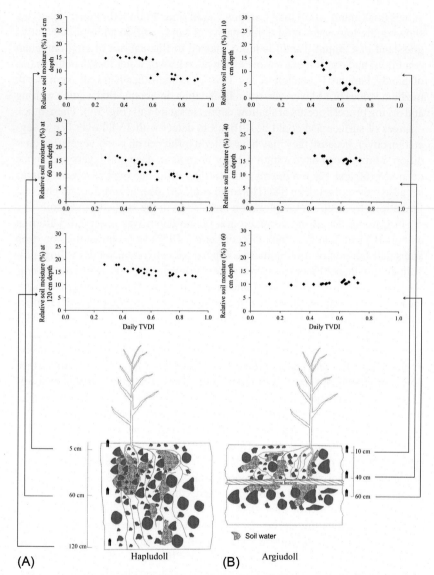

FIG. 5 Diagram of the relationship between vegetation water stress index (TVDI) and root zone soil moisture at different depths in two root systems: (A) Sandy Pampas (sandy soil, La Ydalina station) and (B) Northern Hills (soil with a dense horizon, Campus Tandil station).

decreases. However, vegetation and soil type affects the SM content by the changes in land surface energy and soil porosity (Chang et al., 2012; Chen et al., 2009) and hence some differences should be noted. Although water content was measured at different depths in each soil, a distinct dynamic of vegetation and SM in both soil types can be deduced. In La Ydalina, higher TVDI values (maximum $=0.9$) than Campus Tandil (maximum $=0.7$) were observed, showing the low water-retention capacity of sandy soil. In addition, the native grassland of Campus Tandil is more adapted to fluctuations of environmental conditions and can achieve a more efficient control of water loss than the maize of Sandy Pampas (Canadell et al., 1996). On the other hand, at 5 cm depth medium values of SM were absent in La Ydalina, indicating that a quick wetting and drying process occurs at shallow depths in sandy soil (Fig. 5A). These quick changes of surface SM could be difficult to detect with TVDI over dense vegetation cover, because they may have a slight influence on water vegetation condition. Maize crops have a high density of roots in deep soil layers and hence deep SM (60 and 120 cm depths) has a strong influence over crop water status because they extract deep SM (Holzman et al., 2014b). In addition, no physical limitations for deep root exploration are visible in the studied sandy soil.

In Campus Tandil no correlation at 60 cm depth was observed. Different authors (Canadell et al., 1996; Gregory et al., 1978) have shown that the deep and small fraction of root biomass is crucial for ecosystem water cycling, providing more than 20% of transpired water during dry periods. However, no correlation at 60 cm depth would indicate that the root biomass of grassland is located at medium depth and that the dense horizon of Argiudoll (Bt) implies a physical limitation for soil exploration by roots (Whiteley and Dexter, 1981) (Fig. 5B). Hence, a method of monitoring vegetation water stress based mainly on the canopy temperature has aptitude to reflect water content in soil layers explored by roots and shows limitations to detect highly dynamic soil surface horizons or deep layers not explored by roots. These limitations should be lower in soils with high vertical integration (e.g., uniform texture or poor pedogenetic development) where the vertical gradient in soil water content is low.

Linear adjustments that represent the relationship between TVDI and SM were found at different depths (Table 1). Strong coefficient of determination was observed in most cases, with values higher than 0.70. These results are consistent with values reported in previous works. Using apparent thermal inertia from MODIS in subhumid areas of Thailand, Chang et al. (2012) found Pearson correlation coefficient of 0.84, 0.57, and 0.49 at 10, 100, and 200 cm depths, respectively. Evaluating the errors, they concluded that the approach performed the best for the SM estimation at 100 cm depth. Although in the current study we analyzed data of two stations, no significant differences about coefficient of determination in deep soil layers were observed. On the other hand, the decreasing slope of adjustments in deep soil layers confirms the higher SM stability abovementioned. This stability can be intensified in wet periods when the saturated zone (water table ≈ 1.0–1.5 m depth) can provide moisture to the deep part of the nonsaturated zone.

The obtained results show that the TVDI could be used as a proxy of root zone SM and for mapping of such conditions. Based on the TVDI-subsurface soil water content adjustments, maps of wetness condition were achieved for both agroclimatic zones during wet and dry periods (Fig. 6). A significant

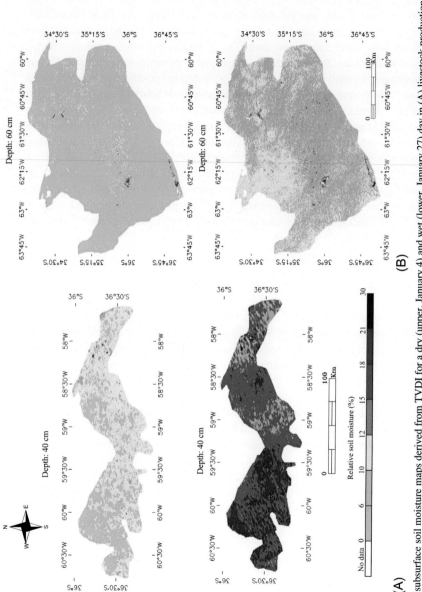

FIG. 6 Daily subsurface soil moisture maps derived from TVDI for a dry (upper, January 4) and wet (lower, January 27) day in (A) livestock production zone of Flooding Pampas and Northern Hills and (B) agricultural zone of Sandy Pampas.

spatial and temporal variability was observed. During the dry period in both areas very low water availability is visible. This pattern is more evident in Sandy Pampas where high water availability is located only near to low areas between dunes reflecting the shallow water table (Fig. 6B). In Northern Hills and Flooding Pampas, SM increases eastward where the influence of ocean and shallow water table increases (Fig. 6A). On the other hand, the humid period is characterized by widespread excessive moisture with irregular pattern according to rainfall distribution, water table depth, and drainage networks integration. Such spatial integration is noticeable in Northern Hills and Flooding Pampas where the water is one of the main factors of spatial homogeneity at landscape scale. In addition, in Sandy Pampas, general high water availability was observed with the highest values located mainly in low areas of dunes and near water bodies. It should be noted that this spatial and temporal variability is crucial for regional crop and livestock production. In Holzman et al. (2014a) we reported the impact of water stress and excess on soybean and wheat yield through the TVDI. Given that irrigation is insignificant at regional scale in the Argentine Pampas, SM changes during critical growth stage (flowering and grain filling) of crops can produce fluctuation of 50% in crop yield in the study area.

4 CONCLUSIONS

In this chapter the ability of a water stress index based on remote sensed LST and VI to estimate subsurface SM was analyzed. Although several works have studied the relationship between surface SM, ET, and optical/thermal data, different factors of soil-vegetation system affecting the subsurface water content estimation were preliminarily analyzed in this study. A high correlation ($R^2 > 0.69$) between the remotely sensed TVDI and soil water content at 10, 40, 60, and 120 cm depths in the great plains of the Argentine Pampas was found. However, the method showed limitations in estimating surface SM over full vegetation and soils of rapid drying, showing the high vertical gradient of water content where the decoupling between the surface and deep soil layers is frequent. In addition, limitations were observed in deep soil layers not explored by roots. In this case, shallow root systems and dense horizons or physical limitation for root penetration regulate root distribution in the soil and therefore the aptitude of the method. On the other hand, the approach is an attractive alternative to identify the spatial connection between deep roots and groundwater in areas with a shallow water table where piezometer measurements are not available.

Although previous calibration with ground-based observations is necessary to obtain soil water content values, no prior SM data are required to apply the method. In this regard, in regions poorly or not covered by in situ stations a multitemporal analysis can be carried out to define baselines of the TVDI parameters that allow the standardization of soil water estimates. It should be noted that this approach is promising for understanding hydrological and ecological processes difficult to analyze through traditional ground-based methods. In this regard, current and future satellite missions with medium spatial resolution like Aqua/Terra and Sentinel missions provide long thermal/optical data series to analyze trends in land processes over great regions and basins with low spatial measurements. In addition, the use of atmospherically corrected images (LST and VI or reflectance products) can reduce the data processing, increasing the adoption of the method by decision makers. In this sense, the agricultural service of Argentina (Oficina de Riesgo Agropecuario, http://www.ora.gov.ar/ tvdi.php) utilizes the method for operationally monitoring regional crop growth water condition. Future studies may analyze spatial interactions associated with SM in different soil and vegetation types and explore alternatives to overcome the data gap in optical/thermal sensors due to cloud cover.

REFERENCES

Akther, M.S., Hassan, Q.K., 2011. Remote sensing based estimates of surface wetness conditions and growing degree days over northern Alberta, Canada. Boreal Environ. Res. 16 (5), 407–416.

Bayala, M.I., Rivas, R.E., 2014. Enhanced sharpening procedures on edge difference and water stress index basis over heterogeneous landscape of sub-humid region. Egypt. J. Remote Sens. Space. Sci. 17 (1), 17–27.

Bhattacharya, B.K., Mallick, K., Nigam, R., Dakore, K., Shekh, A.M., 2011. Efficiency based wheat yield prediction in a semi-arid climate using surface energy budgeting with satellite observations. Agr. Forest. Meteorol. 151 (10), 1394–1408.

Boulet, G., Chehbouni, A., Gentine, P., Duchemin, B., Ezzahar, J., Hadria, R., 2007. Monitoring water stress using time series of observed to unstressed surface temperature difference. Agr. Forest. Meteorol. 146 (3 –4), 159–172.

Canadell, J., Jackson, R.B., Ehleringer, J.B., Mooney, H.A., Sala, O.E., Schulze, E.-D., 1996. Maximum rooting depth of vegetation types at the global scale. Oecologia 108 (4), 583–595.

Carlson, T.N., Gillies, R.R., Perry, E.M., 1994. A method to make use of thermal infrared temperature and NDVI measurements to infer surface soil water content and fractional vegetation cover. Remote Sens. Rev. 9, 161–173.

Carlson, T.N., Gillies, R.R., Schmugge, T.J., 1995. An interpretation of methodologies for indirect measurement of soil water content. Agr. Forest. Meteorol. 1923 (95), 191–205.

Carmona, F., Rivas, R., Ocampo, D., Schirmbeck, J., Holzman, M., 2011. Sensores para la medición y validación de variables hidrológicas a escalas local y regional a partir del balance de energía. Aqua-LAC 3 (1), 26–36.

Chakrabarti, S., Member, S., Bongiovanni, T., Judge, J., Member, S., 2014. Assimilation of SMOS soil moisture for quantifying drought impacts on crop yield in agricultural regions. IEEE J. Sel. Topics Appl. Earth Observ. Remote Sens. 7 (9), 3867–3879.

Chang, T.-Y., Wang, Y.-C., Feng, C.-C., Ziegler, A.D., Giambelluca, T.W., Liou, Y.-A., 2012. Estimation of root zone soil moisture using apparent thermal inertia with MODIS imagery over a tropical catchment in northern Thailand. IEEE J. Sel. Topics Appl. Earth Observ. Remote Sens. 5 (3), 752–761.

Chen, X., Zhang, Æ.Z., Chen, Æ.X., 2009. The impact of land use and land cover changes on soil moisture and hydraulic conductivity along the karst hillslopes of southwest China. Environ. Earth Sci. 59 (4), 811–820.

Chen, C.-F., Son, N.-T., Chang, L.-Y., Chen, C.-C., 2011. Monitoring of soil moisture variability in relation to rice cropping systems in the Vietnamese Mekong Delta using MODIS data. Appl. Geogr. 31 (2), 463–475.

Cho, J., Lee, Y.-W., Lee, H.-S., 2014. Assessment of the relationship between thermal-infrared-based temperature–vegetation dryness index and microwave satellite-derived soil moisture. Remote Sens. Lett. 5 (7), 627–636.

Crow, W.T., Ryu, D., Famiglietti, J.S., 2005. Upscaling of field-scale soil moisture measurements using distributed land surface modeling. Adv. Water Resour. 28 (1), 1–14.

Crow, W., Kustas, W., Prueger, J., 2008. Monitoring root-zone soil moisture through the assimilation of a thermal remote sensing-based soil moisture proxy into a water balance model. Remote Sens. Environ. 112 (4), 1268–1281.

Friedl, M.A., 2002. Forward and inverse modeling of land surface energy balance using surface temperature measurements. Remote Sens. Environ. 79 (2 –3), 344–354.

Gregory, P.J., McGowan, M., Biscoe, P.V., Hunter, B., 1978. Water relations of winter wheat: 1. Growth of the root system. J. Agric. Sci. 91 (1), 91–102.

Han, Y., Wang, Y., Zhao, Y., 2010. Estimating soil moisture conditions of the Greater Changbai Mountains by land surface temperature and NDVI. IEEE Trans. Geosci. Remote Sens. 48 (6), 2509–2515.

Holzman, M.E., Rivas, R., Bayala, M., 2014a. Subsurface soil moisture estimation by VI–LST method. IEEE Geosci. Remote Sens. Lett. 11 (11), 1951–1955.

Holzman, M.E., Rivas, R., Piccolo, M.C., 2014b. Estimating soil moisture and the relationship with crop yield using surface temperature and vegetation index. Int. J. Appl. Earth Obs. Geoinf. 28, 181–192.

Li, F., Crow, W.T., Kustas, W.P., 2010. Towards the estimation root-zone soil moisture via the simultaneous assimilation of thermal and microwave soil moisture retrievals. Adv. Water Resour. 33 (2), 201–214.

Liu, H.Q., Huete, A., 1995. Feedback based modification of the NDVI to minimize canopy background and atmospheric noise. IEEE Trans. Geosci. Remote Sens. 33 (2), 457–465.

Mallick, K., Bhattacharya, B.K., Patel, N.K., 2009. Estimating volumetric surface moisture content for cropped soils using a soil wetness index based on surface temperature and NDVI. Agr. Forest. Meteorol. 149 (8), 1327–1342.

Moran, M.S., Clarke, T.R., Inoue, Y., Vidal, A., 1994. Estimating crop water deficit using the relation between surface-air temperature and spectral vegetation index. Remote Sens. Environ. 49 (3), 246–263.

NASA, EOSDIS, 2013. Reverb – ECHO. Retrieved from, http://reverb.echo.nasa.gov/.

Nemani, R.R., Running, S.W., 1989. Estimation of regional surface resistance to evapotranspiration from NDVI and thermal-IR AVHRR data. J. Appl. Meteorol. 28, 276–284.

Nutini, F., Boschetti, M., Candiani, G., Bocchi, S., Brivio, P., 2014. Evaporative fraction as an indicator of moisture condition and water stress status in semi-arid rangeland ecosystems. Remote Sens. 6 (7), 6300–6323.

Petropoulos, G.P., Griffiths, H.M., Dorigo, W., Xaver, A., Gruber, A., 2014. Surface soil moisture estimation: significance, controls, and conventional measurement techniques. In: Remote Sensing of Energy Fluxes and Soil Moisture Content. Petropoulos, G.P. (Ed.). CRC Press (Taylor and Francis Group), Boca Raton, FL, pp. 29–48.

Petropoulos, G.P., Ireland, G., Barrett, B., 2015. Surface soil moisture retrievals from remote sensing: current status, products & future trends. Phys. Chem. Earth, Parts A/B/C. 44, pp. 36–56.

Rivas, R., Caselles, V., 2004. A simplified equation to estimate spatial reference evaporation from remote sensing-based surface temperature and local meteorological data. Remote Sens. Environ. 93 (1 –2), 68–76.

Sandholt, I., Rasmussen, K., Andersen, J., 2002. A simple interpretation of the surface temperature/vegetation index space for assessment of surface moisture status. Remote Sens. Environ. 79 (2 –3), 213–224.

Stisen, S., Sandholt, I., Nørgaard, A., Fensholt, R., Jensen, K.H., 2008. Combining the triangle method with thermal inertia to estimate regional evapotranspiration — applied to MSG-SEVIRI data in the Senegal River basin. Remote Sens. Environ. 112 (3), 1242–1255.

Wan, Z., Wang, P., Li, X., 2004. Using MODIS land surface temperature and normalized difference vegetation index products for monitoring drought in the southern Great Plains, USA. Int. J. Remote Sens. 25 (1), 61–72.

Wang, C., Qi, S., Niu, Z., Wang, J., 2004. Evaluating soil moisture status in China using the temperature-vegetation dryness index (TVDI). Can. J. Remote. Sens. 30 (5), 671–679.

Whiteley, G.M., Dexter, A.R., 1981. Elastic response of the roots of field crops. Physiol. Plant. 51, 407–417.

Chapter 5

Spatiotemporal Estimates of Surface Soil Moisture from Space Using the T_s/VI Feature Space

G.P. Petropoulos*, G. Ireland*, H. Griffiths*, T. Islam[†,‡], D. Kalivas[§],
V. Anagnostopoulos[¶,‖], C. Hodges** and P.K. Srivastava[††,‡‡]

**Department of Geography & Earth Sciences, Ceredigion, Wales, United Kingdom, [†]NASA Jet Propulsion Laboratory, Pasadena, CA, United States, [‡]California Institute of Technology, Pasadena, CA, United States, [§]Agricultural University of Athens, Athens, Greece, [¶]National Technical University of Athens, Athens, Greece, [‖]InfoCosmos Ltd, Athens, Greece, **Geo Smart Decisions Ltd, Llanidloes, Powys, United Kingdom, [††]NASA Goddard Space Flight Center, Greenbelt, MD, United States, [‡‡]Banaras Hindu University, Banaras, India*

1 INTRODUCTION

Soil moisture content (SMC) at the Earth's surface generally refers to the water content in the upper layer of the soil, expressed usually as either a dimensionless ratio of two masses or two volumes, or given as a ratio of a mass per unit volume. Exact information on the spatiotemporal variation of parameters such as SMC is of key significance because of its control on the various physical processes of the Earth system (Barrett and Petropoulos, 2013). It is a key state variable within the global energy cycle due to its control on the partitioning of available energy at the Earth's surface into latent (LE) and sensible (H) heat exchange across regions where the evaporation regime is, at least intermittently, water limited (as opposed to energy limited) (Vereecken et al., 2014). SMC is also a significant component of the hydrological cycle, governing the partitioning of rainfall into infiltration and runoff, thus affecting stream flow, groundwater recharge, and precipitation. Accurate measurement of soil water dynamics and distribution is also of great importance in a large number of regional- and global-scale applications, such as monitoring of plant water requirements, plant growth and productivity, the management of irrigation and cultivation procedures (Petropoulos et al., 2015), and the monitoring of land degradation and desertification (McCabe and Wood, 2006). Furthermore, the

Satellite Soil Moisture Retrieval. http://dx.doi.org/10.1016/B978-0-12-803388-3.00005-X

importance of SMC is underlined by the fact that a number of global and European organizations are also interested in its dynamics within a number of key framework directives (Piles et al., 2016). One such example is the Global Energy and Water Cycle Experiment (GEWEX), which requires relevant SMC estimates to study the dynamics (including the thermodynamics) of the atmosphere, its interactions with the Earth's surface, and its effects on the global energy and water cycle.

As a result, there is evidently a requirement from many interrelated disciplines for an accurate SMC observational capability across a broad range of scales. A number of quantitative methods have been utilized to analyze the spatiotemporal dynamics and distribution of SMC (Vereecken et al., 2014). At smaller scales, a number of conventional approaches are used to measure SMC directly using ground instrumentation, broadly grouped into point measurements with electromagnetic soil moisture sensors, hydrogeophysical methods, and electrical resistivity tomography (Dorigo et al., 2011). While such approaches may be considered the most direct and accurate way to measure SMC, their use over large areas is limited, mainly due to the expense of maintaining the field equipment and the large spatial variability in SMC. Efforts have also been made at a global scale to establish and maintain operational networks of in situ soil moisture sensors that can provide long-term soil moisture measurements and related meteorological information (Petropoulos et al., 2013a). These networks provide valuable distributed point measurements but are also insufficient to characterize the spatial and temporal variability of soil moisture at large scales. Given these limitations, Earth Observation (EO) technology is recognized as the only viable solution for obtaining estimates of SMC at the spatiotemporal scales and accuracy levels required by many applications (Petropoulos et al., 2014).

More recent approaches have focused on exploiting the synergy between EO data types and land surface process models, including two-layer soil vegetation atmosphere transfer (SVAT) models, to provide improved spatiotemporal estimations of SMC. Such approaches aim to combine the horizontal coverage and spectral resolution of remote sensing data with the vertical coverage and fine temporal continuity of those models, where a more powerful synergistic avenue can be developed to provide improved estimates of SMC (Petropoulos et al., 2015). Commonly termed in the literature as the "triangle" method, it is based on the physical relationships that exist when a satellite-derived land surface temperature (T_s) is plotted against a spectral vegetation index (VI) in combination with a land biosphere model. In summary, within this group of approaches, theoretical boundary lines are derived from the observed inverse relationship between T_s and VI within a scatter plot feature space, determined from the distribution of the remote sensing-based pixel envelope (Gillies et al., 1997). A description of the method can be found in Carlson (2007). Various validation studies have demonstrated the "triangle's" ability to provide estimates of SMC with accuracies within 5% vol vol^{-1} over homogenous areas

(Gillies et al., 1997). The significant prospect of the T_s/VI methods, and of the "triangle" in particular, is further documented by the fact that variants of this technique are being considered at present in the development of operational products of SMC on a global scale (Chauhan et al., 2003; ESA STSE, 2012; Piles et al., 2016). Also, a variant of the "triangle" is already presently deployed over Spain to operationally deliver SMC maps at 1 km spatial resolution from the European Space Agency (ESA)'s own Soil Moisture and Ocean Salinity (SMOS) satellite (Merlin, 2013; Merlin et al., 2012, 2013; Piles et al., 2011).

Although the potential of the "triangle" has been recognized by many authors, there is limited research on validating outputs of this modeling approach at regional to mesoscale, as well as for European ecosystems. Indeed, validation exercises of this technique have been scarce to date, performed mostly using high-resolution airborne or satellite EO data for sites located mostly (if not solely) in US ecosystems. Thus, the use of moderate- to high-resolution sophisticated EO spaceborne sensors, such as those from Advanced Spaceborne Thermal Emission and Reflection Radiometer (ASTER) and the Advanced Along Track Scanning Radiometer (AATSR), in the method implementation would be of key interest. In this context, this study explores the effectiveness of integrating data from both ASTER and AATSR with the SimSphere land biosphere model via the "triangle" to derive spatiotemporal estimates of SMC over different ecosystems in Europe. The predicted SMC were validated against co-orbital in situ measurements acquired from a series of sites representative of a variety of environmental and topographical conditions belonging to the CarboEurope flux network.

2 THE T_S/VI DOMAIN

Numerous studies have documented the triangular (or trapezoidal) shape of the T_s/VI space that emerges from a scatter plot based on a contextual relationship between T_s and VI (Gillies et al., 1997; Carlson, 2007; Fan et al., 2015; Maltese et al., 2015). The emergence of the triangular shape dispersion arises from T_s being less sensitive to water content at the surface in vegetated areas than in areas of exposed soil. Such a scatter plot is characterized by four physical bounds, as shown in Fig. 1.

The so-called dry edge or warm edge is defined by the locus of points of highest temperature but that contain differing amounts of bare soil and vegetation. It is assumed to represent conditions of limited surface soil water content and zero evaporative flux from the soil, characterizing surfaces with the largest water stress for a range of VIs (Tian et al., 2013). Likewise, the left-hand border (the so-called wet edge or cold edge) describes the water availability with respect to the vegetation conditions; that is, representing the maximum soil wetness conditions and potential SMC/ET for a range of VIs (Fan et al., 2015; Maltese et al., 2015). Variation along the "base" of the triangle represents pixels of bare soil and is assumed to reflect the combined

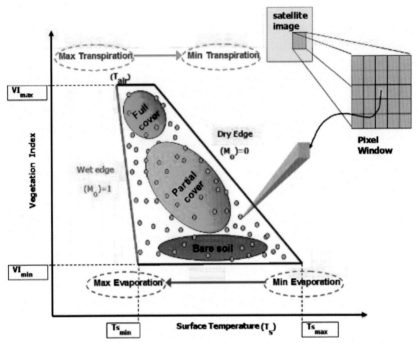

FIG. 1 Summary of the key descriptors and physical interpretations of the T_s/VI feature space "scatter plot." *(Figure adopted from Petropoulos et al. (2009).)*

effects of soil water content variations and topography, while the triangle's apex equates to full vegetation cover (as this is expressed by the highest VI value). Determination of both the dry and wet edges is a critical procedure, in that they provide important boundary conditions to the contextual T_s/VI relationship (Tian et al., 2013). Based on these boundary conditions, points within the triangular space correspond to pixels at varying intervals of VI and surface soil water content, between the extreme boundaries of bare soil and dense vegetation. For data points having the same VI, T_s can range markedly. Thus, for pixels with the same VI, those with minimum T_s represent the case of the strongest evaporative cooling, while those with maximum T_s represent those with the weakest evaporative cooling. Therefore, the dry edge and wet edge determine SMC for pixels within the two limiting edges and subsequently determine SMC estimation. More detailed descriptions on the physical properties encapsulated in the T_s/VI triangle domain as well as the main factors driving the shape of the T_s/VI scatter plot can be found in Petropoulos et al. (2009). The use of the T_s/VI methods in SMC retrievals from EO data is also reviewed in Petropoulos et al. (2015).

3   EXPERIMENTAL SET UP AND DATA SETS

3.1   CarboEurope In Situ Measurements

In situ SMC measurements were acquired from selected European sites representative of different ecosystem conditions belonging to the CarboEurope validated observational network. CarboEurope is part of FLUXNET, the largest global in situ measurements network that coordinates regional and global analysis of micrometeorological fluxes and ancillary parameters (Baldocchi et al., 1996). At CarboEurope, SMC is measured in at least two depths (surface and root zone) at 30-min frequencies using standardized instrumentation across sites. All collected data are quality-controlled, gap-filled, and standard procedures for error corrections are prescribed. All data was obtained from the FLUXNET database (http://fluxnet.ornl.gov/obtain-data) at Level 2 processing, to allow consistency and interoperability. This processing level includes the originally acquired in situ data from which any erroneous data (e.g., ones caused by obvious instrumentation error) have been removed. Additionally, atmospheric in situ data was obtained from the freely distributed University of Wyoming's weather balloon data archive (http://weather.uwyo.edu/upperair/sounding.html). Local profiles of temperature, dew point temperature, wind direction, wind speed, and atmospheric pressure were taken from the nearest possible experimental sites acquired at 06.00 h for initialization of the SVAT model.

Selection of the test sites and identification of the specific image dates was conducted based on a number of principles: (a) each site had to be part of a validated network at the same processing level to allow direct comparisons between results from different sites; (b) atmospheric profile (radiosonde/sounding) data were available and used primarily for initialization of the SVAT model; (c) test sites needed to cover a geographical distribution that represented a variety of climatic, topographic, and environmental conditions to be able to stratify the results by land cover type; and (d) sites needed to be within locations that were cloud free in their instantaneous field of view (IFOV) on the selected days concurrent to ASTER/AATSR overpass.

The study sites used to validate the "triangle" varied depending on the availability of satellite data per sensor. Thus, different sites and image dates were utilized in implementing the "triangle" using from AATSR and ASTER. For the "triangle" implemented utilizing ASTER, in situ measurements were acquired for 16 days between the years 2000 and 2004 from seven CarboEurope sites from two different countries. Table 1 summarizes the characteristics of the experimental sites used in this study. The in situ data, which corresponded to the "triangle" implemented using AATSR data, were acquired for 12 sites from 47 days in total covering the period from 2007 to 2011, allowing for a sufficient database for model parameterization and validation to be developed (Table 1).

TABLE 1 Location and Characteristics of the CarboEurope Flux Tower Sites Used in This Study

Site Name	Geographic Coordinates	Country	Land Cover	Elevation	Climate	Sensor
Aguamarga (ES-Agu)	36.8347/−2.2511	Spain	Open shrubland (OSH)	199 m	Arid steppe cold	AATSR
Amoladeras (ES-Amo)	36.9405/−2.0329	Spain	Open shrubland (OSH)	58 m	Arid steppe cold	AATSR
Espirra (PT-Esp)	38.6394/−8.6018	Portugal	Evergreen broadleaf forest (EBF)	95 m	Warm temperate with dry, hot summer	AATSR
Mitra IV Tojal (PT-Mi2)	38.4765/−8.0246	Portugal	Grassland (GRA)	190 m	Warm temperate with dry, hot summer	AATSR
Collelongo-Selva Piana (IT-Col)	41.8493/13.588	Italy	Deciduous broadleaf forest (DBF)	1560 m	Warm temperate, fully humid with hot summer	AATSR
Lecceto (IT-Lec)	43.3046/11.271	Italy	Evergreen needle-leaf forest (ENF)	314 m	Warm temperate, fully humid with hot summer	AATSR
Nonantola (IT-Non)	44.6898/11.089	Italy	Mixed forest (MF)	20 m	Warm temperate, fully humid with hot summer	AATSR
Bonis (IT-Bon)	39.4778/16.535	Italy	Evergreen needle-leaf forest (ENF)	1170 m	Warm temperate with dry, hot summer	AATSR
Negrisia (IT-Neg)	45.7476/12.447	Italy	Cropland (CRO)	9 m	Warm temperate, fully humid with hot summer	AATSR
Castellaro (IT-Cas)	45.0700/8.7175	Italy	Cropland (CRO)	84 m	Warm temperate, fully humid with hot summer	AATSR

TABLE 1 Location and Characteristics of the CarboEurope Flux Tower Sites Used in This Study—cont'd

Site Name	Geographic Coordinates	Country	Land Cover	Elevation	Climate	Sensor
Renon/Ritten (IT-Ren)	46.5878/11.435	Italy	Evergreen needle-leaf forest (ENF)	1730 m	Snow, fully humid, hot summer	AATSR/ASTER
Malga Arpaco (IT-Mal)	46.1167/11.703	Italy	Grassland (GRA)	1730 m	Polar tundra	AATSR/ASTER
Borgo Cioffi (IT-Bci)	40.5254/14.1250	Italy	Cropland (CRO)	20 m	Temperate, arid	ASTER
Roccarespampani 3 years (IT-Ro3)	42.4078/11.9297	Italy	Deciduous broadleaf forest (DBF)	244 m	Mediterranean, montane	ASTER
Roccarespampani 11 years (IT-Ro1)	42.3897/11.9208	Italy	Deciduous broadleaf forest (DBF)	224 m	Mediterranean, montane	ASTER
Monte Boldone (IT-Mbon)	46.0296/11.0690	Italy	Grassland (GRA)	1550 m	Sub-continental	ASTER
Lavarone (IT-Lav)	45.9553/11.2673	Italy	Mixed forest (MF)	1353 m	Sub-continental	ASTER
Loobos (NL-Loo)	52.1666/05.7436	Netherlands	Evergreen needle-leaf forest (ENF)	25 m	Temperate, oceanic	ASTER
Lelystad (NL-Lel)	52.5242/05.5516	Netherlands	Grassland (GRA)	0 m	Temperate, oceanic	ASTER

3.2 The Advanced Spaceborne Thermal Emission and Reflection Radiometer (ASTER) Imagery

ASTER is a high-spatial resolution imaging spectroradiometer flown on board Terra, the flagship satellite of NASA's Earth Observing System (EOS). The visible and near-infrared (VNIR) sensor provides four bands at 15-m resolution, the shortwave infrared (SWIR) sensor provides six bands at 30-m resolution, and the thermal infrared (TIR) sensor provides five bands at 90-m resolution. The swath width for all sensors is 60 km. Thus, ASTER is able to provide 15 spectral bands with 15- to 90-m resolutions (dependent on band(s)) with a repeat cycle of 16 days. For each of the days used in this study, the higher level (Level 2) validated ASTER image products of Surface Reflectance (AST07) and Surface Kinetic Temperature (AST08) were obtained from the ASTER IMS and/or the EOS Data Gateway. The AST07 product contains daytime atmospherically corrected surface spectral reflectance data for the VNIR and SWIR bands at a spatial resolution of 15 and 30 m, respectively, calculated from the ASTER L1 data. The AST08 product contains daytime and nighttime Kinetic Temperature (T_{kin}) at 90 m spatial resolution derived from the five ASTER TIR channels.

3.3 The Advanced Along Track Scanning Radiometer (AATSR) Imagery

AATSR is a dual-view, multichannel, imaging radiometer with 1 km spatial resolution, a 512 km swath width, a temporal resolution of 1–3 days, and provides measurements of reflected and emitted radiation taken at seven different wavelengths. It is flown on board Envisat, the ESA's satellite. One of the distinct characteristics of AATSR includes the use of the along track scanning technique from the instrument's conical scanning mechanism. This permits acquiring two observations from different observation angles of any given point on the Earth's surface. This dual-view design of the radiometer allows accounting for the direct measurement of the effect of the atmosphere on the observations acquired, resulting in accurate atmospheric correction of the data. In this study, the ATS_NR_2P geophysical product was acquired to provide information on the spatial distribution of F_r and T_s. ATS_NR_2P is a full spatial resolution (~1 km × 1 km) product that contains the values of various geophysical parameters for each pixel. Product parameters over land include the NDVI and surface brightness temperature/radiance (ESA ENVISAT-1 Products Specifications Manual, 2013). These products are generated as orbital strips, subdivided into two grid types, 50 × 50 km squares (equal area) and on a longitude-latitude grid of 0.5° × 0.5°. A total of 47 AATSR images were acquired for each of the experimental sites/days selected previously. All AATSR data were obtained directly from ESA's EOLiSA platform (https://earth.esa.int/web/guest/eoli).

3.4   The SimSphere Land Biosphere Model

SimSphere, was used synergistically with the ASTER/AATSR data for the implementation of the "triangle" method. SimSphere was originally developed by Carlson and Boland (1978). Formerly known as the Penn State University Biosphere-Atmosphere Modeling Scheme (PSUBAMS) (Carlson and Boland, 1978; Lynn and Carlson, 1990), it was considerably modified to its current state by Gillies et al. (1997) and later by Petropoulos et al. (2013b). Since its development, it has diversified and become highly varied in its application use (for a comprehensive overview of the model use, see Petropoulos et al., 2009). It is currently maintained and freely distributed by the Department of Geography and Earth Sciences at Aberystwyth University (http://www.aber.ac.uk/simsphere).

4   METHODOLOGY

4.1   Preprocessing

Preliminary satellite data preprocessing followed commonly applied steps between the two different sensors. The preprocessing of the ASTER data included application of the appropriate scale factors (0.001 and 0.1 for AST07 and AST08 products, respectively) to each data set to obtain the physical values of the represented parameters. Following data scaling, the AST07 VNIR surface spectral reflectance data was resampled to the 90 m spatial resolution of the TIR-retrieved data sets of surface Kinetic Temperature to ensure a common spatial resolution between products. Subsequently, cloud-contaminated pixels were removed from the ASTER images and water masks were created manually to mask out rivers, lakes, and open seawater, to allow clearer interpretation of the triangular T_s/VI domains for only the land surface. Preprocessing for the AATSR included similar steps. Scale factors were initially applied to obtain the true values of the biophysical parameters before surfaces, including clouds, snow, and water, were masked from each image. Subsequently, data quality flags were used as guidance to further exclude ambiguous pixels/data, and finally, data sets were subset to a radius of $\sim$50 km around each validation study site used to develop the T_s/VI scatter plot domain.

4.2   "Triangle" Implementation

Implementing the "triangle" is based on combining a T_s/VI scatter plot (derived using both ASTER and AATSR imagery) with a SVAT model (herein SimSphere) for deriving spatially distributed estimates of SMC. An overview of the method implementation is provided below; more details can be found in Carlson (2007). The same steps were followed for the "triangle" implemented using both ASTER and AATSR imagery. Following preprocessing of the satellite imagery, the normalized difference vegetation index (NDVI) is computed from the red and near-infrared (NIR) spectral bands (Deering and Rouse, 1975). The NDVI is then scaled to an N^* value following the equation

$$N^* = \frac{\text{NDVI} - \text{NDVI}_0}{\text{NDVI}_s - \text{NDVI}_0} \tag{1}$$

where NDVI_0 and NDVI_s are the minimum and maximum values of NDVI at minimum (0%) and maximum (100%) vegetation cover, respectively. These values are generally computed from the scatter plot of the T_s versus the NDVI maps. N^* is then related to the fractional vegetation cover (F_r), following Gillies and Carlson (1995) and Choudhury et al. (1994):

$$F_r = N^{*2} \tag{2}$$

This transformation allows both the SVAT-simulated and the measured T_s from the satellite sensor to be plotted on the same scale. Following computation of the F_r, T_s normalization is then undertaken using the following the equation:

$$T_{scaled} = \frac{T_0 - T_{min}}{T_{max} - T_{min}} \tag{3}$$

where T_{min} and T_{max} are the expected minimum and the maximum T_s for wet vegetated pixels and for the dry, bare soil, respectively, interpolated from the scatter plot bounds and T_s corresponds to the temperature value of any pixel in the scene.

4.3　Coupling EO With the SVAT Model to Retrieve SMC

The satellite observations of T_s (or equally T_{scaled}) and F_r are then coupled with SimSphere to derive the inversion equations, which provide the spatially explicit maps of SMC. Briefly, this is done by first parameterizing SimSphere, although any other SVAT model with similar capabilities can also be used. SimSphere parameterization is done using the time and geographical location as well as the general soil and vegetation characteristics of the study site together with the appropriate atmospheric profile data. The model then is iterated repetitively until the extreme values of F_r and T_s in the T_s/F_r scatter plot between the simulated (modeled with the SVAT) and observed (obtained from the EO-derived data) are matched. Once the model tuning is completed, SimSphere is run repetitively, keeping the time (corresponding to the satellite overpass) the same but varying F_r and SMC over all possible values (0–100% and 0–1, respectively), in increments of 10 and 0.1 respectively, for all possible theoretical combinations of SMC and F_r. The result is a matrix of model outputs: SMC, F_r, T_{scaled} (or equally T_s) for the time of satellite overpass and calculated for each combination of F_r and SMC. Next, this output matrix is used to derive a series of nonlinear (quadratic) equations, relating F_r and T_{scaled} to each of the other variables of interest (i.e., SMC). These equations derived from the SVAT model matrix model outputs have the general form (shown here for the version relating SMC to F_r and T_s and/or T_{scaled}):

$$SMC = \sum_{p=0}^{3} \sum_{q=0}^{3} a_{pq} \left(T_{scaled}^*\right)^p (F_r)^q \qquad (4)$$

where the coefficients $a_{p,q}$ are derived from nonlinear regression between the matrix values of F_r, T_{scaled}, and SMC and p and q vary from 0 to 3. That way, the set of physically based relationships between the various surface-atmosphere parameters, as described by the detailed biophysical descriptions included in the SVAT model and inherent in the matrix outputs, are used to derive a series of simple, empirical relations relating the SMC parameter to just the locations of F_r and T_{scaled} recorded at that location. Because these variables of F_r and T_{scaled} are derivable from the satellite data, these equations are then used to derive the required spatially explicit maps of SMC from the satellite products of F_r and T_{scaled}.

5 RESULTS

To quantify the agreement between the "triangle"-derived estimates and the reference in situ observations, a series of appropriate statistical performance metrics were computed, including the root mean square difference (RMSD), the linear regression fit model coefficient of determination (r), the bias or mean bias error (MBE), and the scatter or mean standard deviation (MSD). Table 2 lists the formulas that express the above statistical terms. The summarized results of the statistical comparisons between the in situ volumetric moisture content (VMC) and the predicted SMC (0–1 SMC is the same as 0–1 VMC) via the "triangle" method for all image dates from both sensors are shown in Table 2. Fig. 2 depicts examples of spatial maps of SMC derived from the "triangle" inversion technique for one case day using ASTER and AATSR T_{kin} and F_r images.

TABLE 2 Summary of the Different Statistical Comparisons of SMC Predicted From the "Triangle" (Both ASTER and AATSR Scenarios) Versus the In Situ Observed Values

Satellite Data Used in "Triangle"	Mean Bias Error/MBE	Mean Standard Deviation/MSD	RMSD	R
ASTER	0.08	0.18	0.19	0.561
AATSR	0.01	0.06	0.06	0.844

Bias, scatter, and RMSD are expressed in vol vol^{-1}, R is unitless.

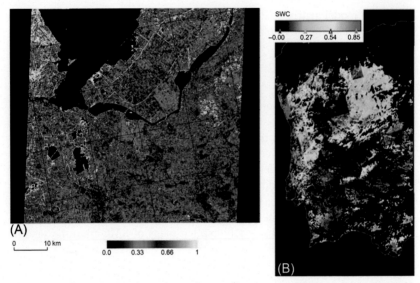

FIG. 2 (A) The ASTER "triangle"-inverted map of instantaneous SMC derived for March 30, 2004, for two of our test sites in The Netherlands. Here, the spatial resolution is 90 m and also clouds and water have been masked out from the scene. (B) A similar map produced from the AATSR data derived for a different date.

Although it was not possible to quantitatively evaluate the spatially distributed maps of SMC, visual inspection of both clearly reflected the large spatial variability of this parameter across topography and land use classes, which could also be the result of variability in the soil properties under each sensors' IFOV.

With regard to the comparisons performed between the point-predicted SMC derived from the ASTER "triangle" and the in situ measurements for all days of analysis (Table 2 and Fig. 3), results showed that the "triangle" estimated SMC with comparable accuracies to previous analogous verification studies of the technique using different sensors and implementation conditions (Gillies and Carlson, 1995; Gillies et al., 1997; Brunsell and Gillies, 2003). A minor overestimation of the in situ measurements by the "triangle"-derived SMC estimates was evident, illustrated by a mean bias of 0.08 vol vol^{-1}. A high scatter of 0.18 vol vol^{-1} also indicated a relatively poor estimation of the in situ data by the model, notably resulting in an RMSD of 0.19 vol vol^{-1} when all days were considered. Interestingly, RMSD was 0.09 vol vol^{-1} above the required accuracy range of 0.10 vol vol^{-1} for the operational retrieval of SMC. A moderate r (0.561) was exhibited for the agreement between both data sets, which was potentially due to the large divergence (RMSD— 0.47 vol vol^{-1}) between the compared SMC data sets over the coniferous forest site for the November 21, 2002, case day, which the "triangle" was apparently unable to closely reproduce.

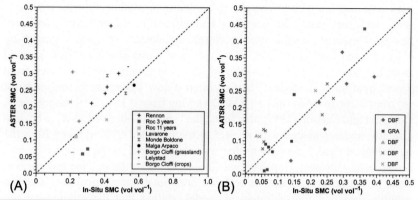

FIG. 3 (A) Agreement between point-predicted SMC (ASTER "triangle") versus the in situ point measurement of the 0–5 cm VMC (% vol) at the time of ASTER overpasses. The dashed line represents the 1:1 line of agreement. (B) Agreement between point-predicted SMC (AATSR "triangle") versus the in situ point measurement of the 0–5 cm VMC (% vol) at the time of AATSR overpasses. The dashed line represents the 1:1 line of agreement.

In terms of the SMC comparisons (Table 2, Fig. 3) between the in situ measurements and the estimates from the "triangle" implemented using AATSR imagery, results showed a significant improvement on the validation results from the ASTER "triangle." Notably, when the AATSR was used in the "triangle", a considerable increase in correlation between both data sets was evident, demonstrated by high r values of 0.844, an improvement of 0.283 on the ASTER results. The AATSR "triangle" estimates also displayed a minor improvement in bias prediction by 0.08 vol vol^{-1} (MBE—0.01 vol vol^{-1}) and a more significant improvement of $\sim$0.12 vol vol^{-1} (MSD—0.06 vol vol^{-1}) for scatter, again signifying an improved model performance. This improved model performance was reflected in a substantial decrease in error of $\sim$0.12 vol vol^{-1}, resulting in an RMSD of 0.06 vol vol^{-1}.

6 DISCUSSION

The validity of the "triangle" in deriving spatially distributed maps of instantaneous SMC has been previously examined using a range of EO sensors with varying degrees of agreement reported with regard to the ability of the technique to derive SMC; yet generally comparable to the results presented herein for both the ASTER and AATSR "triangle" schemes. For example, Carlson et al. (1995) implemented the "triangle" using National Oceanic and Atmospheric Administration (NOAA) Advanced Very High Resolution Radiometer (AVHRR) satellite data over an area of Newcastle in the United Kingdom. They attributed their results to the decoupling of the surface from the deeper substrate in regions of rapid soil drying (where large vertical gradients in soil moisture may exist near the surface) and also to the coarse resolution of the AVHRR sensor, which they stated did not allow the retrieval of the underlying soil processes. However, results are in closer agreement to those reported by Capehart and Carlson (1997), who performed comparisons of SMC derived from the "triangle" method using AVHRR data versus SMC simulated from a soil hydrological model. They reported a low degree of correspondence (r^2 from 0.266 to 0.441) and an RMSD varying from 0.15 to 0.19, respectively. Capehart and Carlson (1997) also found the "triangle"-derived SMC to be consistently underestimated with respect to the modeled SMC derived from the hydrological model, something that was also evident in the comparison presented here. However, Capehart and Carlson (1997) attributed the low correlations to the poor mismatch in both the horizontal and vertical scales of the satellite data and the hydrological model, stating that the satellite-derived T_{kin}/T_{rad} may be responding to the soil water content in a layer much shallower than the minimum resolution of the hydrological model. Gillies et al. (1997), in a verification study of the "triangle," also returned comparable results to those found in this study. In comparisons between VMC measurements and SMC predicted from the "triangle" using high-spatial resolution airborne data (of a spatial resolution similar to LANDSAT TM) authors had reported r^2 and standard errors in SMC estimation varying from 0.29 to 0.79 and from 8.73% to 8.25%, respectively.

Many authors have already documented as main sources of error in the implementation of the "triangle" (Gillies, 1993; Gillies and Carlson, 1995; Carlson et al., 1995; Gillies et al., 1997; Brunsell and Gillies, 2003; Chauhan et al., 2003), the following: uncertainty in the retrievals of F_r and T_{kin} from the remotely sensed data, spatial discrepancy in the resolution differences between the observed and model simulations, errors in SVAT initialization, errors or uncertainties in the regression equations, error introduced by the normalization procedures used to process the F_r and T_{kin} data, mismatch between the observed and simulated "triangles," and subjectivity in regard to the selection of the "triangle's" anchor points.

A number of sensitivity analyses on SimSphere (e.g., Ireland et al., 2015; Petropoulos et al., 2015) have found that F_r has an important contribution to the overall sensitivity of modeled SMC, adding to the rationale of the inclusion of F_r in the "triangle" method inversion process (i.e., the regression equations). Thus, any uncertainty in the derivation of F_r from the remotely sensed data would inherently have an effect on the accuracy of the simulated output by the "triangle," and thus a decrease in correlation between compared data sets. Furthermore, other researchers working on T_s/VI methods have also suggested that more accurate retrieval of T_s would be expected to have highly significant implications as this parameter is scaled in the implementation, and thus improved accuracies (or even inaccuracies) would be enhanced in the subsequent scaled T_s (Gillies and Temesgen, 2000). Spatial discrepancies between the "triangle" inverted parameter estimations and the in situ data can also cause large mismatches when comparing the two data sets. Ideally, the in situ observations used for comparisons would be colocated in both time and space with the satellite overpass. In reality, however, in situ measurements from ground observation networks are generally measured in the range of meters, whereas remotely sensed footprints are often on a much larger scale. Uncertainties or errors thus arise from the lack of representativeness in the remotely sensed data due to scale mismatch, geolocation errors, and noise from surface heterogeneity at the satellite spatial resolution. Moreover, apart from scale mismatch, particularly for SMC, questions arise as to the extent to which point comparisons between the satellite-derived SMC and observed VMC are meaningful, as the soil moisture is known to be highly spatially variable, not only horizontally, but also vertically (Carlson et al., 1995; Chauhan et al., 2003). Capehart and Carlson (1997), for instance, have pointed out that SMC derived from TIR remote sensing data is probably not very useful in understanding the column-average soil water content, but could still be useful in providing some insight into spatial variations in soil texture and hydraulic properties at the surface. Furthermore, SMC measured by in situ stations are normally averaged from the top 0–5 cm of the soil, whereas in the "triangle" implementation estimates the soil water content availability, which is equated to the SMC only if divided by the soil's field capacity. Such discrepancies could also account in part for explaining the discrepancies found between the EO-derived predictions and the in situ data.

7 CONCLUSIONS

This study aimed at exploring the effectiveness of integrating data from both ASTER and AATSR with the SimSphere land biosphere model via the "triangle" method to derive SMC over different ecosystems in Europe. Derived SMC was validated against co-orbital in situ measurements acquired from several sites belonging to the CarboEurope flux tower network. Overall, comparison of the "triangle"-derived estimates with the in situ measured SMC from the ASTER imagery exhibited an RMSD of 0.19 vol vol^{-1}, which although did not reach the required accuracy of 0.10 vol vol^{-1} for operational implementation, was comparable to previous studies. A clear improvement was evident when the AATSR data was used in the "triangle" method, with comparison results for all days of analysis returning an RMSD of 0.06 vol vol^{-1}. Correlation coefficient results demonstrated a good agreement between the in situ and both "triangle" schemes for the estimation of SMC (ASTER r—0.561/AATSR r—0.844), with the AATSR results again outperforming the ASTER. Our results provide strong supportive evidence for the potential value of the "triangle" inversion modelling technique to accurately derive estimates of SMC, and also represent an important step towards operational implementation of this approach.

ACKNOWLEDGMENTS

Authors wish to gratefully acknowledge the financial support provided by the European Commission under the Marie Curie Career Re-Integration Grant "TRANSFORM-EO" project and the High Performance Computing Facilities of Wales (HPCW) under the PREMIER-EO project for the completion of this work. The authors are also grateful to the CarboEurope site managers and the satellite data providers for the provision of the in situ and EO data, respectively, which supported the implementation of this study.

REFERENCES

Baldocchi, D.D., Valentini, R., Running, S.R., Oechel, W., Dahlman, R., 1996. Strategies for measuring and modeling carbon dioxide and waterfluxes over terrestrial ecosystems. Glob. Chang. Biol. 2, 159–168.

Barrett, B., Petropoulos, G.P., 2013. Satellite remote sensing of surface soil moisture. Chapter 4, pages 85–120. In: Petropoulos, G.P., (Ed.), Remote Sensing of Energy Fluxes and Soil Moisture Content. Taylor and Francis, ISBN: 978-1-4665-0578-0.

Bhattacharya, B.K., Mallick, K., Patel, N.K., Parihar, J.S., 2010. Regional clear sky evapotranspiration over agricultural land using remote sensing data from Indian geostationary meteorological satellite. J. Hydrol. 387 (1), 65–80.

Brunsell, N.A., Gillies, R.R., 2003. Scale issues in land-atmosphere interactions: implications for remote sensing of the surface energy balance. Agr. Forest. Meteorol. 117, 203–221.

Capehart, W.J., Carlson, T.N., 1997. Decoupling of surface and near-surface soil water content: a remote sensing perspective. Water Resour. Res. 33 (6), 1383–1395.

Carlson, T.N., Boland, F.E., 1978. Analysis of urban-rural canopy using a surface heat flux/temperature model. J. Appl. Meteorol. 17 (7), 998–1013.

Carlson, T.N., Gillies, R.R., Schmugge, T.J., 1995. An interpretation of methodologies for indirect measurement of soil water content. Agric. For. Meteorol. 77 (3), 191–205.

Carlson, T.N., 2007. An overview of the "triangle method" for estimating surface evapotranspiration and soil moisture from satellite imagery. Sensors 7, 1612–1629.

Choudhury, B.J., Ahmed, N.U., Idso, S.B., Reginato, R.J., Daughtry, C.S.T., 1994. Relations between evaporation coefficients and vegetation indices studied by model simulations. Rem. Sens. Environ. 50 (1), 1–17.

Chauhan, N.S., Miller, S., Ardanuy, P., 2003. Spaceborne soil moisture estimation at high resolution: a microwave-optical/IR synergistic approach. Int. J. Remote Sens. 22, 4599–4646.

Deering, D.J., Rouse, J.W., Haas, R.H., Schell, J.A., 1975. Measuring production of grazing units from Landsat MSS data. In: Proceedings of the 10th International Symposium of remote Sensing of Environment, ERIM, Ann Arbor, Michighan, August 23–25, pp. 1169–1178.

Dorigo, W.A., Wagner, W., Hohensinn, R., Hahn, S., Paulik, C., Xaver, A., Gruber, A., Drusch, M., Mecklenburg, S., van Oevelen, P., Robock, A., Jackson, T., 2011. The International Soil Moisture Network: a data hosting facility for global in situ soil moisture measurements. Hydrol. Earth Syst. Sci. 15, 1675–1698. http://dx.doi.org/10.5194/hess-15-1675-2011.

European Space Agency, 2012. Support to Science Element (2012): A Pathfinder for Innovation in Earth Observation. ESA, p. 41. Available from: http://due.esrin.esa.int/stse/files/document/STSE_report_121016.pdf (accessed on 10.07.13).

European Space Agency, 2013. ENVISAT-1 Products Specifications – AATSR Products Specifications. Available at: https://earth.esa.int/documents/10174/437508/Vol-07-Aats-4C.pdf.

Fan, L., Xiao, Q., Wen, J., Liu, Q., Tang, Y., You, D., ... Li, X., 2015. Evaluation of the airborne CASI/TASI T_s-VI space method for estimating near-surface soil moisture. Remote Sens. 7 (3), 3114–3137.

Gillies, R.R., 1993. A Physically-based Land Sue Classification Scheme Using Remote Solar and Thermal Infrared Measurements Suitable for Describing Urbanisation (PhD thesis). University of Newcastle, UK, 121 p.

Gillies, R.R., Carlson, T.N., 1995. Thermal remote sensing of surface soil water content with partial vegetation cover for incorporation into climate models. J. Appl. Meteorol. 34 (4), 745–756.

Gillies, R.R., Temesgen, B., 2000. In: Thermal Remote Sensing in Land surface Processes. CRC Press, New York, pp. 160–183.

Gillies, R.R., Carlson, T.N., Cui, J., Kustas, W.P., Humes, K.S., 1997. Verification of the "triangle" method for obtaining surface soil water content and energy fluxes from remote measurements of the Normalized Difference Vegetation Index NDVI and surface radiant temperature. Int. J. Remote Sens. 18, 3145–3166.

Ireland, G., Petropoulos, G.P., Carlson, T.N., Purdy, S., 2015. Addressing the ability of a land biosphere model to predict key biophysical vegetation characterisation parameters with Global Sensitivity Analysis. Environ. Model. Software 65, 94–107.

Lambin, E.F., Ehrlich, D., 1996. The surface temperature-vegetation index space for land cover and land-cover change analysis. Int. J. Remote Sens. 17 (3), 463–487.

Lynn, B.H., Carlson, T.N., 1990. A stomatal resistance model illustrating plant vs. external control of transpiration. Agr. Forest. Meteorol. 52 (1), 5–43.

Maltese, A., Capodici, F., Ciraolo, G., Loggia, G.L., 2015. Soil water content assessment: critical issues concerning the operational application of the triangle method. Sensors 15 (3), 6699–6718.

McCabe, M., Wood, E.F., 2006. Scale influences on the remote estimation of evapotranspiration using multiple satellite sensors. Rem. Sens. Environ. 105, 271–285.

Merlin, O., 2013. An original interpretation of the wet edge of the surface temperature-albedo space to estimate crop evapotranspiration (SEB-1S), and its validation over an irrigated area in northwestern Mexico. Hydrol. Earth Syst. Sci. 17 (9), 3623–3637.

Merlin, O., Ruediger, C., Al Bitar, A., Richaume, P., Walker, J.P., Kerr, Y.H., 2012. Disaggregation of SMOS soil moisture in southeastern Australia. IEEE Trans. Geosci. Rem. Sens. 50 (5), 1556–1571.

Merlin, O., Jose Escorihuela, M., Aran Mayoral, M., Hagolle, O., Al Bitar, A., Kerr, Y., 2013. Self-calibrated evaporation-based disaggregation of SMOS soil moisture: an evaluation study at 3 km and 100 m resolution in Catalunya, Spain. Rem. Sens. Environ. 130, 25–38.

Nishida, K., Nemani, R., Running, S., Glassy, J.M., 2003. Remote sensing of land surface evaporation (I) theoretical basis for an operational algorithm. J. Geophys. Res. 41, 493–501.

Petropoulos, G.P., Carlson, T.N., Wooster, M.J., Islam, S., 2009. A review of T_s/VI remote sensing based methods for the retrieval of land surface fluxes and soil surface moisture content. Adv. Phys. Geochem. 33 (2), 1–27.

Petropoulos, G.P., Griffiths, H.M., Dorigo, W., Xaver, A., Gruber, A., 2013a. Surface soil moisture estimation: significance, controls and conventional measurement techniques. In: Petropoulos, G.P. (Ed.), Remote Sensing of Energy Fluxes and Soil Moisture Content. Taylor and Francis, New York, ISBN: 978-1-4665-0578-0, pp. 29–48 (Chapter 2).

Petropoulos, G.P., Konstas, I., Carlson, T.N., 2013b. Automation of SimSphere land surface model use as a standalone application and integration with EO data for deriving key land surface parameters. European Geosciences Union, April 7–12th, 2013, Vienna, Austria.

Petropoulos, G.P., Ireland, G., Srivastava, P.K., Ioannou-Katidis, P., 2014. An appraisal of soil moisture operational estimates accuracy from SMOS MIRAS using validated in-situ observations acquired at a Mediterranean environment. Int. J. Remote Sens. 35 (13), 5239–5250.

Petropoulos, G.P., Ireland, G., Barrett, B., 2015. Surface soil moisture retrievals from remote sensing: evolution, current status, products & future trends. Phys. Chem. Earth 83–84, 36–56. http://dx.doi.org/10.1016/j.pce.2015.02.009.

Piles, M., Camps, A., Vall-Llossera, M., Corbella, I., Panciera, R., Rüdiger, C., Kerr, Y., Walker, J., 2011. Downscaling SMOS-derived soil moisture using MODIS visible/infrared data. IEEE Trans. Geosci. Rem. Sens. 49 (9), 3156–3166.

Piles, M., Petropoulos, G.P., Ireland, G., Sanchez, N., 2016. A novel method to retrieve soil moisture at high spatio-temporal resolution based on the synergy of SMOS and MSG SEVIRI observations. Rem. Sens. Environ.

Sandhold, I., Rasmussen, K., Andersen, J., 2002. A simple interpretation of the surface temperature/vegetation index space for assessment of surface moisture status. Remote Sens. Environ. 79, 213–224.

Tian, J., Su, H., Sun, X., Chen, S., He, H., Zhao, L., 2013. Impact of the spatial domain size on the performance of the T_s-VI triangle method in terrestrial evapotranspiration estimation. Rem. Sens. 5 (4), 1998–2013.

Vereecken, H., Huisman, J.A., Pachepsky, Y., Montzka, C., van der Kruk, J., Bogena, H., Weihermüllera, L., Herbsta, M., Martinez, G., Vanderborght, J., 2014. On the spatio-temporal dynamics of soil moisture at the field scale. J. Hydrol. 516, 76–96.

Chapter 6

Spatial Downscaling of Passive Microwave Data With Visible-to-Infrared Information for High-Resolution Soil Moisture Mapping

M. Piles*,† and N. Sánchez‡

*Universitat Politècnica de Catalunya, IEEC/UPC, Barcelona, Spain, †Barcelona Expert Center, Institute of Marine Sciences, CSIC, Barcelona, Spain, ‡University of Salamanca/CIALE, Villamayor, Spain

1 INTRODUCTION

Theoretical and experimental evidence have proved that L-band passive microwaves are optimal for measuring soil moisture (SM) under up to moderate vegetation densities and under all weather conditions. However, practical constraints on the L-band required antenna size and the altitude of low Earth orbits limit the spatial resolution of the observations to ~40 km. The two first series of new generation satellites dedicated to measuring the Earth's surface SM have L-band radiometers onboard: the ESA's Soil Moisture and Ocean Salinity (SMOS, 2009–17; Font et al., 2010; Kerr et al., 2010); and NASA's Soil Moisture Active Passive (SMAP, 2015–18; Entekhabi et al., 2010) missions. In this context, enhancing the spatial resolution of passive microwave observations is widely anticipated to extend the applicability of the data to regional and local studies.

A variety of multisensor/multiresolution approaches have been proposed for refining the broad resolution of SMOS and SMAP observations. SMAP has a high-resolution radar onboard to enhance the spatial resolution of SM estimates. Hence, the use of temporal covariations of active and passive data is the central approach to SMAP disaggregation (Das et al., 2011). There are other approaches both for Active data (Tomer et al., 2015) and using different schemes (Verhoest et al., 2015). However, operations of active-passive

Satellite Soil Moisture Retrieval. http://dx.doi.org/10.1016/B978-0-12-803388-3.00006-1
109

products ceased abruptly with the failure of the SMAP radar after about 10 weeks of operations. For SMOS, combination with optical/thermal infrared (TIR) data seems to be the most promising strategy (Merlin et al., 2010, 2012; Piles et al., 2011, 2014; Fang and Lakshmi, 2014; Sánchez-Ruiz et al., 2014; Piles et al., 2016). Thermal and infrared sensors can provide complementary information of SM patterns at higher spatial resolutions than radiometers (from tens of meters to several kilometers). The surface reflectance observed by optical sensors can be used to provide an indirect estimate of SM through empirical spectral vegetation indexes (VIs) (Gao et al., 2013; Lobell and Asner, 2002). The common method used by TIR remote sensing to estimate SM is by calculating thermal inertia (Qin et al., 2013; Verstraeten et al., 2006). Yet, the observations from optical/TIR sensors are used to assess SM status but not to provide a quantitative estimate. To take advantage of microwave and visible-to-infrared remote sensing, synergistic techniques can be developed to optimally blend the multisensor information into SM estimates at different spatial resolutions. This idea has motivated a number of studies in recent decades (Chauhan et al., 2003; Pellenq et al., 2003; Kim and Hogue, 2012; Piles et al., 2011, 2014; Sobrino et al., 2012; Merlin et al., 2012; Fang and Lakshmi, 2014; Sánchez-Ruiz et al., 2014). The proposed SM downscaling or disaggregation approaches vary with respect to input ancillary data (e.g., optical data, topography, soil depth, field capacity), the nature of the scale linking model (physical, semiempirical, empirical), and the underlying physical assumptions (i.e., how SM is linked to available fine-scale modeled or observational information). This chapter focuses on the semiempirical SM downscaling methods based on the triangular relationship observed between the land surface temperature (LST) and VI space to characterize regional SM variability. These downscaling approaches have attracted significant attention as they do not require extensive ground-based data or model-based information and are relatively easy to implement. A strong negative correlation exists between the LST and a VI, with the SM condition being the main factor influencing the LST/VI slope (Carlson et al., 1994; Moran et al., 1994). The emergence of the triangle (or trapezoid) shape in the LST/VI feature space is the result of the low sensitivity of LST to SM variation over highly vegetated areas, but its increased sensitivity (and thus greater spatial variation) over areas of bare soil (Carlson, 2007; Petropoulos et al., 2009).

Based on the LST/VI feature space, Piles et al. (2011) proposed a SM downscaling scheme for SMOS that links the SM with surface temperature, a VI, and brightness temperatures using a polynomial regression model. The model combines highly accurate but low spatial resolution SMOS radiometric information with high spatial resolution but low sensitivity visible-to-infrared imagery to SM across spatial scales. An upgraded version of this algorithm is operational at the Barcelona Expert Center (BEC) and provides daily estimates of surface SM at 1 km spatial resolution over the Iberian Peninsula (Piles et al., 2014). These level 4 (L4) maps are available from year 2010 to the present and can be freely accessed through BEC visualization and distribution services (cp34-bec.cmima.csic.es). Global SMOS data as well as NASA's Terra/Aqua

Moderate Resolution Imaging Spectroradiometer (MODIS) data over the Iberian Peninsula are received in near real-time (NRT) at BEC facilities; the downscaling algorithm is triggered twice a day, corresponding to SMOS ascending and descending passes to produce high-resolution SM maps in NRT (delay of <6 h). Since 2012, these maps are routinely used by *Diputació de Barcelona* local fire prevention services in their early warning system to detect extremely dry soil and vegetation conditions posing a risk of fire (Piles et al., 2013; Chaparro et al., 2015), and in the estimation of Gross Primary Productivity (Sánchez-Ruiz et al., 2015). Also, these maps have been used to assess the impact of water stress in recent forest decline episodes (Chaparro et al., 2014).

Following the abovementioned studies for the synergistic use of SMOS and MODIS, other studies have been focused on improving the semiempirical linking model, by exploiting the multiangular and full-polarimetric information to be used (Piles et al., 2012), by exploring the use of VIs from the shortwave infrared (SWIR) (Sánchez-Ruiz et al., 2014), and by exploring the LST and SM relationship dependence on season and time of the day (Pablos et al., 2014). More recently, the feasibility of applying this synergistic approach to estimate high spatiotemporal resolution SM from SMOS and Spinning Enhanced Visible and Infrared Imager (SEVIRI) geostationary data has also been addressed (Piles et al., in press). The downscaling concept has also been validated from airborne platforms, using LST, VI, and microwave-derived information such as L-band brightness temperatures (Martín et al., 2011; Sánchez et al., 2014), surface emissivity (Sobrino et al., 2012), and Global Navigation Satellite System-Reflectometry (GNSS-R) reflectivity (Sánchez et al., 2015). This chapter provides an overview of results showing the synergistic use of passive microwave and VIS/IR for multiscale SM estimation, with case studies using satellite and airborne observations. Section 2 provides a thorough description of the synergistic approach. Section 3 presents results from its evaluation using satellite platforms, and Section 4 shows the applicability of the methodology on airborne platforms. The last section includes an overview of future research lines and recommendations.

2   A SEMIEMPIRICAL MODEL TO CAPTURE THE SYNERGY OF PASSIVE MICROWAVES WITH OPTICAL DATA AT DIFFERENT SPATIAL SCALES

Theoretical and experimental studies have demonstrated that there can be a unique relationship between soil moisture s_m, a vegetation index VI, and land surface temperatures LST on a given region under specific climatic conditions and land surface types. This relationship can be expressed through a regression formula such as (Carlson et al., 1994)

$$s_m = \sum_{i=1}^{n}\sum_{j=1}^{n} a_{ij} \cdot VI_i \cdot LST_j \qquad (1)$$

where n should be chosen so as to give a reasonable representation of the data. SM downscaling algorithms have been developed based on this relationship between SM, VI, and LST (Chauhan et al., 2003; Kim and Hogue, 2012). In the context of SMOS, Piles et al. (2011) showed that the relationship between land surface parameters and SM is strengthened if brightness temperatures T_B are added to the right side of Eq. (1):

$$s_m = \sum_{i=1}^{n} \sum_{j=1}^{n} \sum_{k=1}^{n} a_{ijk} \cdot VI_i \cdot LST_j \cdot T_{B_k} \tag{2}$$

The brightness temperature term includes information on all parameters that dominate the Earth's emission at L-band, in addition to SM; for example, soil roughness, soil texture, soil temperature, vegetation opacity, and vegetation scattering albedo. A downscaling approach for SMOS using MODIS was proposed, based on the assumption that the chosen linking model—a regression formula from Eq. (2)—holds at the two spatial scales under consideration; that is, the coarse scale of passive observations (e.g., 40 km from SMOS) and the fine-scale of optically-derived parameters (e.g., 1 km from MODIS). Parameters in the model are represented in both a low- and a high-spatial resolution grid: VI and LST and are linearly aggregated to 40 km and T_B resampled at 1 km. The model is first applied at low resolution using a coarse scale estimate of SM as a reference (e.g., SMOS L2 or SMOS L3 data); a system of linear equations is set up using all the pixels of a specific morning/afternoon pass and scene, and model fitting coefficients a_{ijk} are obtained by the least squares method. Then, the model is applied at high resolution using the obtained coefficients to get the disaggregated SM.

After the successful launch in 2009 of SMOS, the downscaling algorithm was applied to a set of SMOS images acquired during the commissioning phase over the Murrumbidgee catchment, in southeastern Australia, and validated with the *in situ* data from the OZnet SM monitoring network (Piles et al., 2011). Fig. 1 shows visual results of the application of the proposed downscaling algorithm to SMOS data acquired over OZnet on Feb. 17, 2010 (6:00 am). The different panels show in the top row the 40 km SMOS L2 SM and the 1 km SMOS/MODIS-derived SM using a regression model from Eq. (1), and using a regression model from Eq. (2), and in the bottom row the SMOS T_B at horizontal polarization and 42.5 degrees incidence angle, the 1 km MODIS/Aqua LST and the 1 km MODIS/Terra normalized difference vegetation index (NDVI) (see Camps et al. (2012) for further details). Results evidence that the use of brightness temperatures in the linking model is needed to capture SM variability at high resolution, and to reproduce changes in SM due to rain events occurring between the passes of SMOS and MODIS. Results from comparison with *in situ* data indicate that SM variability is effectively captured at 1 and 10 km scales without a significant degradation of the root mean square error (Piles et al., 2011).

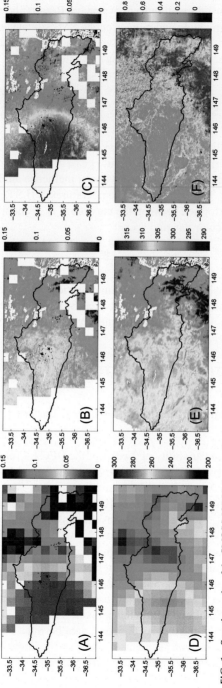

FIG. 1   Sample results of the SMOS/MODIS downscaling approach over the Murrumbidgee catchment, in Southeast Australia, from Feb. 17, 2010 (6:00 am). (A) 40 km SMOS SM (m³/m³); (B) 1 km SMOS-derived SM maps (m³/m³) using the linking model in Eq. (1); (C) 1 km SMOS-derived SM maps (m³/m³) using the linking model in Eq. (2); (D) 40 km SMOS T_{Bh} image (K) at 42.5 degrees; (E) 1 km MODIS/TERRA LST (K); (F) 1 km MODIS/TERRA NDVI. Empty areas in the image correspond to unsuccessful SMOS SM retrievals, or clouds masking MODIS LST measurements. (*Adapted from Camps, A., et al., 2012.*)

In Piles et al. (2012) a variety of model formulations including SMOS polarimetric and multiangular information were evaluated. Results showed that more robust coefficient determination and more accurate SM estimates were obtained when multiangular brightness temperatures at vertical and horizontal polarizations were used. Based on these results, the following linear linking model was proposed to relate the two instruments across spatial scales

$$s_m = a_0 + a_1 \cdot LST_N + a_2 \cdot VI + \sum_{i=1}^{3} a_{3i} \cdot T_{BH\theta_i N} + \sum_{i=1}^{3} a_{4i} \cdot T_{BV\theta_i N} \tag{3}$$

where LST_N are normalized LST, VI is the vegetation index (i.e., the optically derived vegetation descriptor), and $T_{BH\theta_i N}$ and $T_{BV\theta_i N}$ are normalized horizontal and vertically polarized SMOS T_B, respectively, at incidence angles θ_i of 32.5, 42.5, and 52.5 degrees. Normalization of temperatures T (LST_N, $T_{BH\theta_i N}$, and $T_{BV\theta_i N}$) in Eq. (3) is performed between the maximum T_{max} and minimum T_{min} obtained for each particular day and scene:

$$T_N = \frac{T - T_{min}}{T_{min} - T_{max}} \tag{4}$$

The possibility of using the proposed semiempirical model to obtain fine-scale SM maps from microwave missions such as SMOS considerably widens the applicability of the data to fields that, initially, were not considered.

3 HIGH-RESOLUTION SOIL MOISTURE MAPPING FROM SPACE

3.1 Long-Term Validation Over the Central Part of the Duero Basin

The temporal and spatial variability of 2 years of the BEC L4 fine-scale (1 km) SM product was evaluated through comparison with ground-based measurements acquired at the *in situ* Soil Moisture Measurement Stations Network of the University of Salamanca (REMEDHUS) located in the central part of the Duero basin, Spain (Piles et al., 2014). Results showed that the downscaling method improves the spatial representation of SMOS coarse SM estimates (SMOS L2) while maintaining temporal correlation and root mean square differences (RMSD) with ground-based measurements. Fig. 2 shows the temporal evolution of SM time-series (i.e., SMOS L2, BEC L4, *in situ*) over REMEDHUS. It can be seen that area-averaged downscaled estimates match well with *in situ* data (circles are enclosed within the network's SM variability in shaded yellow). Scatter plots of Fig. 3 display the agreement between remotely sensed and *in situ* SM time-series, with segments illustrating the linear fit of seasonal data. Results are shown for a representative station of rainfed cereals, the most common land use in the area, for SMOS L2 (Fig. 3A) and for BEC L4 (Fig. 3B).

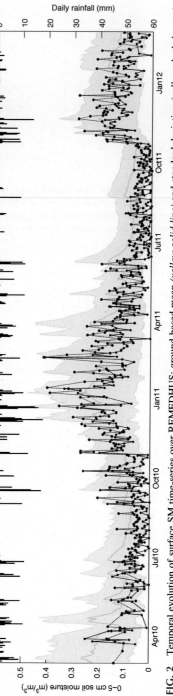

FIG. 2 Temporal evolution of surface SM time-series over REMEDHUS: ground-based mean (*yellow solid line*) and standard deviation (*yellow shaded areas*), SMOS L2 (*black stars*), 40-km aggregated SMOS/MODIS downscaled (*blue circles*). Daily mean rainfall on top. (*Adapted from Piles, M., et al., 2014.*)

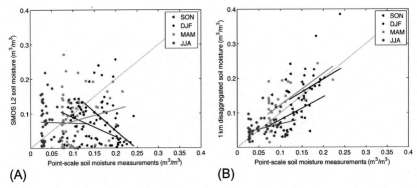

FIG. 3 Results of the seasonal analysis for the hydrological year starting in Sep. 2010 (afternoon passes) over the 07 station (rainfed cereals). (A) Scatter plots of SMOS L2 SM versus point-scale measurements. (B) Scatter plot of 1-km SMOS-MODIS disaggregated SM versus point-scale measurements. Segments are the linear fit of seasonal (3 months) data. *(Adapted from Piles, M., et al., 2014.)*

It can be seen that the slope of the linear correlation is significantly improved in the L4 maps (it is closer to the 1:1 line) and the dynamic range of *in situ* SM measurements is reproduced in the high-resolution maps, including stations with different mean soil wetness conditions (see further results in Piles et al., 2014). This evaluation study supports the use of this downscaling approach to enhance the spatial resolution of SMOS observations over semiarid regions such as the Iberian Peninsula.

A sensitivity study of the relationship between LST and SM was performed using 2 years (2012 and 2013) of *in situ* measurements and spaceborne observations from SMOS and MODIS Terra/Aqua over the REMEDHUS network (Pablos et al., 2014). Using *in situ* SM and LST, results revealed that instantaneous SM exhibits a higher correlation with daily maximum LST than with instantaneous, daily mean, daily median, daily minima, and diurnal cycle LST. This could be explained by the fact that the time of daily maximum LST is also the time of maximum potential evapotranspiration; that is, when there is a high atmospheric demand for water and therefore higher coupling of SM and LST.

In the case of spaceborne observations, for each SMOS overpass (6 and 18 coordinated universal time (UTC)) on a given day there are four available MODIS overpasses: two from Terra (9:30 and 21:30 UTC) and two from Aqua (13:30 and 1:30 UTC). Fig. 4 shows a daily cycle of *in situ* LST with labels indicating the different SMOS and MODIS overpasses. Results of the eight possible combinations of correlations—SMOS morning/afternoon with MODIS Terra/Aqua day and with MODIS Terra/Aqua night—are shown in Fig. 5 for the REMEDHUS network (left), and also extended to the whole Iberian Peninsula (right). Higher correlations are obtained between SMOS SM and MODIS

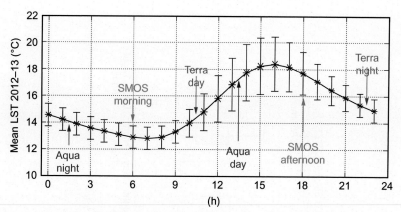

FIG. 4 LST *in situ* annual cycle. Error bars represent hourly mean and standard deviation. Labels indicate the daily SMOS and MODIS Terra/Aqua overpasses.

Aqua/Terra daytime than nighttime LST. These results are consistent with the *in situ* analysis, as daytime LST is closest to daily maximum LST.

This LST-SM sensitivity study supports the use of MODIS Terra/Aqua daytime observations in the SMOS downscaling algorithm. In Piles et al. (2014) SMOS observations were combined with the closest MODIS overpass (Aqua night and Terra day for morning passes, Aqua day and Terra night for afternoon passes) and overall results indicated that better fine-scale SM estimates should be expected from afternoon than from morning orbits. However, if only daytime acquisitions are used, the statistical scores obtained for the same period from both SMOS morning and afternoon orbits are comparable (not shown for brevity).

The linking model in Eq. (3) using SMOS morning/afternoon and MODIS Terra/Aqua daytime observations is implemented at BEC to generate its L4 SM product at 1 km spatial scale for the Iberian Peninsula.

3.2   Exploring the Use of SWIR-Based Vegetation Indices to Disaggregate SMOS Observations to 500 m

Some studies have demonstrated that a direct and significant relationship exists between SM and the spectral response in the visible and near-infrared (VNIR) region. Products such as the NDVI, reflectances, derivatives, or other vegetation indices in the VNIR region have been compared against *in situ* or modeled SM (Farrar et al., 1994; Adegoke and Carleton, 2002; Khanna et al., 2007; Wang et al., 2007; Schnur et al., 2010). However, some research evidenced that the SWIR region could provide better results for detection of soil water content (Lobell and Asner, 2002; Whiting et al., 2004; Finn et al., 2011). Moreover, the SWIR band presents regions of water absorption by plant leaves, and it is well known that SWIR reflectance is negatively related to leaf water content

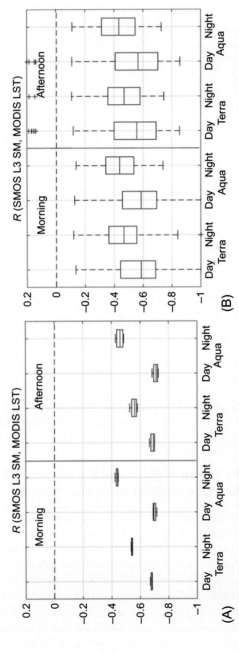

FIG. 5 Correlation between SMOS surface SM (morning and afternoon) and MODIS Terra/Aqua (night and day) LST over (A) REMEDHUS and (B) the Iberian Peninsula. *(Adapted from Pablos, M., et al., 2014.)*

(Fensholt and Sandholt, 2003). The spectral signature of vegetation in near-infrared (NIR) and SWIR bands can thus be associated to the plant water status. Furthermore, there are few studies linking this status, described by the NIR-SWIR indices, to SM in the root zone (Lobell and Asner, 2002; Fensholt and Sandholt, 2003).

The use of high-resolution LST and SWIR-based vegetation indices to disaggregate SMOS observations down to 500 m SM maps were explored in Sánchez-Ruiz et al. (2014). The hypothesis suggested in this research was that SWIR-NIR indices representing the plant water conditions can improve the current SM downscaling approaches based in the synergy between visible, infrared, and microwave observations. In this line, the MODIS-derived normal difference water index (NDWI) is a potential source to estimate vegetation water content due to its availability and temporal resolution (Chen et al., 2005). Indeed, many disaggregation methods using VNIR data as inputs have been applied to improve SM retrievals (Chauhan et al., 2003; Merlin et al., 2005, 2013; Piles et al., 2011; Choi and Hur, 2012; Kim and Hogue, 2012), but little research has been done applying the SWIR-based indices. In Sánchez-Ruiz et al. (2014), these so-called water indices extracted from MODIS products have been used together with MODIS LST and SMOS T_B to improve the spatial resolution of ~40 km SMOS SM estimates using the linking model in Eq. (3).

The indices were calculated as the normalized ratio of the 1240, 1640, and 2130 nm SWIR bands and the 858 nm NIR band from the 8-day 500 m land surface reflectance MODIS product. The three different absorption regions of the SWIR could lead to different results, as was expected after comparing the scatter plot of LST versus vegetation/water indices in a two-dimensional space (Fig. 6). In all cases, SM conditions vary from high values on the bottom of the scatter plot to low values on the top. In this figure, scatter plots of NDVI and NDWI3 showed a wider range of variation of the index, while the scatter plot of NDWI1 showed a nearly flat shape, with values around zero. Therefore changes in SM were expected to be better predicted when using NDVI and NDWI3 in Eq. (3), due to their larger dynamic range.

The retrieved SM maps were compared against *in situ* measurements from REMEDHUS. Results of this comparison showed that using the proposed vegetation/water indices in the algorithm preserves the sensitivity of the SMOS original SM product at 40 km, and that the temporal dynamics of the ground-based SM is captured at 500 m spatial resolution. Although differences between the NDWI1, NDWI2, NDWI3, and NDVI were small, the NDWI1 index resulted in the best proxy of vegetation state to be included in the model linking SMOS and MODIS. This indicates that, contrary to what we expected, using a VI with a higher dynamic range is not having an important effect in the linking model. Also, this experiment suggests that the VI provides enough fine-scale information for the downscaling; that is, SM can be estimated at the spatial scale of the VI even if the LST information is at a lower spatial resolution.

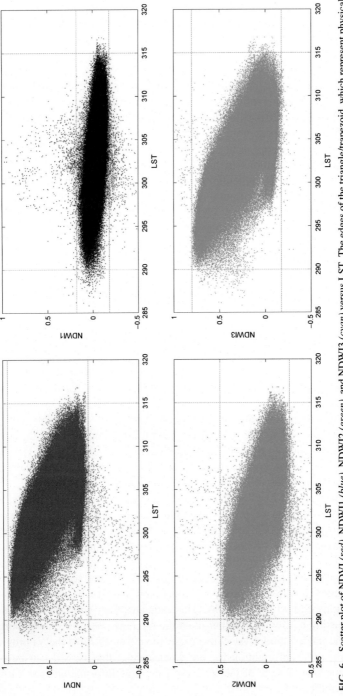

FIG. 6 Scatter plot of NDVI (*red*), NDWI1 (*blue*), NDWI2 (*green*), and NDWI3 (*cyan*) versus LST. The edges of the triangle/trapezoid, which represent physical limits, are denoted by dashed gray lines. Data are from May 16, 2011, and correspond to images of the whole Iberian Peninsula. (*Adapted from Sánchez-Ruiz, S., et al., 2014.*)

4 AIRBORNE FIELD EXPERIMENTS

4.1 Airborne Platform for Simultaneous Thermal, VNIR Hyperspectral and Microwave L-Band Acquisions: Proof-of-Concept and Soil Moisture Estimation at Very High Spatial Resolution

A test flight campaign over the area of Gimenells (Lleida, 41°39′21.34″N, 0°24′21.25″E) was performed on Mar. 31, 2011 (Martín et al., 2011). Fig. 7 shows the Institut Cartogràfic de Catalunya (ICC) airplane Cessna 208 Caravan where two photogrammetric sensors were used to simultaneously acquire TIR and VNIR hyperspectral data together with L-band brightness temperatures.

This first field experiment served as a proof-of-concept to show the feasibility of acquiring simultaneous observations from the Compact Airborne Spectrographic Imager (CASI), the Thermal Airborne Spectrographic Imager (TASI), and the Airborne RadIomEter at L-band (ARIEL-2). The ARIEL-2

FIG. 7 Thermal (TASI) and VNIR (CASI) hyperspectral sensors (bottom right) and radiometer installation (bottom left) onboard ICC airplane (top).

radiometer has been designed and manufactured by the *Universitat Politècnica de Catalunya* and is the second version of the ARIEL prototype, which was developed for small remote controlled aircrafts (Acevo-Herrera et al., 2010).

ARIEL-2 brightness temperatures, CASI NDVI, and TASI LST are shown in Fig. 8A–C, respectively. These data were employed to estimate SM at the spatial resolution of the radiometer (200 m, Fig. 8D) by inversion of a forward model (Martín et al., 2011), and at the spatial resolution of the optical sensors (2 m, Fig. 8E) by using a simplification of the linking model in Eq. (3), in which single look-angle (nadir) T_B at horizontal polarization were used.

A second airborne experiment with this sensor combination took place in the REMEDHUS area in which the application of the algorithm both at very high spatial and spectral resolution was explored (Sánchez et al., 2014). As an alternative to broadband multispectral remote sensing sensors, which cover relatively wide ranges of the electromagnetic spectrum, hyperspectral sensors collect images with high spectral resolution that could provide indirect information of SM (Grandjean et al., 2010; Finn et al., 2011), although the best spectral range for detecting SM has yet to be determined. Moreover, little research has investigated the potential of combining microwave brightness temperatures and hyperspectral solar reflectance; with this aim we carried out a field experiment in which hyperspectral observations were combined with brightness temperature and LST to obtain fine-scale SM estimations (Sánchez et al., 2014). Airborne hyperspectral observations were acquired with the same two sensors: TASI 600 (8–11.5 μm, 32 bands) and CASI 550 (400–1000 nm, 72 bands). Radiometric data was acquired by ARIEL-2, which was provided by the Universitat Politècnica de Catalunya.

The performance of the triangle method, as stated by Carlson (2007), is more effective with higher-resolution imagery such as that from Landsat or aircraft radiometers because the triangle is more easily resolved. Fig. 9 shows a subzone of the NDVI and LST images of the study area containing irrigated and rainfed crops and the associated scatter plot between NDVI-CASI and LST-TASI. The different response of irrigated areas versus rainfed and stubble-covered areas is clearly related to their soil water content. The link between the scatter plot and the imagery confirmed the potential of the NDVI-LST space for separating the SM content associated with the different land uses and agreed with the shape and the interpretation of the conceptual LST-NDVI space described in previous studies (Lambin and Ehrlich, 1996; Sandholt et al., 2002).

The hyperspectral optical and thermal data sets were first analyzed and processed to select the best hyperspectral features to be included in the SM retrieval procedure. In this case, the processing of hyperspectral bands was designed to determine which bands captured the different SM content of each sample. The lambda versus lambda R^2 model (Thenkabail et al., 2004, 2013) was applied to provide a data-mining technique to highlight wavebands with unique information content related to SM. The analysis showed that the optimal spectral regions ranged from 620 to 678 nm and 730 to 926 nm for comparison with

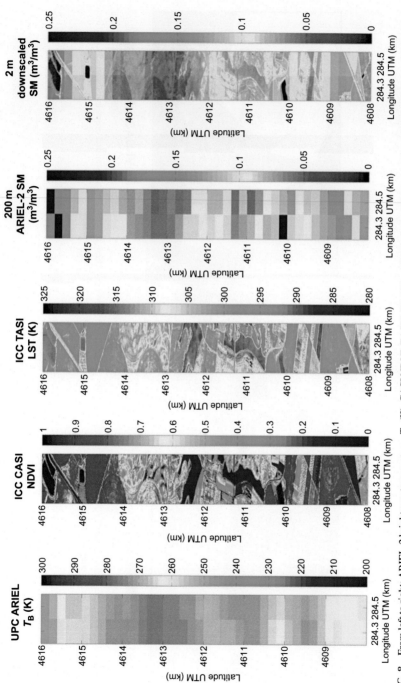

FIG. 8 From left to right: ARIEL-2 brightness temperatures T_B (K), CASI NDVI, TASI LST, 200-m SM from ARIEL-2, 2-m downscaled SM from the combination of ARIEL-2, CASI, and TASI observations, using a simplification of Eq. (3), which only uses single look-angle T_B at horizontal polarization.

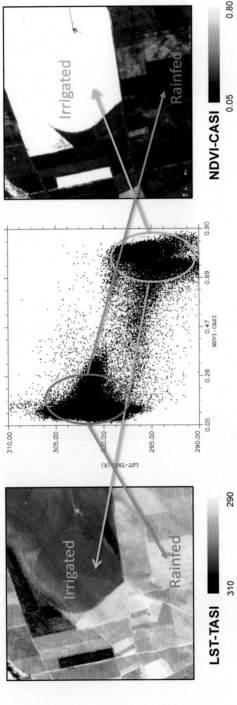

FIG. 9 LST-TASI image (left), NDVI-CASI image (right), and scatter plot of an example area where the irrigated and rainfed crops are separately distributed (center). *(Adapted from Sánchez, N., et al., 2014)*

SM and provided nonredundant information at the same time. The test of the hyperspectral normalized indices in those bands *versus* the SM showed a very good direct correlation with *in situ* SM (R between 0.75 and 0.80), and thus they were selected as vegetation proxies in the linking model. Later, the resulting SM from this model was compared with ground measurements taken in the area during the flight ($n = 63$ samples), which covered a fairly wide range of SM conditions. The comparison showed an $R > 0.76$ and an RMSD below or equal to 0.07 m^3/m^{-3}, which could be considered very satisfactory.

This work showed that a hyperspectral-based airborne field experiment is able to provide SM estimations with good accuracy and detailed spatial resolution. The application of the linking model at very high spatial resolution resulted in SM estimates that agreed well with *in situ* measurements. The fine spatial detail of the resulting estimates enhanced the representativeness of the *in situ* SM measurements for validating the results. Spatial resolution improved the validation and provided the spatial framework to disaggregate SM at very high spatial resolution (3.5 m). Hence, this method could replace intensive field campaigns or high-cost data collection at different spatial scales, and could be particularly promising for precision agriculture.

4.2   Airborne GNSS-R and Landsat 8 for Soil Moisture Estimation

Reflectivity from GNSS-R could be potentially used to monitor vegetation and soil variables related to the water content. While the synergy between thermal, optical, and passive microwave observations is well known for the estimation of SM and vegetation parameters, the use of remote sensing sources based on the GNSS remains unexplored. The most recent experiments are based on the attenuation and scattering that vegetation cover produces into the GNSS signal before it impinges on the ground and after it is reflected to the receiving antenna. Due to the novelty of this technique, the performance of airborne reflectometers for vegetation/soil applications needs further research, and remains a challenge of increasing interest, especially with the approval of a new dedicated GNSS-R spaceborne mission, i.e. the NASA CYGNSS (Ruf et al., 2014), expected to be launched in Oct. 2016, and the ESA GEROS-ISS experiment (Wickert et al., 2013). With the aim of filling a gap in this field, an innovative sensor developed by the Universitat Politècnica de Catalunya based on GNSS-R was tested for SM estimation, the Light Airborne Reflectometer for GNSS-R Observations (LARGO) (Alonso-Arroyo et al., 2014) (see Fig. 10). The objective was to evaluate the combined use of GNSS-R observations, with a time-collocated Landsat 8 image for SM retrieval under semiarid climate conditions (Sánchez et al., 2015). Also, we tested if the combined use of optical and thermal data with the GNSS-R signal improves the sensitivity to SM obtained with each of the sensors separately. A total of two RGB-based indices and five NIR- and SWIR-based indices were tested as proxies of vegetation and SM status, together with the land temperature, all of them from Landsat 8. The

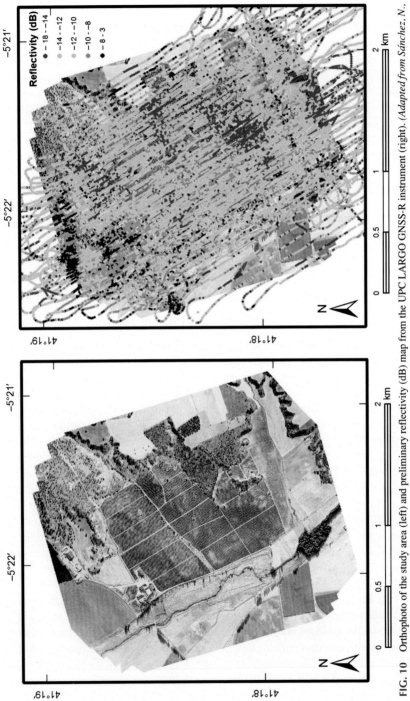

FIG. 10 Orthophoto of the study area (left) and preliminary reflectivity (dB) map from the UPC LARGO GNSS-R instrument (right). *(Adapted from Sánchez, N., et al., 2015)*

brightness temperature of the algorithm in Eq. (3) was replaced here by the reflectivity observed with LARGO, binned between 30 and 50 degrees and between 10 and 30 degrees incidence angles. The optical and thermal data sets provided the spatial framework needed to adjust the GNSS-R observations and the resulting SM estimates to a fixed grid of 30 m spatial resolution.

Taking these data sets separately, and considering the particular weather and vegetation conditions of the experiment, LARGO reflectivity did not show distinctive sensitivity to the main vegetation covers present in the area. However, a noticeable relationship to SM was detected. This relationship was strongly reinforced if the reflectivity was merged to the surface temperature and some vegetation/water indices, all of them retrieved from Landsat 8 bands. Indeed, encouraging correlations were found with the ground measurements ($R > 0.60$) when applying the linking model, much higher than taking each variable separately. It was also found that, for the same SM and roughness conditions, reflectivity from larger incidence angles is higher than for lower ones, and consequently less affected by noise. The joint use of GNSS-R reflectivity, water/vegetation indices, and thermal maps from Landsat 8 not only allows capturing SM spatial gradients under very dry soil conditions, but also holds great promise for accurate SM estimation.

5 FUTURE LINES AND RECOMMENDATIONS

Taking advantage of new satellite-based information on SM (e.g., SMOS, SMAP), LST (e.g., MODIS, SEVIRI), and vegetation (e.g., Landsat-8, Sentinel 2), the performance and robustness of the LST/VI relationship to SM under different atmospheric conditions, terrain, and land covers should be evaluated using different temperatures (instantaneous, thermal inertia, maxima, and minima) and vegetation proxies (fraction of vegetation cover, leaf index, water indices, time-integrated indices). Also, the scale invariance on the LST/VI-SM relationship that is assumed in the downscaling needs to be confirmed and evaluated to determine possible uncertainty ranges in the SM estimates. We anticipate that physical and meteorological heterogeneity of the surface can lead to significant differences in scale, due to subfootprint variability.

The abovementioned studies should be based on satellite observations, but also incorporate *in situ* observations as well as modeling estimates as a benchmark. New metrics to assess the accuracy of multiscale SM estimates have been developed (McColl et al., 2014; Merlin et al., 2015) and should be used to evaluate the efficiency of the synergistic techniques and how they improve or degrade the accuracy and precision of the resulting SM. However, the scaling issue when comparing satellite estimates to point scale measurements still remains. The effects of this inherent scale difference between *in situ* and satellite data could be minimized if strategies of hydrological modeling are used to replace the discrete ground measurements (Polcher et al., 2015).

The benefits of high spatial and temporal SM mapping from the combination of passive microwaves and LST/VI geostationary data is under review, as well as the impact of the relationship between the instantaneous surface SM and the LST at different temporal scales, to decide on a new strategy for synergistic SM retrievals. Also, the fusion of SMOS SM with MODIS LST and water/vegetation indices can be applied to develop agricultural drought products at high spatial and temporal resolutions. These products may condense the soil and temperature conditions while including the lagged response of vegetation.

In the last few decades, there has been a trend toward smaller satellites for imaging the Earth. Small satellites reduce total mission costs and continued technological development is expected to further increase their sensing capabilities. In this context, synergistic techniques can leverage the strengths and synergies of the different sensors onboard small satellites.

ACKNOWLEDGMENTS

The work presented on this paper was supported by the European Community Seventh Framework Programme, FP7-Space-2013-1 (Project E-GEM-ID 607126), the Spanish MINECO (Projects AYA2012-39356-C05 and ESP2015-67549-C3), and the European Regional Development Fund (ERDF).

REFERENCES

Acevo-Herrera, R., Aguasca, A., Bosch-Lluis, X., Camps, A., Martínez-Fernández, J., Sánchez-Martín, N., Pérez-Gutiérrez, C., 2010. Design and first results of an UAV-borne L-band radiometer for multiple monitoring purposes. Remote Sens. 2, 1662–1679.

Adegoke, J.O., Carleton, A.M., 2002. Relations between soil moisture and satellite vegetation indices in the U.S. Corn Belt. Am. Meteorol. Soc. 3, 395–405.

Alonso-Arroyo, A., Camps, A., Monerris, A., Rudiger, C., Walker, J.P., Forte, G., Pascual, D., Park, H., Onrubia, R., 2014. The light airborne reflectometer for GNSS-R observations (LARGO) instrument: initial results from airborne and rover field campaigns. In: Proc. IEEE Int. Geosci. Remote Sens. Symp, pp. 4054–4057.

Camps, A., Font, J., Corbella, I., Vall-Llossera, M., Portabella, M., Ballabrera-Poy, J., González, V., Piles, M., Aguasca, A., Acevo, R., Bosch, X., Duffo, N., Fernández, P., Gabarró, C., Gourrion, J., Guimbard, S., Marín, A., Martínez, J., Monerris, A., Mourre, B., Pérez, F., Rodríguez, N., Salvador, J., Sabia, R., Talone, M., Torres, F., Pablos, M., Turiel, A., Valencia, E., Martínez-Fernández, J., Sánchez, N., Pérez-Gutiérrez, C., Baroncini-Turricchia, G., Rius, A., Ribó, S., 2012. Review of the CALIMAS team contributions to European Space Agency's soil moisture and ocean salinity mission calibration and validation. Remote Sens. 4, 1272–1309.

Carlson, T.N., Gillies, R.R., Perry, E.M., 1994. A method to make use of thermal infrared temperature and NDVI measurements to infer surface soil water content and fractional vegetation cover. Rem. Sens. Rev. 9 (1–2), 161–173.

Carlson, T., 2007. An overview of the 'triangle method' for estimating surface evapotranspiration and soil moisture from satellite imagery. Sensors 7, 1612–1629.

Chaparro, D., Vayreda, J., Martínez-Vilalta, J., Vall-llossera, M., Banqué, M., Camps, A., Piles, M., 2014. SMOS and climate data applicability for analyzing forest decline and forest fires.

In: Geoscience and Remote Sensing Symposium (IGARSS), 2014 IEEE International, 13–18 July, pp. 1069–1072.

Chaparro, D., Vall-llossera, M., Piles, M., Camps, A., Rudiger, C., 2015. Low soil moisture and high temperatures as indicators for forest fire occurrence and extent across the Iberian Peninsula. In: Geoscience and Remote Sensing Symposium (IGARSS), 2015 IEEE International, Milan, pp. 3325–3328.

Chauhan, N., Miller, S., Ardanuy, P., 2003. Spaceborne soil moisture estimation at high resolution: a microwave-optical/IR synergistic approach. Int. J. Remote Sens. 22, 4599–4622.

Chen, D., Huang, J., Jackson, T.J., 2005. Vegetation water content estimation for corn and soybeans using spectral indices derived from MODIS near- and short-wave infrared bands. Remote Sens. Environ. 98, 225–236.

Choi, M., Hur, Y., 2012. A microwave-optical/infrared disaggregation for improving spatial representation of soil moisture using AMSR-E and MODIS products. Remote Sens. Environ. 124, 259–269.

Das, N.N., Entekhabi, D., Njoku, E., 2011. Algorithm for merging SMAP radiometer and radar data for high resolution soil moisture retrieval. IEEE Trans. Geosci. Remote Sens. 49, 1504–1512.

Entekhabi, D., Njoku, E., O'Neill, P., Kellogg, K., Crow, W., Edelstein, W., Entin, J., Goodman, S., Jackson, T., Johnson, J., Kimball, J., Piepmeier, J., Koster, R., Martin, N., McDonald, K., Moghaddam, M., Moran, S., Reichle, R., Shi, J., Spence, M., Thurman, S., Tsang, L., Zyl, J.V., 2010. The Soil Moisture Active Passive (SMAP) mission. Proc. IEEE 98, 704–716.

Fang, B., Lakshmi, V., 2014. Soil moisture at watershed scale: remote sensing techniques. J. Hydrol. 516, 258–272.

Farrar, T.J., Nicholson, S.E., Lare, A.R., 1994. The influence of soil type on the relationships between NDVI, rainfall and soil moisture in semiarid Botswana. II. NDVI response to soil moisture. Remote Sens. Environ. 50, 121–133.

Fensholt, R., Sandholt, I., 2003. Derivation of a shortwave infrared water stress index from MODIS near- and shortwave infrared data in a semiarid environment. Remote Sens. Environ. 87, 111–121.

Finn, M.P., Lewis, M., Bosch, D., Giraldo, M., Yamamoto, K., Sullivan, D.G., Kincaid, R., Luna, R., Allam, G.K., Kvien, C., Williams, M.S., 2011. Remote sensing of soil moisture using airborne hyperspectral data. GISci. Remote Sens. 48, 522–540.

Font, J., Camps, A., Borges, A., Martín-Neira, M., Boutin, J., Reul, N., Kerr, Y., Hahne, A., Mecklenburg, S., 2010. SMOS: the challenging sea surface salinity measurement from space. Proc. IEEE 98, 649–665.

Gao, Z., Xu, X., Wang, J., Yang, H., Huang, W., Feng, H., 2013. A method of estimating soil moisture based on the linear decomposition of mixture pixels. Math. Comp. Model 58, 606–613.

Grandjean, G., Cerdan, O., Richard, G., Cousin, I.P.L., Tabbagh, B., Van Wesemael, B., Stevens, A., Lambot, S., Carre, F., 2010. Digisoil: an integrated system of data collection technologies for mapping soil properties. In: Viscarra Rossel, R.A., McBratney, A., Minasny, B. (Eds.), Proximal Soil Sensing. Springer, New York, pp. 89–101.

Kerr, Y., Waldteufel, P., Wigneron, J.-P., Delwart, S., Cabot, F., Boutin, J., Escorihuela, M.-J., Font, J., Reul, N., Gruhier, C., Juglea, S., Drinkwater, M.R., Hahne, A., Martin-Neira, M., Mecklenburg, S., 2010. The SMOS mission: new tool for monitoring key elements of the global water cycle. Proc. IEEE 98, 666–687.

Khanna, S., Palacios-Orueta, A., Whiting, M.L., Ustin, S.L., Riaño, D., Litago, J., 2007. Development of angle indexes for soil moisture estimation, dry matter detection and land-cover discrimination. Remote Sens. Environ. 109, 154–165.

Kim, J., Hogue, T.S., 2012. Improving spatial soil moisture representation through integration of AMSR-E and MODIS products. IEEE Trans. Geosci. Remote Sens. 50, 446–460.

Lambin, E.F., Ehrlich, D., 1996. The surface temperature-vegetation index space for land cover and land-cover change analysis. Int. J. Remote Sens. 17, 463–487.

Lobell, D.B., Asner, G.P., 2002. Moisture effects on soil reflectance. Soil Sci. Soc. Am. J. 66, 722–727.

Martín, F., Marchan, J.F., Aguasca, A., Vall-llossera, M., Corbera, J., Camps, A., Piles, M., Pipia, L., Tarda, A., Villafranca, A.G., 2011. Airborne soil moisture determination using a data fusion approach at regional level. In: Proc. IEEE Int. Geosci. Remote Sens. Symp, pp. 3109–3112. 24–29.

McColl, K.A., Vogelzang, J., Konings, A.G., Entekhabi, D., Piles, M., Stoffelen, A., 2014. Extended triple collocation: estimating errors and correlation coefficients with respect to an unknown target. Geophys. Res. Lett. 41 (17), 6229–6236.

Merlin, O., Chehbouni, A.G., Kerr, Y., Njoku, E.G., Entekhabi, D., 2005. A combined modeling and multispectral/multiresolution remote sensing approach for disaggregation of surface soil moisture: application to SMOS configuration. IEEE Trans. Geosci. Remote Sens. 43, 2036–2050.

Merlin, O., Bitar, A.A., Walker, J.P., Kerr, Y.H., 2010. An improved algorithm for disaggregating microwave-derived soil moisture based on red, near-infrared and thermal-infrared data. Remote Sens. Environ. 114, 2305–2316.

Merlin, O., Rudiger, C., Bitar, A.A., Richaume, P., Walker, J.P., Kerr, Y.H., 2012. Disaggregation of SMOS soil moisture in Southeastern Australia. IEEE Trans. Geosci. Remote Sens. 50, 1556–1571.

Merlin, O., Escorihuela, M.J., Mayoral, M.A., Hagolle, O., Albitar, A., Kerr, Y., 2013. Self-calibrated evaporation-based disaggregation of SMOS soil moisture: an evaluation study at 3 km and 100 m resolution in Catalunya, Spain. Remote Sens. Environ. 130, 25–38.

Merlin, O., Malbéteau, Y., Notfi, Y., Bacon, S., Khabba, S.E.-R., Jarlan, L., 2015. Performance metrics for soil moisture downscaling methods: application to DISPATCH data in central Morocco. Remote Sens. 7, 3783–3807.

Moran, M.S., Clarke, T.R., Inoue, Y., Vidal, A., 1994. Estimating crop water deficit using the relation between surface-air temperature and spectral vegetation index. Rem. Sens. Environ. 49 (3), 246–263.

Pablos, M., Piles, M., Sanchez, N., Gonzalez-Gambau, V., Vall-llossera, M., Camps, A., Martinez-Fernandez, J., 2014. A sensitivity study of land surface temperature to soil moisture using in-situ and spaceborne observations. In: Proc. IEEE Int. Geosci. Remote Sens. Symp, pp. 3267–3269.

Pellenq, J., Kalma, J., Boulet, G., Saulnier, G.M., Wooldridge, S., Kerr, Y., Chehbouni, A., 2003. A disaggregation scheme for soil moisture based on topography and soil depth. J. Hydrol. 276, 112–127.

Petropoulos, G., Carlson, T.N., Wooster, M.J., Islam, S., 2009. A review of T_s/VI remote sensing based methods for the retrieval of land surface energy fluxes and soil surface moisture. Prog. Phys. Geogr. 33 (2), 224–250.

Piles, M., Camps, A., Vall-Llossera, M., Corbella, I., Panciera, R., Rudiger, C., Kerr, Y.H., Walker, J., 2011. Downscaling SMOS-derived soil moisture using MODIS visible/infrared data. IEEE Trans. Geosci. Remote Sens. 49 (9), 3156–3166.

Piles, M., Vall-llossera, M., Laguna, L., Camps, A., 2012. A downscaling approach to combine SMOS multi-angular and fullpolarimetric observations with MODIS VIS/IR data into high resolution soil moisture maps. In: Proc. IEEE Int. Geosci. Remote Sens. Symp, pp. 1247–1250.

Piles, M., Vall-llossera, M., Camps, A., Sánchez, N., Martínez-Fernández, J., Martínez, J., González-Gambau, V., 2013. On the synergy of SMOS and Terra/Aqua MODIS: high resolution soil moisture maps in near real-time. In: Proc. IEEE Int. Geosci. Remote Sens. Symp, pp. 3423–3426.

Piles, M., Sánchez, N., Vall-llossera, M., Camps, A., Martínez-Fernández, J., Martínez, J., González-Gambau, V., 2014. A downscaling approach for SMOS land observations: two year evaluation of high resolution soil moisture maps over the Iberian Peninsula. IEEE J. Sel. Topics Appl. Earth Observ. Remote Sens. 7, 3845–3857.

Piles, M., Petropoulos, G., Sánchez, N., González-Zamora, A., Ireland, G., in press. Towards improved spatio-temporal resolution soil moisture retrievals from the synergy of SMOS and MSG SEVIRI spaceborne observations. Rem. Sens. Environ. http://dx.doi.org/10.2016/j.rse.2016.02.048.

Piles, M., Petropoulos, G., Ireland, G., Sanchez, N., 2016. A novel method to retrieve soil moisture at high spatio-temporal resolution based on the synergy of SMOS and MSG SEVIRI observations. Rem. Sens. Environ.

Polcher, J., Piles, M., Gelati, E., Barella-Ortiz, A., Tello, M., 2015. Comparing surface-soil moisture from the SMOS mission and the ORCHIDEE land-surface model over the Iberian Peninsula. Remote Sens. Environ. 174, 69–81

Qin, J., Yang, K., Lu, N., Chen, Y., Zhao, L., Han, M., 2013. Spatial upscaling of in-situ soil moisture measurements based on MODIS-derived apparent thermal inertia. Remote Sens. Environ. 138, 1–9.

Ruf, C., Chang, P., Clarizia, M.P., Jelenak, Z., Ridley, A., Rose, R., 2014. CYGNSS: NASA Earth venture tropical cyclone mission. In: Meynart, R., Neeck, S.P., Shimoda, H. (Eds.), Proceedings of SPIE, Sensors, Systems, and Next-Generation Satellites XVIII. SPIE, Amsterdam, The Netherlands, p. 924109.

Sánchez, N., Piles, M., Martínez-Fernández, J., Vall-llosera, M., Pipia, L., Camps, A., Aguasca, A., Pérez-Aragüés, F., 2014. Hyperspectral optical, thermal and microwave L-band observations for soil moisture retrieval at very high spatial resolution. Photogramm. Eng. Remote Sens. 80 (8), 745–755.

Sánchez, N., Alonso-Arroyo, A., Martínez-Fernández, J., Piles, M., González-Zamora, Á., Camps, A., Vall-llosera, M., 2015. On the synergy of airborne GNSS-R and Landsat 8 for soil moisture estimation. Remote Sens. 7, 9954–9974.

Sánchez-Ruiz, S., Piles, M., Sánchez, N., Martínez-Fernández, J., Vall-llossera, M., Camps, A., 2014. Combining SMOS with visible and near/shortwave/thermal infrared satellite data for high resolution soil moisture estimates. J. Hydrol. 49, 3156–3166.

Sánchez-Ruiz, S., Moreno, A., Martínez, B., Piles, M., Maselli, F., Carrara, A., Gilabert, M.A., 2015. Impact of water stress on GPP estimation from remote sensing data in Mediterranean ecosystems. In: Proc. IV Recent Advances on Quantitative Remote Sensing, pp. 338–343.

Sandholt, I., Rasmussen, K., Andersen, J., 2002. A simple interpretation of the surface temperature/vegetation index space for assessment of surface moisture status. Remote Sens. Environ. 79, 213–224.

Schnur, M.T., Xie, H., Wang, X., 2010. Estimating root zone soil moisture at distant sites using MODIS NDVI and EVI in a semi-arid region of southwestern USA. Ecol. Inform. 5, 400–409.

Sobrino, J.A., Franch, B., Mattar, C., Jiménez-Muñoz, J.C., Corbari, C., 2012. A method to estimate soil moisture from Airborne Hyperspectral Scanner (AHS) and ASTER data: application to SEN2FLEX and SEN3EXP campaigns. Remote Sens. Environ. 117, 415–428.

Thenkabail, P.S., Enclona, E.A., Ashton, M.S., Van Der Meer, B., 2004. Accuracy assessments of hyperspectral waveband performance for vegetation analysis applications. Remote Sens. Environ. 91, 354–376.

Thenkabail, P.S., Mariotto, I., Gumma, M.K., Middleton, E.M., Landis, D.R., Huemmrich, F.K., 2013. Selection of hyperspectral narrowbands (HNBs) and composition of hyperspectral two-band vegetation Indices (HVIs) for biophysical characterization and discrimination of crop types using field reflectance and Hyperion/EO-1 data. IEEE J. Sel. Topics Appl. Earth Observ. Remote Sens. 6, 1–13.

Tomer, S.K., Al Bitar, A., Sekhar, M., Zribi, M., Bandyopadhyay, S., Sreelash, K., Sharma, A.K., Corgne, S., Kerr, Y., 2015. Retrieval and multi-scale validation of soil moisture from multi-temporal SAR data in a semi-arid tropical region. Remote Sens. 7 (6), 8128–8153.

Verhoest, N.E.C., van den Berg, M.J., Martens, B., Lievens, H., Wood, E.F., Pan, M., Kerr, Y.H., Al Bitar, A., Tomer, S.K., Drusch, M., Vernieuwe, H., De Baets, B., Walker, J.P., Dumedah, G., Pauwels, V.R.N., 2015. Copula-based downscaling of coarse-scale soil moisture observations with implicit bias correction. IEEE Trans. Geosci. Rem. Sens. 53(6): 3507–3521.

Verstraeten, W.W., Veroustraete, F., van der Sande, C.J., Grootaers, I., Feyen, J., 2006. Soil moisture retrieval using thermal inertia, determined with visible and thermal spaceborne data, validated for European forests. Remote Sens. Environ. 101, 299–314.

Wang, X., Xie, H., Guan, H., Zhou, X., 2007. Different responses of MODIS-derived NDVI to root-zone soil moisture in semi-arid and humid regions. J. Hydrol. 340, 12–14.

Whiting, M.L., Li, L., Ustin, S.L., 2004. Predicting water content using Gaussian model on soil spectra. Remote Sens. Environ. 89, 535–552.

Wickert, J., Andersen, O.B., Beyerle, G., Chapron, B., Cardellach, E., D'Addio, S., Foerste, C., Gommenginger, C., Gruber, T., Helm, A., Hess, M., Hoeg, P., Jaeggi, A., Jakowski, N., Kern, M., Lee, T., Martin-Neira, M., Montenbruck, O., Pierdicca, N., Rius, A., Rothacher, M., Shum, C., Zuffada, C., 2013. GEROS-ISS: innovative GNSS reflectometry/occultation payload onboard the International Space Station for the Global Geodetic Observing System. *American Geophysical Union*, #G51A-0871.

Microwave Soil Moisture Retrieval Techniques

Section III

Microwave Soil Moisture
Retrieval Techniques

Chapter 7

Soil Moisture Retrieved From a Combined Optical and Passive Microwave Approach: Theory and Applications

C. Mattar*, A. Santamaría-Artigas*, J.A. Sobrino[†]
and J.C. Jiménez-Muñoz[†]
*University of Chile, Santiago, RM, Chile, [†]Universitat de València, Valencia, Spain

1 INTRODUCTION

Surface soil moisture (SM) retrieved from new L-band remote sensing satellites such as the Soil Moisture and Ocean Salinity (SMOS) and the Soil Moisture Active and Passive (SMAP) missions generate a new scenario in the current trends to analysis of the global water cycle. In the case of SMOS, an explorer mission of the European Space Agency (ESA), the 5 years of L-band radiometry information (Kerr et al., 2001, 2012) create a challenge to develop and adapt combined approaches for the better estimation of SM at regional and global scales. The combined optical-passive microwave method is one of these approaches, which consist of the adaptation of the tau-omega model (Ulaby et al., 1986) to estimate the surface SM.

The tau-omega model estimates the surface's brightness temperature by assuming two emission layers related to the soil and vegetation. This model was adapted for several land cover types gathered in the L-band microwave emission of the biosphere (L-MEB) model, which is the core of the SMOS mission for the retrieval of SM values (Wigneron et al., 2007). Several new approaches to improve the estimates generated from the SMOS mission have been published, and the combined optical-passive microwave approach is one of them.

The use of the synergic optical-passive microwave approach at regional scales could generate interesting consequences for the improvement of surface SM retrievals used for hydrological and forecast modeling. However, the theoretical concepts of the combined approach need to be analyzed in its

Satellite Soil Moisture Retrieval. http://dx.doi.org/10.1016/B978-0-12-803388-3.00007-3

assumptions and the possible application to different land cover types at global scale. Thus, the objective of this chapter is to describe the theoretical optical-passive microwave approach, its application at in situ and regional levels, and provide further perspective to adapt thermal infrared (TIR) information for the synergistic approach.

2 RADIATIVE TRANSFER EQUATION

The basis of radiative transfer in the passive microwave has a relation to the Rayleigh-Jens approximation, in which the brightness temperature is equivalent to the land leaving radiance emitted in the microwave domain. The contribution of different brightness temperature totalizes the brightness temperature observed by a remote sensor (TB_p). These terms can be summarized in the general radiative equation proposed by Kerr and Njoku (1990):

$$TB_p = TB_{at}\uparrow + \exp(-\tau_{at}\uparrow)TB_{suf} + \exp(-\tau_{at}\uparrow)\left(TB_{at}\downarrow + \exp(-\tau_{at}\downarrow)TB_{sky}\right)r_{suf}$$

$$(1)$$

where suf, at, and sky represent the contribution from the surface, atmosphere, and sky respectively, and r_{suf} is the surface reflectivity. Both TB_{at} and τ_{at}, the brightness temperature and the atmospheric opacity respectively, can be upwelling ($\uparrow$) or downwelling ($\downarrow$). These terms are in relation to the incidence angle (θ) and the polarization (p). Fig. 1 shows a general scheme of the brightness temperature captured by a sensor.

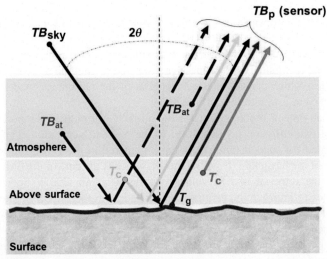

FIG. 1 Radiative contributions of brightness temperature at sensor level. TB_{sky} is the cosmic emission, TB_{at} is the atmospheric emission, T_c is the surface emission (ie, vegetation), and T_g is the land emission. Both T_c and T_g contributes to TB_{suf}.

A key parameter in this equation is the surface reflectivity r_{suf}, which is the integral of the dispersion coefficient above all the directions. This element influences the brightness temperature of the surface TB_{suf}, the land surface emissivity (LSE) ($\varepsilon = 1 - r_{\text{suf}}$), and therefore the brightness temperature for a given polarization in relation to the effective temperature of the ground. Eq. (1) denotes the general components that constitute the emitted energy from a given surface through the atmosphere path to the sensor, summarizing all the existing energy sources in this radiative process. However, Eq. (1) can be rearranged to consider the energy emitted by the surface and vegetation layer above its surface. In this way, the vegetation layer represents a certain opacity (τ) and single scattering albedo (ω). This single combination of the opacity and the dispersion is called the tau-omega (τ-ω) model, which is an established method for SM retrievals and where the optical-passive microwave approach can demonstrate an interesting synergy.

2.1 Tau-Omega Model

The τ-ω model considers the global emission from two layers (soil and vegetation) through four main terms: (1) direct emission from vegetation attenuated by the atmosphere, (2) reflected emission from the vegetation by the soil and attenuated by the same vegetation layer and the atmosphere, (3) the soil emission attenuated by the vegetation and the atmosphere, and (4) the emission of the surface (soil + vegetation) attenuated by the atmosphere. These terms can be expressed in the following equation to describe the brightness temperature (TB) emitted by a surface:

$$TB(\theta, p) = (1 - \omega)(1 - \gamma)(1 + \gamma\Gamma_s(\theta, p))T_V^E$$
$$+ \gamma(1 - \Gamma_s(\theta, p))T_S^E + TB_\theta^{\text{sky}\downarrow}\Gamma_s(\theta, p)\gamma^2(\theta, p) \tag{2}$$

where T_S^E and T_V^E are the soil and vegetation effective temperature respectively, $TB_\theta^{\text{sky}\downarrow}$ represents the contribution of the atmospheric and cosmic downwelling irradiance. Γ_s is the soil reflectivity, ω is the single scattering albedo of the surface, and γ is the vegetation attenuation factor, which can be expressed in terms of the vegetation optical depth (τ):

$$\gamma(\theta, p) = \exp\left(-\frac{\tau(\theta, p)}{\cos\theta}\right) \tag{3}$$

Eqs. (2) and (3) are influenced by polarization (p) and incidence angle (θ). Eq. (2) is also dependent on ω, and on the two main temperatures (ie, soil (T_S^E) and vegetation (T_V^E)) that contribute to the surface emission. These parameters are essential for the correct interpretation and use of tau-omega models, and are as well the basis for generating certain approximations that allows reliable SM estimates by using a combined optical-passive microwave approach. The effective temperature, the single scattering albedo, and the vegetation optical depth are fundamentals for this type of combined approach and they are described in the following sections.

2.2 Effective Temperature, Single Dispersion and Vegetation Optical Depth

According to the Rayleigh-Jens approximation, emitted energy from the soil in the microwave domain is proportional to the thermodynamic temperature and the brightness temperature, and can be expressed as the product of the emissivity and the effective temperature (Holmes et al., 2006). The whole soil layer contributes to the soil thermal emission. From the point of origin to the soil surface, the intensity is attenuated by the intervening soil, whose absorption is related to the moisture content. The net intensity at the soil surface, called the effective temperature, is a superposition of the intensities emitted at various depths within the soil (Choudhury et al., 1982). The theoretical formulation of the effective temperature requires fine vertical profile information on both SM and soil temperature. A few test sites provide a sufficiently fine measurement of the vertical profiles in the soil. To estimate the effective temperature with minimum soil profile information, several parameterizations have been developed for use with L-band radiometry. For instance, Choudhury et al. (1982) showed that for short time periods, the effective temperature can be described as a linear function of the soil temperature at two depths. Wigneron et al. (2001) developed a two-depths parameterization making it suitable for seasonal studies, by taking into account the influence of SM on the attenuation of microwave energy. Holmes et al. (2006) developed a similar parameterization that provided slightly improved results than the abovementioned approach but is a bit more complex as it requires the additional use of a simulation model (Wigneron et al., 2007). A comprehensive comparison between all these methods was published in Wigneron et al. (2008) and most recently, Lv's approach (Lv et al., 2014) has also published a new effective temperature retrieval method, although further work needs to be carried out to compare all these methods at global scale.

The effective soil and vegetation temperature can be assumed equal to a single value $T_{gc} \approx T_g \approx T_c$ (Mattar et al., 2012). In particular, the effects of temperature gradients within the vegetation canopy are not accounted for. With an overpass around dawn these gradients should be minimized and T_c can be expected to be close to the air temperature, while T_g can be estimated from atmospheric models outputs (Hornbuckle and England, 2005). For instance, In L-MEB, an estimate of an effective composite temperature T_{gc} (including both soil and vegetation media) is dependent on the vegetation optical depth as

$$T_{gc} = AT_c + (1-A)T_g \qquad (4)$$

where

$$A = B(1 - \exp(-\tau)) \Rightarrow A \leq 1 \qquad (5)$$

considering A and B as coefficients of the vegetation structure effect that depend on the vegetation structure. The rationale of these equations is detailed in Wigneron et al. (2007). As τ increases, both attenuation of soil emission and vegetation emission increase, making the effective temperature T_{gc} closer to the vegetation effective temperature. For bare soil conditions (ie, for $\tau = 0$), T_{gc} is equal to T_g. On the other hand, in the case of single scattering albedo, the values of ω registered and documented in the literature are low. For instance, Wigneron et al. (2004a) demonstrated values of $\omega < 0.05$ and slightly higher values for maize crops (ie, 0.1). However, the single scattering albedo at large scales can be averaged over a coarse footprint such as in passive microwave data and it is not necessary to account for a prominent distinction between different canopies.

The vegetation optical depth is key in the τ-ω because it determines the state of vegetation cover over the ground and also the water intercepted by this vegetation, contributing to the total emission of soil and vegetation layer. Several works have demonstrated the linear relation between the vegetation optical depth and the vegetation water content (VWC) (Jackson and Schmugge, 1991; van de Griend and Wigneron, 2004) by using a constant parameter suitable for a wide range of crops:

$$\tau = bVWC \qquad (6)$$

Where VWC is the vegetation water content and b is a parameter that changes according to the wavelength and in the case of L-band, it is equivalent to 0.12 ± 0.03 (Wigneron et al., 2007). However, as it is difficult to provide estimates of VWC at global scale, the contribution of the standing vegetation to τ was parameterized as a function of the leaf area index (LAI) demonstrating an intrinsic relation between τ and LAI. Despite this linkage, the estimation of VWC is not simple because it changes according to the spatial scale. Moreover, three different components comprise the total vegetation optical depth, which are the optical depth of the stand vegetation (τ_{SP}), the vegetation optical depth linked to the litter (τ_L), and the water content intercept by the both vegetation types described before (τ_{LP}). In this sense, the vegetation optical depth can be described as

$$\tau = \tau_{SP} + \tau_L + \tau_{LP} \qquad (7)$$

A comprehensive description of these components is detailed in Wigneron et al. (2007). Despite the complexity to estimate the optical depth as a function of its different components, there are some interesting approximations that can be used as an indicator of the optical depth and also to contribute to the SM retrievals.

2.3 A Combined Optical-Passive Microwave Approach

The combination of optical-passive microwave data has been used in several works to estimate the snow cover mapping (Salomonson et al., 1995), flooding risk (Chakraborty et al., 2014; Lacava et al., 2015), vegetation monitoring (Shi et al., 2008; Jones et al., 2011) and land surface modeling (Wigneron et al., 2007; Dente et al., 2014). The combination of these data sources can contribute to the SM estimation by using different electromagnetic domains in single semi-empirical approaches.

To simplify Eq. (2) without introducing significant errors in the result (Jackson et al., 1995), it was assumed that the effective soil and vegetation temperatures were equal ($T_S^E = T_V^E = T_C$ = effective surface temperature). Moreover, the upward atmospheric emission can be neglected for ground-based measurements at L-band. The downward brightness temperature $TB_\theta^{sky\downarrow}$ is also very small after reflection on the ground and the attenuation through the vegetation (Pellarin et al., 2003) can be neglected. Furthermore, the scattering effects also can be neglected, which is generally a good approximation at L-band (Wigneron et al., 2004a), so ω can be considered equal to zero and the new approach of the τ-ω model can be written as

$$TB(\theta, p) = T_c \left(1 - \Gamma_s(\theta, p)\gamma^2(\theta, p) \right) \tag{8}$$

where the surface emissivity $e(\theta, p)$ is defined by the ratio between the polarized brightness temperature and the ground temperature as $e(\theta, p) = TB(\theta, p)/T_c$, so it can be arranged in terms of soil reflectivity as

$$e(\theta, p) = 1 - \Gamma_s(\theta, p)\gamma^2(\theta, p) \tag{9}$$

remarking that the surface reflectivity is equivalent to the surface emissivity $\Gamma(\theta, p) = 1 - e(\theta, p)$. When replacing the surface reflectivity in Eq. (9),

$$\Gamma(\theta, p) = \Gamma_s(\theta, p)\gamma^2(\theta, p) \tag{10}$$

Using Eq. (3) and taking the logarithm function of both terms, Eq. (5) can be rewritten as

$$\cos(\theta)\log(\Gamma_s(\theta, p)) - 2\tau(\theta, p) = \cos(\theta)\log(\Gamma(\theta, p)) \tag{11}$$

As soil reflectivity $\Gamma_s(\theta, p)$ is often considered close to a linear function of surface SM (w_s) and it is rather low for very dry soils (Schmugge et al., 1994), the soil reflectivity can be assumed proportional to surface SM according to

$$\Gamma_s(\theta, p) \approx Ap(\theta)w_s^a \tag{12}$$

where $Ap(\theta)$ is a coefficient depending on the sensor configuration. The value $Ap(\theta)$ is site dependent and implicitly accounts for all the soil characteristics that determine soil emission: mainly soil texture and structure, surface roughness, and so forth. Many studies in the field of passive microwave remote sensing are based on SM relationships derived from single configuration measurements of the surface emission (Jackson et al., 1999; Chanzy et al., 1997). These latter approaches are appropriate in areas with a low contribution from the vegetation to the surface emission and low roughness effects (Saleh et al., 2006a). It is important to note that the linear approximation in Eq. (7) is a crude approximation of the reflectivity curve, which is more typically S-shaped and does not go through the origin, especially at H polarization (Wigneron et al., 2004b). So Eq. (11) can be reordered as the relation between SM, surface reflectivity, and vegetation optical depth:

$$\log(w_s) = c_0 \log(\Gamma(\theta, p)) + c_1(\theta)\tau(\theta, p) + c_2(\theta, p) \tag{13}$$

where c_0, c_1, and c_2 are regression coefficients. Optical depth can be generally assumed as independent of incidence angle like $\tau(\theta, p) \cong \tau(p)$ as described in Wigneron et al. (2004b). The optical depth is intrinsically related to the VWC and it can be parametrized by using the LAI. This relation has been analyzed for several crops by Calvet et al. (2011) and considered key in the L-MEB model to process the SMOS Level 2 products (Kerr et al., 2010). However, the reason for combining LAI and vegetation optical depth have been analyzed in other works. Some of them are linked to the feasibility to adapt reliable LAI products from remote sensing and for soil vegetation atmosphere transfer (SVAT) modeling (Wigneron et al., 2002). In the case of L-MEB (Wigneron et al., 2007), the LAI product derived from a Moderate Resolution Imaging Spectroradiometer (MODIS) is used as an input to estimate the level 2 parameters. Moreover, semiempirical relationships were also evidenced for a well-instrumented test site describing a parametric equation between optical depth and LAI (Saleh et al., 2006a, 2006b).

The use of LAI can be also related to the normalized difference vegetation index (NDVI), as both vegetation indicators represent in general terms the dominance of vegetation cover for a given surface. In the case of LAI the structural parameter is related to the vegetation optical depth, although the NDVI, as a measure of the vegetation vigor, can be related to the LAI but not directly to the optical depth. Nevertheless, several works considered that NDVI might also be linked to the optical depth and its spatial-temporal variations at global scale. A global scale analysis related to the spatial and temporal patterns of vegetation optical depth, LAI, and NDVI was published by Jones et al. (2011). Furthermore, Jackson et al. (2004) demonstrated the relationship between NDVI and VWC, which are key for single parametrization between NDVI-LAI and VWC. Lawrence et al. (2014) analyzed the relationship of vegetation optical depth over several crops in central North America combining the MODIS and SMOS data. More specifically, in the context of SM estimation by combined optical-passive microwave data, Parrens et al. (2012) used LAI at regional scale to estimate the SM using bipolarized and biangular passive microwave information. Mattar et al. (2012) demonstrated at the surface monitoring of soil reservoir experiment (SMOSREX) test site (de Rosnay et al., 2006) that the use of NDVI improved the estimation of SM by using a combined optical-passive microwave approach. Similar results were published by Miernieki et al. (2014) at the Valencia Anchor Site (VAS) (Wigneron et al., 2012; Schwank et al., 2010) (Wigneron et al., 2011) where the use of NDVI from remote sensing improves the SM estimates over vineyards by using different approaches. In most of these works, the use of a vegetation indicator derived from optical data such as LAI or NDVI improved the SM estimation over land covers where a phenology curve is evidenced.

At in situ levels or test site, the NDVI and LAI can be estimated by several instruments and different methods describing the phenology stage of the surface and the crop. Thus, well-instrumented sites can be adapted to calibrate and validate the use of combined optical-passive microwave approach such as SMOS-REX in France and ELBARRA in Spain, where the optical measurements are complemented by passive-microwave radiometry, SM measurements, and weather station data. Nevertheless, the use of combined optical-passive microwave at regional or global scale are still ongoing and deserve special attention owing to the current optical data available from different optical sensors such as Landsat Thematic Mapper (TM)/Enhanced Thematic Mapper Plus (ETM+)/ Operational Land Imager (OLI), MODIS, or Meteosat Second Generation-Spinning Enhanced Visible Infrared Imager (MSG-SEVIRI), which are an examples of heliosynchronous and geostationary missions and different passive microwave missions such as SMOS or SMAP.

Based on the potential of using a combined optical-passive microwave approach at global scale, it is relevant to describe a possible relation that can be applied to both in situ and global scales so that the vegetation optical depth can be parametrized in terms of a vegetation indicator such as NDVI, Enhanced Vegetation Index (EVI), or LAI. However, LAI is theoretically linked to the vegetation optical depth, but the most used vegetation index is NDVI; thus, the vegetation optical depth can be analyzed in terms of both vegetation indicators as

$$\tau = f(VEG) \cong aVEG \tag{14}$$

where a is a constant that accounts mainly for the effect of the vegetation structure and VEG is the vegetation index. Using Eqs. (13) and (14), SM can be expressed as a function of the microwave reflectivity and a vegetation index VEG, which can be assumed as LAI or NDVI:

$$\log(w_s) = a_0 \log(\Gamma(\theta, p)) + a_1 VEG + a_2(\theta, p) \tag{15}$$

where a_0, a_1, and a_2 are the regression coefficients and VEG is the vegetation content of a given surface. The above equation was initially developed for one polarization and one incidence angle. However, it is valid for both horizontal and vertical polarizations and for all incidence angles. So, equations obtained at different polarizations and incidence angles can be summed. It is likely that the statistical regression will be more representative whether several angles and both polarizations are accounted for in the retrieval approach. For instance, an adaptation of Eq. (15) presenting two different angles (denoted by indexes "1" and "2"), both H and V polarizations, and replacing the reflectivity by the ratio between the brightness and surface temperature ($\Gamma = 1 - TB/T_c$) is detailed as follows:

$$
\begin{aligned}
\log(w_s) = a + b &\left(\log \left(1 - \frac{TB_{\theta1,V}}{T_c} \right) \right) + c \left(\log \left(1 - \frac{TB_{\theta2,V}}{T_c} \right) \right) \\
+ d &\left(\log \left(1 - \frac{TB_{\theta1,H}}{T_c} \right) \right) + e \left(\log \left(1 - \frac{TB_{\theta2,H}}{T_c} \right) \right) \\
+ f &VEG
\end{aligned}
\tag{16}
$$

where a, b, c, d, e, and f are the regression coefficients. Eq. (16), considered as an optical-passive microwave approach, will be described and tested at in situ and regional scales in the following sections.

2.4 Case Study—Optical Passive Microwave at in-situ level

In situ measurements of SM, surface temperature, passive microwave temperature, and visible/near-infrared reflectance acquired in the framework of the SMOSREX (de Rosnay et al., 2006) were used in this case study. SMOSREX is a part of the Interdisciplinary Field Experiment on Radiometry (PIRRENE) program and is located at the National Office of Aerospace Study and Research (ONERA) test site near the town of Mauzac in France (43°23′8.74″N; 1°17′32.63″E; 188 m. a. s. l.). SMOSREX integrates a number of studies in the field of L-band passive microwaves, from the development of emission models, to the assimilation of L-band data for the estimation of SM in the root zone. The experiment started in 2001 and became fully operational on Jan. 2003.

At the SMOSREX site (see Fig. 2), L-band radiometric observations were obtained by the L-band radiometer for estimating water in soils (LEWIS) with an accuracy of 0.2 K and a field of view of 13.5 degrees (Lemaître et al., 2004). LEWIS was installed at the top of a 13.7 m vertical structure over a fallow where natural grasses grow. An automatic scanning was made eight times per day at five incidence angles (20, 30, 40, 50, and 60 degrees) over two adjacent areas: a fallow field and a bare soil field. In routine mode, LEWIS monitors the brightness temperature of the fallow field at an incidence angle of 40 degrees. Only observations made over the fallow area were considered because the main objective of the combined optical-passive microwave is to include vegetation parameters. Thus, a vegetation index such as the NDVI was derived from infrared and near-infrared reflectance measurements carried out at the incidence angle of 40 degrees using two Cimel optical radiometers from Jul. 2003 to Dec. 2007. More details about the reflectance measurements can be obtained in (Albergel et al., 2010). For this case study data acquired between Jan. 1, 2004, and Dec. 31, 2007 were used. Recently, the LEWIS instrument was relocated to Grenoble after 13 years of measurements at the SMOSREX site.

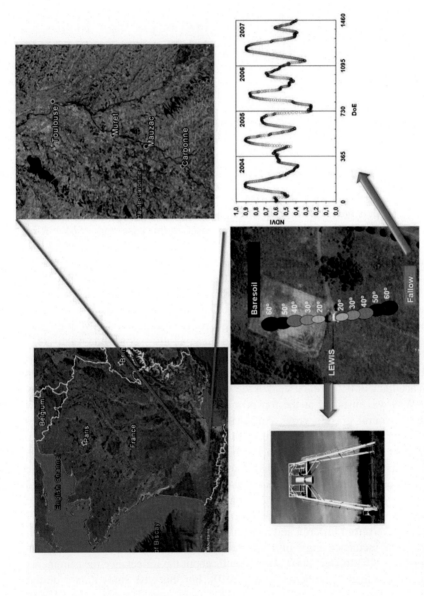

FIG. 2   SMOSREX study area. The LEWIS instrument and NDVI phenology are also shown.

TABLE 1 Coefficients of Determination r^2 ($p < 0.05$) Obtained in the Calibration of the Regression Equation Considering NDVI and Setting VEG Equal to Zero for Each Year of Calibration

Calibration Year	N	r^2 (VEG = 0)	r^2 (VEG = NDVI)
2004	516	0.861	0.888
2005	979	0.654	0.869
2006	800	0.794	0.875
2007	1243	0.584	0.788

Adapted from Mattar, C., Wigneron, J., Sobrino, J.A., Novello, N., Calvet, J.C., Albergel, C., Richaume, P., 2012. A combined optical–microwave method to retrieve soil moisture over vegetated areas. IEEE Trans. Geosci. Remote Sens. 50(5), 1404–1413.

Table 1 shows the results obtained in the calibration of the regression equation (16) from the SMOSREX in situ data. Better results were obtained in terms of r^2, when the NDVI index was included in the regression equation. This can be partially explained by the fact the NDVI brings information of the vegetation dynamics in the regression equation. In the SM retrieval process, the vegetation effect can be better corrected from biangular and bipolarization observations than from monoangular and bipolarization observations (Wigneron et al., 1995). From the obtained results, it seems that the information on the vegetation dynamics brought by the NDVI index is more useful in terms of statistical calibration (eg, r^2). Indeed, larger differences in terms of r^2 were obtained without using NDVI (VEG = 0). For instance, the r^2 at using VEG = 0 varied from 0.861 to 0.584 using the year 2004 or 2007, respectively, for calibration.

Illustration of the retrieval results are given in Fig. 3, which shows the retrieved SM for years 2005–2007 using 2004 as the calibration year. Improved results using the NDVI index (root mean square deviation (RMSE) = 0.064 m^3 m^{-3} vs. 0.075 m^3 m^{-3}) could generally be obtained during these periods. Conversely, SM tends to be overestimated in winter ("wet" season). Besides, outliers (ie, large discrepancies between retrieved and measured SM) were obtained for wet soil conditions mainly when VEG is set equal to zero.

The regression equations were calibrated using 1 year (see the section discussed previously) and then evaluated using the three other years. Conversely to the previous section, better results were generally obtained for the monoangular configuration, which seems more robust when it is used in a retrieval mode. As for the previous section, in most cases (except for the year 2005 for calibration), the RMSE between observed and estimated SM is lower when the NDVI index

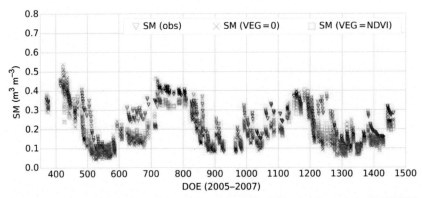

FIG. 3 Retrieved SM estimations for validation years 2005–2007 using the over the SMOSREX site. Measurements of SM are represented by red triangles, the estimated SM retrieved by using the NDVI in green squares and setting VEG equal to zero in blue crosses. *(Adapted from Mattar, C., Wigneron, J., Sobrino, J.A., Novello, N., Calvet, J.C., Albergel, C., Richaume, P., 2012. A combined optical–microwave method to retrieve soil moisture over vegetated areas. IEEE Trans. Geosci. Remote Sens. 50(5), 1404–1413.)*

is used in the regression equations in comparison to the case where NDVI is set equal to zero. The minimum values of the RMSE (0.051 and 0.053 $m^3\ m^{-3}$) were evidenced for the VEG = NDVI configuration, and for calibration years 2005 and 2007. Maximum values of RMSE (0.091 and 0.095 $m^3\ m^{-3}$) were obtained for the VEG = 0 configuration and for calibration years 2004 and 2005.

2.5 Case Study—Optical Passive Microwave at Regional Scale

2.5.1 Study Area

The Optical Passive Microwave (OPM) approach was tested over the central and southern region of Chile. The study area used to test the OPM approach is located between 28–43.5°S, and 69.5–74.5°W. This region shows a wide range of climatic conditions and land covers. The northern zone has climatic conditions of the semiarid type (monthly mean temperatures ranging between 11°C and 19°C and precipitation lower than 12 mm month^{-1} in the winter season), and is mostly covered by sparse vegetation and shrublands. The central zone, which is the most agriculturally productive zone in the country, is covered by several types of crops, forests, and bare soil areas, its climatic conditions are of Mediterranean type (monthly mean temperatures ranging between 9°C and 21°C and precipitations up to 86 mm month^{-1} in the winter season). Finally, the southern zone, covered mostly by forests and grasslands, has temperate and oceanic climatic conditions (monthly mean temperatures ranging between 6°C and 14°C and precipitations greater than 230 mm month^{-1} in the winter season).

2.5.2 Data

Different data sources were used to test the OPM approach over regional scales for the period between Jan. 1, 2010, and Dec. 31, 2012. The microwave emission of the surface was obtained from bipolarized L-band brightness temperature data from the SMOS (Kerr et al., 2010) level 3 product (L3TB—RE02). Temperature and volumetric water content information of the first layer of soil (0–7 cm) was obtained from ERA-Interim's STL1 and VSWL1 products, respectively (Dee et al., 2011). The NDVI from MODIS product MOD13Q1 (Huete et al., 2002) and the LAI from MODIS product MOD15 (Myneni et al., 2002) were used as possible indicators of the vegetation condition. Finally, to evaluate the performance of the OPM approach, land cover class information was obtained from the ECOCLIMAP database (Masson et al., 2003). Land cover classes present in the study area were: crops (C), closed shrublands (C), deciduous forest (DF), evergreen forest (EF), grass-lands (G), mixed forest (MF), open shrublands (OS), rocks (R), woodlands (W), and woody shrublands (WS).

2.5.3 Results

The spatial distribution of the determination coefficient obtained from the calibration of the OPM approach when setting VEG = 0, VEG = LAI, and VEG = NDVI for years 2010, 2011, and 2012 is shown in Fig. 4.

On average, the use of vegetation indicators in the calibration of the OPM approach generates improvements in the r^2. For the northern zone of the study area, which is mainly covered by open, closed, and woody shrublands, these improvements were greater when the NDVI was considered as a vegetation indicator. The central zone, covered by woody shrublands, crops, and evergreen forests, didn't show significant differences in the improvements of the r^2 when either the NDVI or LAI was used. Finally, for the southern zone of the study area covered by different forest classes, the biggest improvements were obtained when the LAI was included in the calibration of the OPM approach. Fig. 5 shows the variations by land cover class and calibration year of the r^2 obtained from the calibration of the OPM approach when setting VEG = 0, VEG = LAI, and VEG = NDVI.

For every land cover class of the study area, the r^2 increased when a vegetation indicator was included in the OPM approach. Table 2 shows the average determination coefficient obtained from the calibration of the OPM approach when setting VEG = 0, VEG = LAI, and VEG = NDVI by land cover class and calibration year.

The crops class showed the highest r^2 values of all classes for all the calibration years and approaches (ie, VEG = 0, VEG = LAI, and VEG = NDVI). Shrub-lands classes showed the highest improvements in terms of r^2 when the NDVI was used in the calibration of the OPM approach. Forest classes however, showed the highest improvements when the LAI was considered as the vegetation index.

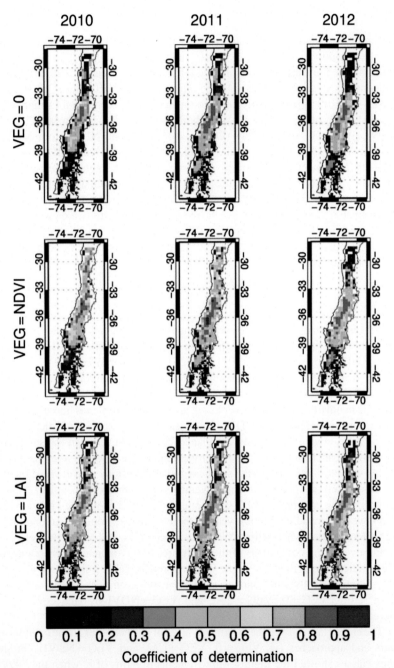

FIG. 4 Spatial distribution of the determination coefficient obtained from the calibration of the OPM approach when setting VEG=0 *(top row)*, VEG=NDVI *(middle row)*, and VEG=LAI *(bottom row)* for years 2010, 2011, and 2012.

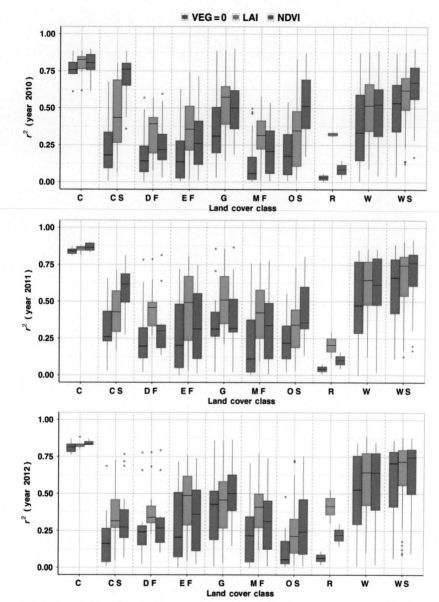

FIG. 5 Box plots by land cover class of the determination coefficient obtained from the calibration of the OPM approach when setting VEG=0, VEG=LAI, and VEG=NDVI for years 2010 (top), 2011 (middle), and 2012 (bottom).

TABLE 2 Average Determination Coefficient by Land Cover Class and Calibration Year Obtained From the OPM Approach With VEG = 0, VEG = NDVI, and VEG = LAI

	Calibration Year								
	2010			2011			2012		
				Average Determination Coefficient					
Land Cover	*VEG = 0*	*LAI*	*NDVI*	*VEG = 0*	*LAI*	*NDVI*	*VEG = 0*	*LAI*	*NDVI*
Closed shrublands	0.22	0.44	0.71	0.32	0.43	0.59	0.17	0.35	0.32
Crops	0.76	0.79	0.79	0.84	0.85	0.87	0.81	0.83	0.84
Deciduous forest	0.20	0.34	0.26	0.27	0.45	0.33	0.28	0.40	0.32
Evergreen forest	0.19	0.37	0.29	0.27	0.45	0.35	0.29	0.46	0.35
Grasslands	0.36	0.54	0.50	0.37	0.52	0.41	0.38	0.44	0.50
Mixed forest	0.13	0.30	0.22	0.22	0.41	0.33	0.23	0.40	0.32
Open shrublands	0.21	0.33	0.51	0.23	0.34	0.43	0.11	0.26	0.27
Rocks	0.03	0.32	0.08	0.04	0.20	0.10	0.06	0.41	0.22
Woodlands	0.37	0.49	0.50	0.49	0.60	0.59	0.49	0.58	0.56
Woody shrublands	0.50	0.59	0.64	0.59	0.66	0.70	0.58	0.63	0.64

The rest of the classes also showed higher r^2 values when a vegetation indicator (either NDVI or LAI) was included in the calibration of the OPM approach.

2.5.4 Discussion

The use of vegetation indices improves the calibration of the OPM approach for all the land cover classes analyzed in this case study. It seems that while both vegetation indicators improve the calibration of the OPM approach, the NDVI performs better for shrublands classes and LAI for forests classes. This behavior might be related to the method used to estimate each vegetation indicator, which for the NDVI is a simple normalized difference between the near-infrared and infrared bands reflectance and for the MODIS LAI is a more complex algorithm that compares observed reflectances with ones modeled for a range of canopy structures and soil patterns.

The application of the optical-passive microwave approach at in situ or regional level is conditioned by the land cover characteristics such as its phonological stage. The selection of NDVI, LAI, or any other vegetation indicator depends of the land cover type proportion. Despite the fact the use of a vegetation indicator always can perform the SM retrievals better than setting vegetation contribution equal to zero, the effects of LAI and NDVI on SM retrievals must still be analyzed.

The retrieval of SM from optical-passive microwave approaches relies on several assumptions, such as the effective temperature, which is difficult to determine. Nevertheless, further comparison of the sensibility errors of the SM retrievals need to be performed. Another source of error is the SM used to calibrate and validate the combined method. In the case of in situ, the land cover is known and the effects of the SM depth can be identified at test site level. However, at regional scale there is a lack in terms of the SM product used to calibrate and validate the optical-passive microwave approach. For instance, ERA-Interim provides SM at better spatial and temporal resolution than SMOS, although its soil depth is larger than SMOS SM retrievals (ie, 0–7 cm in ERA and 0–4 cm in SMOS). This difference could affect the validation of the SM retrieval as the quantity of water contained in the soil is different. Despite these problems, Santamaría-Artigas et al. (2016) showed a good performance when estimating SM at regional scale by using the combined approach.

The period of calibration was demonstrated as a drawback for the optical-passive microwave approach in Miernieki et al. (2014) because the coefficient needed to be calibrated before. Nevertheless, the temporal variation of the coefficients deserves more analysis, ideally for a period of L-band brightness temperature global measurements. Thus, the spatiotemporal regression coefficient analysis could be performed by using long-term time series that ensures the determination of coefficients and the degree of variation at regional and global scale over different land covers.

Finally, a new combination can be developed to consider different surface features including the mid-infrared and TIR. However, to include the land surface temperature (LST) or emissivity in the combined approach a theoretical background needs to be presented as a link to the main features of different parts of the electromagnetic spectra.

2.6 Toward a Thermal Infrared Contribution in the Optical-Passive Microwave Approach

The use of TIR parameters in the combined optical-passive microwave approach is an interesting option that can improve the SM retrievals. Although this chapter focuses on the combination between microwave and Visible and Near Infrared (VNIR) data (vegetation indices), it is worth mentioning the role of TIR data in the SM characterization. This is the purpose of this section, where we briefly discuss in a qualitative point of view how TIR remote sensing can be used to improve retrieval techniques based only on vegetation indices. For a quantitative discussion and detailed description of algorithms the reader is referred to Sobrino et al. (2013).

It is known that thermal remote sensing plays an important role in the research related to the analysis and modeling of land surface energy fluxes and land surface processes (Quattrochi and Luvall, 2000). The main geo/biophysical variables to be retrieved from the measured TIR signal are the LST and LSE. Precisely, these two variables provide useful information about the SM status.

On the one hand, the existence of a robust relationship between the water status of vegetated surfaces, their canopy development, and the corresponding LST has been evidenced (Carlson et al., 1994). It is not possible to directly infer the soil water content of vegetated surfaces from LST, but the latter can be taken under given circumstances as a proxy of soil water status. For example, empirical relationships between soil water content and LST were developed in Fernandez et al. (2006) to be applied over an agricultural area. Although a dependence of the empirical coefficients on the crop type was observed, the overall accuracy of the methodology was around 15%, and the spatial pattern of the retrieved soil water content captured well the expected variations of this parameter.

One of the possible linkages between LST and the SM estimated by using a combined optical-passive microwave is also through the interactions of the effective temperature. The effective temperature can be defined as the interaction of two layers of temperature. Nevertheless, Wigneron et al. (2008) described the possible interaction between surface temperature obtained from TIR and the effective temperature obtaining a good agreement, although that work was carried out at site scale. O'Neill et al. (2008) proposed a new ratio between surface and effective temperature by using airborne L-band measurements finding strong correlations. Similar results were found in Hasan et al. (2014) using LST to estimate the effective temperature by using the Wigneron et al. (2001) approach.

On the other hand, the LSE responds to changes in surface soil properties, and it has been demonstrated that LSE retrieved from TIR data is a better indicator of land cover change than the traditional NDVI (French et al., 2008; French and Inamdar, 2010). However, the quantitative relationship between LSE and SM is currently poorly understood. Only a few studies analyzed the LSE and SM dependence in the TIR by using laboratory and/or remote sensing measurements. Urai et al. (1997) measured the thermal emissivity of a desert sand sample under controlled SM content. An increase in the emissivity was observed with increasing SM content in the 8–10 μm range, whereas almost no changes were observed in the 11–13 μm range. It is important to note that in some cases this phenomenon was not clear because of the temperature estimation error in the sample. Similarly, an increase in the thermal emissivity with SM was also found by Mira et al. (2010), especially for sandy soils in the 8–9 μm range due to water adhering to soils grains. Hulley et al. (2010) analyzed the effects of SM on surface emissivity using field and satellite measurements over a desert sand sample. Results derived from satellite retrievals showed an increase in surface emissivity with increasing SM after a rainfall event. However, laboratory measurements evidenced that the SM changes were uncorrelated with changes in emissivity during the drying process.

Sobrino et al. (2012) estimated SM at airborne and satellite levels by combining remotely sensed images and in situ measurements using empirical relationships between SM, NDVI, and LSE. Results obtained from analysis of the in situ data showed that inclusion of emissivity in the SM algorithm improved the retrievals, especially in the case of the emissivity at around 9 μm. However, this improvement was reduced when the SM algorithm was applied to airborne and satellite data.

In most of the previous studies the increase in LSE with increasing SM was mainly observed in sand samples and at the 8–9 μm range, where a significant depression in emissivity is observed because of the presence of quartz. Spectral regions at higher wavelengths (10–12 μm) do not typically show significant spectral features in the emissivity spectrum, so the relationship between emissivity and SM in this range is hardly observed. We conclude that more analysis is still required to better understand the relationship between LSE and SM. Therefore, a clear quantitative relationship between SM and LSE should still be considered with caution.

It is also important to highlight that LSE can be used as an indicator of SM only when it is retrieved from multispectral TIR data. Most of the traditional techniques for LSE retrieval were based on VNIR data because of the unavailability of multispectral TIR sensors (eg, Sobrino et al., 2008). However, with the launch of the Advanced Spaceborne Thermal Emission and Reflection Radiometer on board the NASA/Terra platform, new possibilities were opened for retrieval of LSE from multispectral TIR data at local to regional scales (Gillespie et al., 1998). The use of MODIS and SEVIRI sensors for this purpose has also been proposed (Hulley and Hook, 2011; Jiménez-Muñoz et al., 2014), thus providing an opportunity to explore the SM-LSE relationship at global scale. The partial evidences of relationships between SM-LST-LSE described in this section, and recent and future data sets, open new possibilities for the exploration of novel approaches for a better estimation of SM.

3 CONCLUSIONS

This chapter showed the potential of the combined optical-passive microwave approach to retrieve SM at in situ and regional scales. Vegetation indexes such as LAI or NDVI are a potential alternative for better estimation of SM. However, the use of this synergic approach deserves more analysis in terms of spatiotemporal variations and further thermal influences to retrieve surface SM. New remote sensing technology will be launched in the future concerning the integration of several ranges of the electromagnetic spectra for better understanding of the biophysical process occurred in the Earth's surface. These technologies also need to be accompanied by a new synergic approach focusing in the combined use of remote sensing information and modeling.

ACKNOWLEDGMENTS

The authors would like to thank the projects Fondecyt-Initial (CONICYT/ref.: 11130359) and the Ministerio de Ciencia e Innovación (CEOS-Spain, AYA2011-29334-C02-01).

REFERENCES

Albergel, C., Calvet, J.C., Mahfouf, J.F., Rüdinger, C., Barbu, A.L., Lafont, S., Roujean, J.L., Walker, J.P., Crapeau, M., Wigneron, J.P., 2010. Monitoring of water and carbon fluxes using a land data assimilation system: a case study for southwestern France. Hydrol. Earth. Syst. Sci. 14, 1109–1124.

Calvet, J.-C., Wigneron, J.-P., Walker, J.P., Karbou, F., Chanzy, A., Albergel, C., 2011. Sensitivity of passive microwave observations to soil moisture and vegetation water content: L-band to W-band. IEEE Trans. Geosci. Rem. Sens. 49 (4), 1190–1199.

Carlson, T.N., Gillies, R.R., Perry, E.M., 1994. A method to make use of thermal infrared temperature and NDVI measurements to infer soil water content and fractional vegetation cover. Remote Sens. Rev. 52, 45–59.

Chakraborty, R., Ferrazzoli, P., Rahmoune, R., 2014. Application of passive microwave and optical signatures to monitor submerging of vegetation due to floods. Int. J. Remote Sens. 35 (16), 6310–6328.

Chanzy, A., Schmugge, T.J., Calvet, J.C., Kerr, Y., van Oevelen, P., Grosjean, O., Wang, J.R., 1997. Airborne microwave radiometry on a semi-arid area during Hapex-Sahel. J. Hydrol. 188–189, 285–309.

Choudhury, B.J., Schmugge, T.J., Mo, T., 1982. A parameterization of effective soil temperature for microwave emission. J. Geophys. Res. 87 (c2), 1301–1304.

Dee, D.P., Uppala, S.M., Simmons, A.J., Berrisford, P., Poli, P., Kobayashi, S., Andrae, U., Balmaseda, M.A., Balsamo, G., Bauer, P., et al., 2011. The ERA-interim Reanalysis: configuration and performance of the data assimilation system. Q. J. Roy. Meteorol. Soc. 137, 553–597.

De Rosnay, P., Calvet, J.C., Kerr, Y., Wigneron, J.P., Lemaître, F., Escorihuela, M.J., Muñoz Sabater, J., Saleh, K., Barrié, J., Bouhours, G., Coret, L., Cherel, G., Dedieu, G., Durbe, G., Fritz, N., Froissard, F., Hoedjes, J., Kruszewski, A., Lavenu, F., Suquia, D., Waldteufel, P., 2006. SMOSREX: a long term field campaign experiment for soil moisture and land surface processes remote sensing. Remote Sens. Environ. 102, 377–389.

Dente, L., Ferrazzoli, P., Su, Z., van der Velde, R., Guerriero, L., 2014. Combined use of active and passive microwave satellite data to constrain a discrete scattering model. Remote Sens. Environ. 155, 222–238.

Fernandez, G., Palladino, M., Urso, G., Moreno, J., Jimenez, J.C., 2006. Exploring soil water content dynamics from thermal observations within the SEN2FLEX experiment. In: Proceedings of SEN2FLEX. ESA Publications Division, Noordwijk, The Netherlands. (published in CD-ROM), WPP-271.

French, A.N., Inamdar, A., 2010. Land cover characterization for hydrological modelling using thermal infrared emissivities. Int. J. Remote Sens. 31 (14), 3867–3883.

French, A.N., Schmugge, T.J., Ritchie, J.C., Hsu, A., Jacob, F., Ogawa, K., 2008. Detecting land cover change at the Jornada Experimental Range, New Mexico with ASTER emissivities. Remote Sens. Environ. 112 (4), 1730–1748.

Gillespie, A.R., Rokugawa, S., Hook, S., Matsunaga, T., Kahle, A.B., 1998. A temperature and emissivity separation algorithm for Advanced Spaceborne Thermal Emission and Reflection Radiometer (ASTER) images. IEEE Trans. Geosci. Rem. Sens. 36, 1113–1126.

Hasan, S., Montzka, C., Rüdiger, C., Ali, M., Bogena, H.R., Vereecken, H., 2014. Soil moisture retrieval from airborne L-band passive microwave using high resolution multispectral data. ISPRS J. Photogramm. Rem. Sens. 91, 59–71.

Holmes, T., de Rosnay, P., de Jeu, R., Wigneon, J.P., Kerr, Y., Calvet, J.C., Escorihuela, M.J., Saleh, K., Lemaitre, F., 2006. A new parameterization of the effective temperature for L-band radiometry. Geophys. Res. Letters. 33. http://dx.doi.org/10.1029/2006GL025724.

Hornbuckle, B.K., England, A.W., 2005. Diurnal variation of vertical temperature gradients within a field of maize: implications for satellite microwave radiometry. IEEE Geosci. Remote Sens. Lett. 2 (1), 74–77.

Huete, A., Didan, K., Miura, T., Rodríguez, E.P., Gao, X., Ferreira, L.G., 2002. Overview of the radiometric and biophysical performance of the MODIS vegetation indices. Rem. Sens. Environ. 83, 195–213.

Hulley, G.C., Hook, S.J., Baldridge, A.M., 2010. Investigating the effects of soil moisture on thermal infrared land surface temperature and emissivity using satellite retrievals and laboratory measurements. Remote Sens. Environ. 114, 1480–1493.

Hulley, G.C., Hook, S.J., 2011. Generating consistent land surface temperature and emissivity products between ASTER and MODIS data for earth science research. IEEE Trans. Geosci. Rem. Sens. 49, 1304–1315.

Jackson, T.J., Schmugge, T.J., 1991. Vegetation effects on the microwave emission of soils. Remote Sens. Environ. 36, 203–212.

Jackson, T.J., Le Vine, D.M., Swift, C.T., Schmugge, T., Schiebe, F.R., 1995. Large area mapping of soil moisture using the ESTAR passive microwave radiometer in Washita92. Remote Sens. Environ. 54 (1), 27–37.

Jackson, T.J., LeVine, D.M., Hsu, A.Y., Oldak, A., Starks, P.J., Swift, C.T., et al., 1999. Soil moisture mapping at regional scales using microwave radiometry: the Southern Great Plains Hydrology Experiment. IEEE Trans. Geosci. Remote Sens. 37, 2136–2150.

Jackson, T.J., Chen, D., Cosh, M., Li, F., Anderson, M., Walthall, C., Dariaswamy, P., Ray Hunt, E., 2004. Vegetation water content mapping using Landsat data derived normalized difference water index for corn and soybeans. Remote Sens. Environ. 92, 475–482.

Jiménez-Muñoz, J.C., Sobrino, J.A., Mattar, C., Hulley, G., Göttsche, F., 2014. Temperature and emissivity separation from MSG/SEVIRI data. IEEE Trans. Geosci. Rem. Sens. 52 (9), 5937–5951.

Jones, M.O., Jones, L.A., Kimball, J.S., McDonald, K.C., 2011. Satellite passive microwave remote sensing for monitoring global land surface phenology. Remote Sens. Environ. 115 (4), 1102–1114.

Kerr, Y., Njoku, E.G., 1990. A semiempirical model for interpreting microwave emission from semiarid land surfaces as seen from space. IEEE Trans. Geosci. Remote Sens. 28, 384–393.

Kerr, Y., Waldteufel, P., Wigneron, J.P., Martinuzzi, J.M., Font, J., Berger, M., 2001. Soil moisture retrieval from space. The Soil Moisutre and Ocean Salinity (SMOS) mission. IEEE Tran. Geosci. Rem. Sens. 39 (8), 1729–1735.

Kerr, Y.H., Waldteufel, P., Richaume, P., Wigneron, J.P., Ferrazzoli, P., Mahmoodi, A., Al Bitar, A., Cabot, F., Gruhier, C., Juglea, S.E., Leroux, D., Mialon, A., Delwart, S., 2012. The SMOS soil moisture retrieval algorithm. IEEE Trans. Geosci. Rem. Sens. 50 (5), 1384–1403.

Kerr, Y.H., Waldteufel, P., Ricahume, P., Davenport, I., Ferrazzoli, P., Wigneron, J.P., 2010. SMOS level 2 processor soil moisture algorithm theoretical basis document (ATBD). SM-ESL (CBSA), CEBIO, Toulouse, SO-TN-ESL-SM-GS-0001, Issue 3.d. 13/06/2010.

Lacava, T., Brocca, L., Coviello, I., Faruolo, M., Pergola, N., Tramutoli, V., 2015. Integration of optical and passive microwave satellite data for flooded area detection and monitoring. Chapter 126. In: Lollino, G. et al., (Eds.), Engineering Geology for Society and Territory, vol. 3. Springer International Publishing, Switzerland, pp. 631–635. http://dx.doi.org/10.1007/978-3-319-09054-2_126.

Lawrence, H., Wigneron, J.-P., Richaume, P., Novello, N., Grant, J., Mialon, A., et al., 2014. Comparison between SMOS Vegetation Optical Depth products and MODIS vegetation indices over crop zones of the USA. Remote Sens. Environ. 140, 396–406.

Lemaître, L., Poussière, J.C., Kerr, Y., Dejus, M., Durbe, R., de Rosnay, P., Calvet, J.C., 2004. Design and test of the ground based L-band radiometer for estimating water in soils (LEWIS). IEEE Trans. Geosci. Remote Sens. 42 (8), 1666–1676.

Lv, S., Wen, J., Zeng, Y., Tian, H., Su, H., 2014. An improved two-layer algorithm for estimating effective soil temperature in microwave radiometry using in situ temperature and soil moisture measurements. Remote Sens. Environ. 152, 356–363.

Masson, V., Champeaux, J., Chauvin, F., Meriguet, C., Lacaze, R., 2003. A global database of land surface parameters at 1-km resolution in meteorological and climate models. J. Clim. 16 (9), 1261–1282.

Mattar, C., Wigneron, J., Sobrino, J.A., Novello, N., Calvet, J.C., Albergel, C., Richaume, P., 2012. A combined optical–microwave method to retrieve soil moisture over vegetated areas. IEEE Trans. Geosci. Remote Sens. 50 (5), 1404–1413.

Miernieki, M., Wigneron, J.P., López-Baeza, E., Kerr, Y., De Jeu, R., De Lennoy, G., Jackson, T., et al., 2014. Comparison of SMOS and SMAP soil moisture retrieval approaches using tower-based radiometer data over a vineyard field. Remote Sens. Environ. 154, 89–110.

Mira, M., Valor, E., Caselles, V., Rubio, E., Coll, C., Galve, J.M., Niclòs, R., Sánchez, J.M., Boluda, R., 2010. Soil moisture effect on thermal infrared (8–13 μm) emissivity. IEEE Trans. Geosci. Remote Sen. 48 (5), 2251–2260.

Myneni, R., et al., 2002. Global products of vegetation leaf area and fraction absorbed PAR from year one of MODIS data. Remote Sens. Environ. 83, 214–231.

O'Neill, P.M., Dehaan, R., Walker, J.P., Hue, I., 2008. Improving the Airborne Remote Sensing of Soil Moisture: Estimating Soil Effective Temperature. Charles Sturt University, 29pp.

Parrens, M., Zakharova, E., Lafont, S., Calvet, J.C., Kerr, Y., Wagner, W., Wigneron, J.P., 2012. Comparing soil moisture retrievals from SMOS and ASCAT over France. Hydrol. Earth Syst. Sci. 16, 423–440.

Pellarin, T., Wingeorn, J.P., Calvet, J.C., Ferrazzoli, P., Douville, H., Lopez-Baeza, E., Pulliainen, J., Simmonds, L.P., Walteufel, P., 2003. Two-year global simulation of L-band brightness temperatures over land. IEEE Trans. Geosci. Remote Sens. 41 (9), 2135–2139.

Quattrochi, D.A., Luvall, J.C., 2000. Thermal Remote Sensing in Land Surface Processes. CRC Press, Boca Raton, FL.

Saleh, K., Wigneron, J.P., de Rosnay, P., Calvet, J.C., Kerr, Y., 2006a. Semi-empirical regressions at L-band applied to surface soil moisture retrievals over grass. Remote Sens. Environ. 101, 415–426.

Saleh, K., Wigneron, J.P., de Rosnay, P., Calvet, J.C., Esocrihuela, M.J., Kerr, Y., Waldteufel, P., 2006b. Impacto of rain interception by vegetation and mulch on the L-band emission of natural grass. Remote Sens. Environ. 101, 127–139.

Santamaría-Artigas, A., Mattar, C., Wigneron, J.-P., 2016. Application of a Combined Optical–Passive Microwave Method to Retrieve Soil Moisture at Regional Scale Over Chile. http://dx.doi.org/10.1109/JSTARS.2015.2512926.

Salomonson, V.V., Hall, D.K., Chien, J.Y.L., 1995. Use of passive microwave and optical data for large-scale snow cover mapping. In: Proceedings of the 2nd Topical Symposium on Combined Optical Microwave Earth and Atmosphere Sensing, Atlanta, GA, April 3–6, pp. 35–37.

Schmugge, T., Jackson, T.J., 1994. Mapping soil moisture with microwave radiometers. Meteorol. Atmos. Phys. 54 (1–4), 213–223.

Schwank, M., Wiesmann, A., Werner, C., Mätzler, C., Weber, D., Murk, A., et al., 2010. ELBARA II, an L-band radiometer system for soil moisture research. Sensors MDPI 10, 584–612.

Shi, J.C., Jackson, T., Tao, J., Du, J., Bindlish, R., Lu, L., Chen, K.S., 2008. Microwave vegetation indices for short vegetation covers from satellite passive microwave sensor AMSR-E. Remote Sens. Environ. 112 (12), 4285–4300.

Sobrino, J.A., Jiménez-Muñoz, J.C., Sòria, G., Romaguera, M., Guanter, L., Moreno, J., Plaza, A., Martínez, P., 2008. Land surface emissivity retrieval from different VNIR and TIR sensors. IEEE Trans. Geosci. Remote Sens. 46 (2), 316–327.

Sobrino, J.A., Franch, B., Mattar, C., Jiménez-Muñoz, J.C., Corbari, C., 2012. A method to estimate soil moisture from Airborne Hyperspectral Scanner (AHS) and ASTER data: application to SEN2FLEX and SEN3EXP campaigns. Remote Sens. Environ. 117, 415–428.

Sobrino, J.A., Mattar, C., Jiménez-Muñoz, J.C., Franch, B., Corbari, C., 2013. On the synergy between optical and TIR observations for the retrievals of soil moisture content: exploring different approaches. In: Petropoulos, G.P. (Ed.), Remote Sensing of Energy Fluxes and Soil Moisture Content. CRC Press, Boca Raton, FL, ISBN 978-1-4665-0578-0, pp. 363–390. http://dx.doi.org/10.1201/b15610-19.

Ulaby, F.T., Moore, R.K., Fung, A.K., 1986. Microwave Remote Sensing – Active and Passive. Artech House, Norwood, MA. vol. III.

Urai, M., Matsunaga, T., Ishii, T., 1997. Relationship between soil moisture content and thermal infrared emissivity of the sand sampled in Muus Desert, China. Remote Sens. Soc. Japan 17 (4), 322–331.

Van de Griend, A., Wigneron, J.P., 2004. On the measurements of microwave vegetation properties: some guidelines for a protocol. IEEE Trans. Geosci. Remote Sens. 42 (10), 2277–2289.

Wigneron, J.P., Chanzy, A., Calvet, J.C., Bruguier, N., 1995. A simple algorithm to retrieve soil moisture and vegetation biomass using passive microwave measurements over crop fields. Remote Sens. Environ. 51, 331–341.

Wigneron, J.P., Laguerre, L., Kerr, Y., 2001. Simple modeling of the L-band microwave emission from rough agricultural soils. IEEE Trans. Geosci. Remote Sens. 39 (8), 1697–1707.

Wigneron, J.P., Chanzy, A., Calvet, J.C., Olioso, A., Kerr, Y., 2002. Modeling approaches to assimilation L-band passive microwave observations over land surfaces. J. Geophys. Res. 107, D14. http://dx.doi.org/10.1029/2001JD000958.

Wigneron, J.P., Pardé, M., Waldteufel, P., Chanzy, A., Kerr, Y., Schmidl, S., Skou, N., 2004a. Characterizing the dependence of vegetation parameters on crop type, view angle and polarization at L-band. IEEE Trans. Geosci. Remote Sens. 42 (2), 416–425.

Wigneron, J.P., Calvet, J.C., de Rosnay, P., Kerr, Y., Waldteufel, P., Saleh, K., Escorihuela, M.J., Kruszewski, A., 2004b. Soil moisture retrievals from biangular L-band passive microwave observations. IEEE Trans. Geosci. Remote Sens. 1 (4), 277–281.

Wigneron, J.P., Kerr, Y., Waldteufel, P., Saleh, K., Escorihuela, M.J., Richaume, P., Ferrazzoli, P., de Rosnay, P., Gurney, R., Calvet, J.C., Grant, J.P., Guglielmetti, M., Hornbuckle, B., Mätzler, C., Pellarin, T., Schwank, M., 2007. L-band microwave emission of the biosphere (L-MEB) model: description and calibration against experimental data sets over crop fields. Remote Sens. Environ. 107, 639–655.

Wigneron, J.-P., Chanzy, A., de Rosnay, P., Rüdiger, Ch., Calvet, J.-C., 2008. Estimating the effective soil temperature at L-band as a function of soil properties. IEEE Tran. Geosci. Rem. Sens. 46 (3), 797–807.

Wigneron, J.-P., Schwank, M., Baeza, E.L., Kerr, Y., Novello, N., Millan, C., Moisy, C., Richaume, P., Mialon, A., Al, A., Bitar, A., Cabot, F., Lawrence, H., Guyon, D., Calvet, J.-C., Grant, J.P., Casal, T., de Rosnay, P., Saleh, K., Mahmoodi, A., Delwart, S., Mecklenburg, S., et al., 2012. First evaluation of the simultaneous SMOS and ELBARA-II observations in the Mediterranean region. Rem. Sens. Environ. 124, 26–37.

Chapter 8

Nonparametric Model for the Retrieval of Soil Moisture by Microwave Remote Sensing

D.K. Gupta*, R. Prasad*, P.K. Srivastava[†,‡] and T. Islam[§,¶]
Indian Institute of Technology (BHU), Varanasi, India, †NASA Goddard Space Flight Center, Greenbelt, MD, United States, ‡Banaras Hindu University, Banaras, India, §NASA Jet Propulsion Laboratory, Pasadena, CA, United States, ¶California Institute of Technology, Pasadena, CA, United States

1 INTRODUCTION

The knowledge about spatial and temporal distribution of soil moisture on the top ground surfaces plays an important role in understanding climate changes such as exchange of water and heat energy between the land surface through evaporation and plant transpiration (Petropoulos et al., in press; Piles et al., in press). Nowadays, microwave remote sensing is a powerful tool to monitor the soil moisture available in the top Earth surfaces all the time. Microwave sensors can be more effective than optical sensors due to their adequate capability to acquire observations at all times and in all weather conditions.

The microwave is sensitive toward dissimilarity between the dielectric constant of liquid water ($\sim$80) and dry soil ($\sim$4). Dobson et al. (1985), Hallikainen et al. (1985), and Wang and Schmugge (1980) as well as other researchers studied the dielectric properties of soil-water mixture. As the water content increases in the soil sample, the dielectric constant of soil-water mixture increases. The values of complex dielectric constant of soil-water mixture varies from 4 to 40. The range of complex dielectric constant of soil-water mixture may be differentiable at microwave frequencies.

Researchers including (Chauhan, 1997; Engman and Chauhan, 1995; Njoku and Li, 1999; Schmugge et al., 1986; Ulaby et al., 1981; Wang et al., 1983) conducted experiments using monostatic geometry of the radar system to develop empirical, semiempirical, and physical models for the estimation of soil surface parameters by active and passive microwave remote sensing using different platforms such as ground-based, airborne, and spaceborne. A limited number

Satellite Soil Moisture Retrieval. http://dx.doi.org/10.1016/B978-0-12-803388-3.00008-5

of bistatic experiments have been performed (Khadhra et al., 2012; Singh et al., 1996). This study may be important in the microwave remote sensing community to understand the interaction bistatic radar with soil surface parameters toward its retrieval.

There are many useful modeling approaches based on soft computing algorithms such as artificial neural network (ANN), support vector regression, fuzzy logic, and others, which have been reported for the estimation of soil moisture and soil surface parameters (Oh et al., 1992; Saleh et al., 2006; Singh, 2005; Singh et al., 1996). These models have the advantages of providing reasonable results in most cases. The ANN learns the highly nonlinear input-output relationship through a process called training. Previous researchers (Chai et al., 2009; Del Frate et al., 2003; Dharanibai and Alex, 2009; Jiang and Cotton, 2004) have widely used ANN for the estimation of surface soil moisture.

The objective of the present investigation is to study the capability of the radial basis function artificial neural network (RBFANN) model for the estimation of soil moisture of soil rough surfaces using bistatic scatterometer data at the X-band. Linear regression analysis is carried out between bistatic scattering coefficients and soil moisture to find the suitable incidence angle for the estimation of soil moisture at Transmit horizontal-Receive horizontal (HH)- and Transmit vertical-Receive vertical (VV)-polarization. A RBFANN model is calibrated using input data sets (bistatic scattering coefficients) and output data sets (soil moisture content in rough soil surface). The calibrated RBFANN model is used for the estimation of soil moisture of rough soil surfaces.

2 MATERIAL AND METHODS

2.1 Instrumentation Setup and Observations

The bistatic scatterometer setup is designed to take observations for different soil moisture contents at the top of bare soil surface for angular range 20–70 degrees in steps of 5 degrees at HH- and VV-polarizations in the specular direction at azimuthal angle ($\phi = 0$). The bistatic scatterometer system can be categorized into two sections. The first section is called the transmitter and second section is called the receiver. The transmitter sends the electromagnetic waves while the receiver receives the electromagnetic waves after interaction from scatter. In this study the transmitter part consist of a pyramidal dual-polarized X-band horn antenna, a waveguide to an N-female coaxial adaptor, and a PSG high-power signal generator (E8257D, 10 MHz to 20 GHz). The receiver part consists of a pyramidal dual polarized X-band horn antenna, a waveguide to an N-female coaxial adaptor, an EPM-P series power meter (E4416A), and a peak and average power sensor (E9327A, 50 MHz to 18 GHz). The gain of these antennas were approximately 20 dB, whereas, their half-power beam width were found at 18 and 20 degrees for E and H plane, respectively. The 90 degrees E-H twisters were used to change the polarization HH to VV and vice versa.

FIG. 1 Photograph of bistatic scatterometer system at measurement site.

Fig. 1 shows the real photo of the experimental setup of the bistatic scatterometer system used in this study. The bistatic scatterometer system had the facility to change the incidence angle from 0 to 90 degrees but the observations were taken only from 20 to 70 degress at steps of 5 degrees. The height and distance from the center of the target can also vary to adjust the focus of both antennas at the center of the target. The antennas were placed in the far field region from the center of the target to minimize the near-field interactions. Table 1 shows the specifications of the bistatic scatterometer system used for bistatic measurements. The system was calibrated by noting the signals returned from an aluminum plate placed on the top of the target.

An outdoor test bed of bare soil surface (4 m × 4 m) was specially prepared to carry out bistatic scatterometer measurements at X-band beside the Department of Physics, Indian Institute Technology (B.H.U.), Varanasi, India. The surface roughness was taken constant during the entire observations to study the microwave response of soil moisture content only. The root mean square height (σ) and correlation length of the test soil surface were 1.61 cm and 11.63 cm, respectively. The bare soil test bed was flooded with water for 20–24 h to have a large range of soil moisture constants before making observations. The gravimetric moisture content of soil is defined as the ratio of the weight of water present in soil to the weight of dry soil. It is expressed as a percentage of soil moisture content. Five random soil samples were collected in an aluminum soil container up to depths of 5 cm from the soil surface. These soil samples were dried in an oven at 100°C for 24 h. The samples were weighed before and after drying to compute the gravimetric moisture content. The average of gravimetric moisture content of all five soil samples were taken to calculate the percentage of soil moisture content of the soil surface.

TABLE 1 Specification of Scatterometer System

RF generator		E8257D, PSG High Power Signal Generator, 10 MHz to 20 GHz (Agilent Technologies)
Power meter		E4416A, EPM-P Series Power meter, 10 MHz to 20 GHz (Agilent Technologies)
Power sensor		Peak and average power sensor (E9327A, 50 MHz–18 GHz) (Agilent Technologies)
Frequency (GHz)		10 ± 0.05 (X-band)
Beam Width	E plane (°)	17.3118
	H plane (°)	19.5982
Band width (GHz)		0.8
Antenna gain (dB)		20
Cross-polarization isolation (dB)		40
Antenna type		Dual-polarized pyramidal horn
Calibration accuracy (dB)		1
Platform height (meter)		3
Incidence angle (°)		20° (nadir) – 70°

2.2 Radial Basis Function Artificial Neural Network

The RBFANN consists of three different feed-forward layers, namely the input layer, the hidden layer, and the output layer. The hidden layer neurons are implemented with radial basis functions (e.g., Gaussian function). The output layer neurons are implemented with linear summation functions as in a multi-layer perceptron.

All the hidden layer neurons take inputs from all input layer neurons. The hidden layer neurons are activated with radial basis function having center and spread parameters. The selective function decreases rapidly with smaller value of width, whereas it decreases slowly with a larger value of width. The center of the radial basis function for ith neuron at the hidden layer is a vector c_i whose size is as the input vector x. The radial distance d_i, between the input vector x and the center of the radial basis function c_i, are computed for each ith neurons in the hidden layer as

$$d_i = \|x - c_i\|, i = 1, 2, 3, \ldots, h \qquad (1)$$

Using the Euclidean distance, the output for each ith neurons of the hidden layer were computed by applying the radial basis function $\phi(r)$ to the Euclidean distance,

$$H_i = \phi(d_i, \sigma_i) \tag{2}$$

The calculation made between the input spaces to hidden space is nonlinear whereas the hidden space to output space is linear.

The jth output is computed as

$$y_j = f_j(\boldsymbol{x}) = w_{oj} + \sum_{i=1}^{h} w_{ij} H_i \tag{3}$$

where j denotes 1, 2, 3, ..., m, y_j is the output of jth neuron, w_{ij} is the weight vector for jth neuron, and H_i is the output from ith hidden layer neuron.

The value of the center and width depend on the pattern to be used for optimization. Generally, the width should be larger than the minimum distance and smaller than the maximum distance between the input neurons and the center of the RBFANN spread to get better generalization (Narendra et al., 1998). The center and weights were found by using the orthogonal least squares algorithm (Chen et al., 1991). Fig. 2 shows the architecture of the RBFANN model.

2.3   Performance Indices

Several performance indices such as %Bias, root mean squared error (RMSE), and Nash-Sutcliffe efficiency (NSE) are used for evaluating the performance of

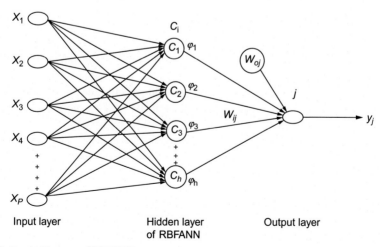

| Input layer | Hidden layer of RBFANN | Output layer |

FIG. 2   Architecture of RBFANN.

the RBFANN model. The percentage bias (%Bias) measures the average tendency of the estimated values to be larger or smaller than their observed values. The optimum value of %Bias is 0.0 and the smaller value of %Bias indicates that accurate model prediction.

$$\%\text{Bias} = 100 \times \left[\frac{\sum (y_i - x_i)}{\sum x_i} \right] \tag{4}$$

where x_i is the observed and y_i is the simulated variable.

RMSE is frequently used to measure the differences between estimated values by a RBFANN model or an estimator and the observed values.

$$\text{RMSE} = \sqrt{\frac{1}{n} \sum_{i=1}^{n} (y_i - x_i)^2} \tag{5}$$

where n is the number of observations.

The NSE is based on the sum of absolute squared differences between the estimated and observed values normalized by the variance of the observed values during the study. The NSE was calculated using the given formula,

$$\text{NSE} = 1 - \frac{\sum_{i=1}^{n} (y_i - x_i)^2}{\sum_{i=1}^{n} (x_i - \bar{x}_i)^2} \tag{6}$$

3 RESULTS AND DISCUSSION

3.1 Assessment of Data Sets

Table 2 shows the linear regression results between bistatic scattering coefficients and soil moisture content at different incidence angle ranges of 20–70 degrees in steps of 5 degrees for HH- and VV-polarization. These results may be used for the assessment of data sets and selection of the suitable incidence angle for the estimation of soil moisture at HH- and VV-polarization. The values of coefficient of determination (R^2) are found higher at lower incidence angles than higher incidence angles. The bistatic scattering coefficients are more sensitive to the moisture content at lower incidence angles than higher incidence angles. The higher value of R^2 is found to be 0.868 and 0.886 at 25 degrees incidence angle for HH- and VV-polarization, respectively. The 25 degrees incidence angle scattering coefficients are taken for the estimation of soil moisture by a nonparametric model (RBFANN). The 75% data sets are used for the calibration and 25% of the data sets are kept for the validation of model.

TABLE 2 Linear Regression Results Between Scattering Coefficient and Soil Moisture

Angle (Degrees)	HH-Polarization			VV-Polarization		
	R^2	SE	SEE	R^2	SE	SEE
20	0.864	0.561	0.386	0.872	1.765	1.174
25	0.868	0.645	0.436	0.886	1.810	1.125
30	0.851	0.691	0.501	0.878	1.863	1.201
35	0.837	0.662	0.506	0.883	1.935	1.219
40	0.846	0.597	0.440	0.844	1.868	1.392
45	0.821	0.755	0.609	0.829	2.088	1.644
50	0.809	0.668	0.563	0.823	1.814	1.460
55	0.808	0.564	0.476	0.840	1.947	1.470
60	0.775	0.386	0.360	0.831	1.684	1.313
65	0.816	0.318	0.262	0.812	1.459	1.216
70	0.810	0.253	0.212	0.847	1.631	1.199

3.2 Estimation of Soil Moisture Using the RBFANN

The RBFANN are calibrated for the estimation of soil moisture content from the moist soil surfaces in the angular ranges of 20–70 degrees for HH- and VV-polarization at the X-band. The observed data sets (bistatic scattering coefficient and soil moisture content) are interpolated into 68 data sets at the incidence angle 25 degrees for HH- and VV-polarization.

The parameters of the RBFANN model need to be optimized for the stable result at HH- and VV-polarization. So the preliminary analysis of parameters of ANN techniques is necessary before using it for any computation. The value of spread is an important parameter for training the RBFANN model. The value of spread is also optimized by simulating the models with different values of spread from 1 to 15 steps 0.5 and continuously watching the RMSE at each value of spread between observed value of soil moisture and estimated value of soil moisture by model. At HH-polarization, the RBFANN model shows the constant RMSE over each value of spread. This means the performance of the RBFANN model may be optimized at the spread value 1. In the case of VV-polarization, the RBFANN model gives the larger value of RMSE until the 6 spread value afterwards is constant until the spread value is 15. The spread value may consider 7 for the RBFANN model.

The performance during the calibration and validation of the RBFANN model for the estimation of soil moisture is evaluated in terms of %Bias, RMSE, and NSE. Fig. 3A and B depict the scatter plot between estimated soil moisture by the RBFANN model and experimentally observed soil moisture using calibration data sets. At HH-polarization, the %Bias showed overperformance in comparison to VV-polarization. The higher value of RMSE is found at HH-polarization in comparison to VV-polarization. The highest value of NSE was found by the RBFANN model at VV-polarization than at HH-polarization.

Fig. 4A and B depict the scatter plot between estimated soil moisture by the RBFANN model and experimentally observed soil moisture for the validation data sets. The value of %Bias showed the overperformance at VV-polarization than at HH-polarization. The higher value of RMSE between observed and estimated values of soil moisture is found at HH-polarization in comparison to VV-polarization. The highest value of NSE is found at VV-polarization in comparison to HH-polarization.

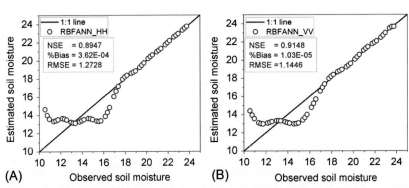

FIG. 3 Scatter plot with 1:1 equiv. Line during the calibration of RBFANN model, (A) at HH-polarization, (B) at VV-polarization.

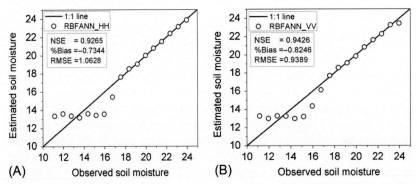

FIG. 4 Scatter plot with 1:1 equiv. Line during the validation of RBFANN model, (A) at HH-polarization, (B) at VV-polarization.

4 CONCLUSIONS

The suitable incidence angle for the estimation of soil moisture of the bare soil surface using bistatic scatterometer data is found to be 25 degrees for both HH- and VV-polarization. The estimation of soil moisture from bare soil surfaces using bistatic scatterometer data by the RBFANN model are found close to the observed data of soil moisture at both like polarizations. The RBFANN may be the good estimator for the estimation of soil moisture. The performance of the RBFANN model is found little better at VV-polarization than HH-polarization for the estimation of soil moisture from bare soil surfaces using multiincidence and dual-polarized bistatic scatterometer data.

REFERENCES

Chai, S.-S., Walker, J.P., Makarynskyy, O., Kuhn, M., Veenendaal, B., West, G., 2009. Use of soil moisture variability in artificial neural network retrieval of soil moisture. Remote Sens. 2, 166–190.

Chauhan, N.S., 1997. Soil moisture estimation under a vegetation cover: combined active passive microwave remote sensing approach. Int. J. Remote Sens. 18, 1079–1097.

Chen, S., Cowan, C.F., Grant, P.M., 1991. Orthogonal least squares learning algorithm for radial basis function networks. IEEE Trans. Neural Netw. 2, 302–309.

Del Frate, F., Ferrazzoli, P., Schiavon, G., 2003. Retrieving soil moisture and agricultural variables by microwave radiometry using neural networks. Remote Sens. Environ. 84, 174–183.

Dharanibai, G., Alex, Z., 2009. ANN technique for the evaluation of soil moisture over bare and vegetated fields from microwave radiometer data. Indian J. Radio Space 38, 283–288.

Dobson, M.C., Ulaby, F.T., Hallikainen, M.T., El-Rayes, M.A., 1985. Microwave dielectric behavior of wet soil—Part II: dielectric mixing models. IEEE Trans. Geosci. Remote Sens. GE-23, 35–46.

Engman, E.T., Chauhan, N., 1995. Status of microwave soil moisture measurements with remote sensing. Remote Sens. Environ. 51, 189–198.

Hallikainen, M.T., Ulaby, F.T., Dobson, M.C., El-Rayes, M.A., Lil-Kun, W., 1985. Microwave dielectric behavior of wet soil—Part 1: empirical models and experimental observations. IEEE Trans. Geosci. Remote Sens. GE-23, 25–34.

Jiang, H., Cotton, W.R., 2004. Soil moisture estimation using an artificial neural network: a feasibility study. Can. J. Remote Sens. 30, 827–839.

Khadhra, K.B., Boerner, T., Hounam, D., Chandra, M., 2012. Surface parameter estimation using bistatic polarimetric X-band measurements. Prog. Electromagn. Res. B 39, 197–223.

Narendra, K., Sood, V., Khorasani, K., Patel, R., 1998. Application of a radial basis function (RBF) neural network for fault diagnosis in a HVDC system. IEEE Trans. Power Syst. 13, 177–183.

Njoku, E.G., Li, L., 1999. Retrieval of land surface parameters using passive microwave measurements at 6–18 GHz. IEEE Trans. Geosci. Remote Sens. 37, 79–93.

Oh, Y., Sarabandi, K., Ulaby, F.T., 1992. An empirical model and an inversion technique for radar scattering from bare soil surfaces. IEEE Trans. Geosci. Remote Sens. 30, 370–381.

Piles, M., Petropoulos, G.P., Ireland, G., Sanchez, N., in press. A novel method to retrieve soil moisture at high spatio-temporal resolution based on the synergy of SMOS and MSG SEVIRI observations. Rem. Sens. Environ.

Petropoulos, G.P., Ireland, G., Barrett, B., in press. Surface soil moisture retrievals from remote sensing: evolution, current status, products & future trends. Phys. Chem. Earth. http://dx.doi.org/10.1016/j.pce.2015.02.009.

Saleh, K., Wigneron, J.-P., de Rosnay, P., Calvet, J.-C., Kerr, Y., 2006. Semi-empirical regressions at L-band applied to surface soil moisture retrievals over grass. Remote Sens. Environ. 101, 415–426.

Schmugge, T., O'Neill, P.E., Wang, J.R., 1986. Passive microwave soil moisture research. IEEE Trans. Geosci. Remote Sens. GE-24, 12–22.

Singh, D., 2005. A simplistic incidence angle approach to retrieve the soil moisture and surface roughness at X-band. IEEE Trans. Geosci. Remote Sens. 43, 2606–2611.

Singh, D., Mukherjee, P., Sharma, S., Singh, K., 1996. Effect of soil moisture and crop cover in remote sensing. Adv. Space Res. 18, 63–66.

Ulaby, F.T., Moore, R.K., Fung, A.K., 1981. Microwave Remote Sensing: Microwave Remote Sensing Fundamentals and Radiometry. Addison-Wesley Publishing Company, Advanced Book Program/World Science Division, Boston, MA.

Wang, J.R., Schmugge, T.J., 1980. An empirical model for the complex dielectric permittivity of soils as a function of water content. IEEE Trans. Geosci. Remote Sens. GE-18, 288–295.

Wang, J.R., O'Neill, P.E., Jackson, T.J., Engman, E.T., 1983. Multifrequency measurements of the effects of soil moisture, soil texture, and surface roughness. IEEE Trans. Geosci. Remote Sens. GE-21, 44–51.

Chapter 9

Temperature-Dependent Spectroscopic Dielectric Model at 0.05–16 GHz for a Thawed and Frozen Alaskan Organic Soil

V. Mironov and I. Savin

Kirensky Institute of Physics, Krasnoyarsk, Russia

1 INTRODUCTION

The most promising results in the remote sensing of soil moisture, freeze/thaw state, and temperature over land surfaces have been obtained using microwave radiometry and radar techniques. Contemporary space missions using these techniques, such as AQUA (Aqua Earth-observing sattelite mission), GCOM-W (Global Change Observation Mission-Water), SMOS (Soil Moisture and Ocean Salinity), SMAP (Soil Moisture Active and Passive), RADARSAT (Radar Satellite), and ALOS PALSAR (Advanced Land Observing Satellite Phased Array type L-band Synthetic Aperture Radar), employ radiometers and radars that function in the frequency range from 1.2 to 89 GHz. Dielectric models of topsoil are a crucial element of retrieval algorithms that are used to obtain geophysical characteristics of the land surface, such as soil moisture, temperature, and freeze/thaw state, with these techniques.

Recent research in microwave radiometry and radar remote sensing has been focused on the Arctic region (Bircher et al., 2012; Jones et al., 2007; Mironov et al., 2013a; Rautiainen et al., 2012; Watanabe et al., 2012). The topsoil in both boreal forests and Arctic tundra generally contains organic-rich soils, which requires that dielectric models for these soils be available to develop retrieval algorithms. A dielectric model for the organic-rich Arctic soil in the area of North Slope, Alaska, was proposed by (Mironov et al., 2010). This model provides predictions of the complex dielectric constant (CDC) for one type of thawed and frozen organic-rich soil as a function of frequency, temperature, and moisture ranging from 1.0 to 16 GHz, 25°C to −30°C, and 0.05 to 1.1 g/g, respectively.

Satellite Soil Moisture Retrieval. http://dx.doi.org/10.1016/B978-0-12-803388-3.00009-7

In this chapter, an enhanced temperature-dependent multirelaxation spectroscopic dielectric model (TD MRSDM) for the organic-rich Arctic soil in the area of North Slope, Alaska, was developed. Compared to the model in Mironov et al. (2010), the frequency range was extended from 1–16 GHz to 0.05–16 GHz, and some additional dielectric relaxations of soil water were detected in the frequency range of 0.05–1.0 GHz. As a result, a new approach to processing the measured dielectric data was applied, and the theoretical dielectric model of moist soil used in Mironov et al. (2010) was modified to address the multirelaxation characteristics of the measured dielectric data of the soil samples as was done in Mironov et al. (2013b) and Mironov and Savin (2015) for thawed and frozen mineral and organic soils. The temperature range for the frozen soil samples was also substantially extended from −30°C to −7°C (Mironov et al., 2010) to −30°C to −1°C, which allowed taking into account the intensive phase transitions of soil water between −1°C and −7°C that were observed in a cooling run.

2 SOIL SAMPLES AND MEASUREMENT PROCEDURES

To develop the dielectric model, we used samples of Arctic soil that were collected (De Roo et al., 2006) in shrub tundra, Alaska, and earlier analyzed in Mironov et al. (2010). The percentages of organic matter and mineral solids in the soil are the following: 80–90% organic matter, 4.5% tiff, 7.5–8.2% quartz, 0.75% plagioclase, 0.75–1.5% mica, and 0.75% smectite. Before performing the dielectric measurements, the samples were ground using a coffee grinder. The techniques of soil samples processing prior to and in the course of dielectric measurements are presented in Mironov et al. (2010).

To conduct dielectric measurements, the soil sample was placed into a cell formed by a section of coaxial waveguide with the cross section of 7/3 mm, the latter ensuring that only the TEM mode is in the measured frequency range. The length of sample placed in the cell and its volume were equal to 17 mm and 0.529 cm^3, respectively. When filling the measurement cell, the soil was compacted with a cylinder pestle. The cell was blocked on both sides with Teflon washers, which prevented the sample from changing in volume. The cell was connected to the ZVK Rohde & Schwarz vector network analyzer to measure the frequency spectra of the S11, S22, S12, and S21 elements of the scattering matrix S over the frequency range from 50 MHz to 16 GHz.

The isothermal measurements were ensured with the use of an SU-241 Espec chamber of heat and cold with accuracy 0.5°C. To control the isothermal measurements, a combined system consisting of the chamber and the network analyzer was developed, using the RS-232 interface and a set of built-in commands. This system allows for setting a sequence of temperatures at which the spectra of the scattering matrices are measured isothermally. After the temperature control system switches the chamber to a next assigned temperature point and this temperature is established inside the chamber, the system starts controlling the root-mean-square deviations between the S12 spectra subsequently measured every minute. When the value of the root-mean-square deviation decreases to below 0.01, the system switches the chamber to the next assigned temperature point, and the process of establishing temperature equilibrium between the sample and the chamber at the assigned temperature point repeats. An interval of time to transfer from one measurement temperature to another is maximal at the largest moisture of the soil sample, taking the following values: (1) 10 min in the case of thawed soil; (2) 40 and 20 min in the case of frozen soil for the temperature intervals of $-15°C < T < -1°C$ and $-30°C < T < -15°C$, respectively.

The algorithm developed in Mironov et al. (2013c) was applied to retrieve the spectra of the CDC of moist sample using the measured values of S12 or S21. This algorithm provides the real and imaginary parts of the CDC, with the errors less than 10%.

3 CONCEPT OF A MULTIRELAXATION SPECTROSCOPIC DIELECTRIC MODEL

As in Mironov et al. (2010), we analyze the CDC of moist soil, ε_s^*, in terms of the reduced complex refractive index (CRI):

$$\left(n_s^* - 1\right)/\rho_d = \left(\sqrt{\varepsilon_s^*} - 1\right)/\rho_d = (n_s - 1)/\rho_d + i\kappa_s/\rho_d \tag{1}$$

where $n_s = Re\sqrt{\varepsilon_s^*}$ and $\kappa_s = Im\sqrt{\varepsilon_s^*}$ are the refractive index (RI) and the normalized attenuation coefficient (NAC), respectively. The NAC is considered to be a proportion of the standard attenuation coefficient to the free-space propagation constant. As observed in Eq. (1), the reduced CRI of a given substance is equal to a difference between the CRI of the substance and the CRI of air normalized by the density of the substance. For the reduced CRI of moist soil, we use the refractive mixing dielectric model as given in Mironov et al. (2010):

$$\frac{n_s - 1}{\rho_d(m_g)} = \begin{cases} \dfrac{n_m - 1}{\rho_m} + \dfrac{(n_b - 1)}{\rho_b}m_g, m_g \leq m_{g1}; \\[2ex] \dfrac{n_m - 1}{\rho_m} + \dfrac{(n_b - 1)}{\rho_b}m_{g1} + \dfrac{(n_t - 1)}{\rho_t}\left(m_g - m_{g1}\right), m_{g1n} \leq m_g \leq m_{g2}; \\[2ex] \dfrac{n_m - 1}{\rho_m} + \dfrac{(n_b - 1)}{\rho_b}m_{g1} + \dfrac{(n_t - 1)}{\rho_t}\left(m_{g2} - m_{g1}\right) + \dfrac{n_{u,i} - 1}{\rho_{u,i}}\left(m_g - m_{g2}\right), m_g \geq m_{g2} \end{cases} \tag{2}$$

$$\frac{\kappa_s}{\rho_d(m_g)} = \begin{cases} \dfrac{\kappa_m}{\rho_m} + \dfrac{\kappa_b}{\rho_b}m_g, m_g \leq m_{g1}; \\[2ex] \dfrac{\kappa_m}{\rho_m} + \dfrac{\kappa_b}{\rho_b}m_{g1} + \dfrac{\kappa_t}{\rho_t}\left(m_g - m_{g1}\right), m_{g1} \leq m_g \leq m_{g2}; \\[2ex] \dfrac{\kappa_m}{\rho_m} + \dfrac{\kappa_b}{\rho_b}m_{g1} + \dfrac{\kappa_t}{\rho_t}\left(m_{g2} - m_{g1}\right) + \dfrac{\kappa_{u,i}}{\rho_{u,i}}\left(m_g - m_{g2}\right), m_g \geq m_{g2} \end{cases} \tag{3}$$

where m_{g1} and m_{g2} are the maximum gravimetric fractions of bound water and total bound water (which consists of bound water and transient bound water), respectively.

According to Eqs. (2) and (3), three ranges of soil moisture, namely, $m_g \leq m_{g1}$, $m_{g1} \leq m_g \leq m_{g2}$, and $m_g \geq m_{g2}$, are clearly distinguished. In the first range ($m_g \leq m_{g1}$), only the bound water component can form in the soil when liquid water is added to an initially dry soil. Therefore, the first range can be identified as the range of bound water. In the second range ($m_{g1} \leq m_g \leq m_{g2}$), an aggregate of both the bound and transient bound components is present; the transient bound water component forms in the soil from liquid water being added to the soil in excess of the maximum bound water fraction, m_{g1}. The second range can therefore be considered as the range of transient bound water. In the third range ($m_g \geq m_{g2}$), an aggregate of all three components (bound, transient bound, and unbound) of soil water are present, and only the unbound water component forms in the soil from the liquid water in excess of the maximum total bound water fraction, m_{g2}. Thus, the third range can be considered as the range of unbound water, which can exist in the form of liquid water or wet ice in thawed and frozen soil, respectively.

The maximum bound and total bound water fractions are determined at the breakpoints in the reduced CRI as a function of moisture, which occur because the reduced RIs and NACs of the bound (($(n_b^* - 1)/\rho_b$ and κ_b/ρ_b), transient bound (($(n_t - 1)/\rho_t$ and κ_t/ρ_t), and unbound (($(n_u - 1)/\rho_u$ and κ_u/ρ_u) soil water components (see Eqs. (2), (3)) are different. The parameter m_{g1} separates the range of bound water from that of transient bound water, and m_{g2} separates the range of transient bound water from that of unbound water.

The subscripts s, d, m, b, t, u, and i, which are related to n, κ, and the density ρ, refer to the moist soil, dry soil, solid component of soil, bound water, transient bound water, unbound-liquid water, and wet ice, respectively. In addition, we assume that $\rho_b = \rho_t = \rho_u = 1$ g/cm^3 in a thawed soil and that $\rho_b = \rho_t = 1$ g/cm^3 and $\rho_i = 0.917$ g/cm^3 in a frozen soil. This assumption is applied to the densities of the soil water components only in the framework of the proposed soil dielectric model and will be confirmed when the dielectric model is validated as a whole.

According to Eq. (1), the RI, n_p, and NAC, κ_p, with the subscript p, which indicate the bound ($p = b$), transient bound ($p = t$), unbound-liquid ($p = u$), and wet ice ($p = i$) soil water components, can be expressed through the dielectric constant (DC), ε_p', and the loss factor (LF), ε_p'', as follows:

$$n_p \sqrt{2} = \sqrt{\sqrt{\left(\varepsilon_p'\right)^2 + \left(\varepsilon_p''\right)^2} + \varepsilon_p'}, \quad \kappa_p \sqrt{2} = \sqrt{\sqrt{\left(\varepsilon_p'\right)^2 + \left(\varepsilon_p''\right)^2} - \varepsilon_p'} \quad (4)$$

Similar to Mironov et al. (2013b), we express the DC and the LF of the components of soil water in Eq. (4) using the equations for the Debye multiple relaxations (Kremer et al., 2002) of nonconductive liquids, which account for only bias electric currents:

$$\varepsilon_p' = \frac{\varepsilon_{0pL} - \varepsilon_{0pM}}{1 + \left(2\pi f \tau_{pL}\right)^2} + \frac{\varepsilon_{0pM} - \varepsilon_{0pH}}{1 + \left(2\pi f \tau_{pM}\right)^2} + \frac{\varepsilon_{0pH} - \varepsilon_{\infty pH}}{1 + \left(2\pi f \tau_{pH}\right)^2} + \varepsilon_{\infty pH},$$

$$\varepsilon_p'' = \frac{\varepsilon_{0pL} - \varepsilon_{0pM}}{1 + \left(2\pi f \tau_{pL}\right)^2} 2\pi f \tau_{pL} + \frac{\varepsilon_{0pM} - \varepsilon_{0pH}}{1 + \left(2\pi f \tau_{pM}\right)^2} 2\pi f \tau_{pM} + \frac{\varepsilon_{0pH} - \varepsilon_{\infty pH}}{1 + \left(2\pi f \tau_{pH}\right)^2} 2\pi f \tau_{pH}$$

(5)

where f is the wave frequency; ε_{0pL}, ε_{0pM}, and ε_{0pH} are the low-frequency limits of the DCs that correspond to the respective relaxations; and $\varepsilon_{\infty pH}$ is a high-frequency limit for the DC of the dipole relaxation. The subscripts H, M, and L refer to the high-frequency, middle-frequency, and low-frequency relaxations, respectively. The high-frequency relaxation is a dipole relaxation, whereas the middle-frequency and low-frequency relaxations are assumed to be the interfacial (Maxwell-Wagner) relaxations that arise due to periodic recharge of soil water layers under the influence of an alternating electromagnetic field. The parameters τ_{pL}, τ_{pM}, and τ_{pH} are the times of the respective relaxations. All of these parameters should be related to the bound ($p = b$), transient bound ($p = t$), unbound-liquid ($p = u$), and wet ice ($p = i$) components of the soil water. In the case of bound water, a three-relaxation form of Eq. (5) is used. In the case of transient bound water, a two-relaxation equation is applied, which follows from Eq. (5) with $\varepsilon_{0tL} = \varepsilon_{0tM}$. In the case of unbound water, a single relaxation equation is used, which follows from Eq. (5) with $\varepsilon_{0uL} = \varepsilon_{0uM} = \varepsilon_{0uH}$.

The loss factors, ε_p'', that are determined by Eq. (5) do not include a term that accounts for the ohmic conductivity of the soil water components. Nevertheless, keeping in mind that only the bias currents account for the DC of moist soil, ε_s', we can express this value in the form

$$\varepsilon_s' = n_s^2 - \kappa_s^2 \qquad (6)$$

where Eqs. (2)–(4) are used to calculate the RI, n_s, and the NAC, κ_s.

At the same time, the LF of moist soil, ε_s'', can be represented as the sum of two terms that account for the bias currents, ε_{sb}'', and the conductivity currents, ε_{sc}'', that run through the moist sample. The terms that account for the bias and ohmic conductivity currents can be expressed as $\varepsilon_{sb}'' = 2n_s\kappa_s$ and $\varepsilon_{sc}'' = \sigma_s/2\pi f \varepsilon_r$, respectively. Here, n_s and κ_s are calculated from Eqs. (4) and (5), σ_s is the specific conductivity of the moist soil, and $\varepsilon_r = 8.854$ pF/m is the DC of the free space. We now represent the specific electrical conductivity of the moist soil, σ_s, as the sum of the specific conductivities, σ_p, of all of the components of soil water ($p = b$, t, u, i), which are weighted by their relative volumetric fractions, W_p; that is, $\sigma_{sc} = W_b\sigma_b + W_t\sigma_t + W_{u,i}\sigma_{u,i}$. By definition, the volumetric fraction, W_p ($p = b$, t, u, i), is expressed as $W_p = V_p/V$, where V is the sample volume, and V_p is the volume of water in the soil that relates to a specific component p. V and V_p can be expressed through the respective masses and densities as $V = M_d/\rho_d$ and $V_p = M_p/\rho_p$. Consequently, the volumetric fraction W_p can be expressed in the form $W_p = m_{g,p}(\rho_d/\rho_p)$, where $m_{g,p}$ is the gravimetric moisture that relates to a specific soil water component p. As a result, the expression for the LF of moist soil can be written as

$$\varepsilon_s'' = \begin{cases} 2n_s\kappa_s + \rho_d(m_g)(m_g/\rho_b)\sigma_b/2\pi f\varepsilon_r, & 0 \leq m_g \leq m_{g1}; \\ 2n_s\kappa_s + \rho_d(m_g)\left[(m_{g1}/\rho_b)\sigma_b + \left[(m_g - m_{g1})/(\rho_t)\right]\sigma_t\right]/2\pi f\varepsilon_r, \\ \quad m_{g1} \leq m_g \leq m_{g2}; \\ 2n_s\kappa_s + \rho_d(m_g)\left[(m_{g1}/\rho_b)\sigma_b + \left[(m_{g2} - m_{g1})/(\rho_t)\right]\sigma_t + \left[(m_g - m_{g2})/\rho_{u,i}\right]\sigma_{u,i}\right] \\ \quad /2\pi f\varepsilon_r, \quad m_g \geq m_{g2} \end{cases}$$

(7)

Eqs. (1)–(7) show that the DC and LF spectra at a given soil temperature as a function of the input variables (dry soil density, ρ_d, gravimetric soil moisture, m_g, and wave frequency, f) can be calculated using the following set of parameters: $(n_m - 1)/\rho_m$, κ_m/ρ_m, m_{g1}, m_{g2}, ε_{0pQ}, $\varepsilon_{\infty pH}$, τ_{pQ}, σ_p, which are related to (i) the bound ($p = b$), transient bound ($p = t$), unbound-liquid ($p = u$), and wet ice ($p = i$) components of the soil water and (ii) the high-frequency ($Q = H$), middle-frequency ($Q = M$), and low-frequency ($Q = L$) relaxations of the soil water components. These parameters can be regarded as characteristics of the MRSDM. The reduced RI, $(n_m - 1)/\rho_m$, and the NAC, κ_m/ρ_m, are mineralogical characteristics of the soil solids. The maximum gravimetric fractions of the bound water, m_{g1}, and of the total bound water, m_{g2}, are hydrological characteristics of the bulk soil. The low limit and high limit DCs, ε_{0pQ} and $\varepsilon_{\infty pH}$, in conjunction with the values of the relaxation time, τ_{pQ}, are relaxation characteristics of the soil water molecules. Finally, the conductivities, σ_p, are electrical characteristics of the soil water solutions. These parameters should vary with temperature. In the next section, we outline the methodology for retrieving all of the parameters of the MRSDM.

4 RETRIEVING THE PARAMETERS OF THE MULTIRELAXATION SPECTROSCOPIC DIELECTRIC MODEL

Laboratory measurements of the CDC spectra were used to retrieve the parameters of the multirelaxation dielectric model of moist soil. The measured reduced RI and NAC are shown in Fig. 1 with the results of fitting the theoretical models (Eqs. (2), (3)) to the measured data. The measured RI and NAC can be calculated using Eq. (1) and the data shown in Fig. 1 and Table 1.

The results in Fig. 1 show that models (Eqs. (2), (3)) are satisfactory. The RI and NAC measurements are piecewise linear in certain moisture ranges, which indicates the contributions of the particular components of water. The maximum fractions m_{g1} and m_{g2} are determined as the transition points from one linear segment of the fit to another (Fig. 1). These values were retrieved by fitting the theoretical models (Eqs. (2), (3)) to the measured reduced RI and NAC

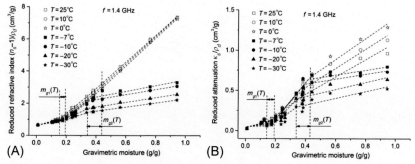

FIG. 1 Behavior of the reduced characteristics for (A) the RI, $(n_s - 1)/\rho_d$, and (B) NAC, κ_s/ρ_d, versus gravimetric moisture at a fixed frequency of 1.4 GHz at temperatures from 25°C to -30°C. The dashed lines indicate fits of Eqs. (2) and (3) to the measured data.

values as a function of soil moisture at various temperatures. Thus, the obtained maximum fractions m_{g1} and m_{g2} depend on the temperature, particularly in the case of frozen soil samples. These dependences were fitted with polynomials and exponential functions and yielded the following equations:

$$
\begin{aligned}
m_{g1} &= 0.21 - 0.002T, \quad 0°C \le T \le 25°C; \\
m_{g1} &= 0.149 + 0.072 \exp(T/13.4), \quad -30°C \le T \le -1°C; \\
m_{g2} &= 0.431 + 0.003T, \quad 0°C \le T \le 25°C; \\
m_{g2} &= 0.322 + 0.14 \exp(T/12.3), \quad -30°C \le T \le -1°C.
\end{aligned}
\tag{8}
$$

By fitting the theoretical models (Eqs. (2), (3)) to the reduced RI and NAC measurements as a function of soil moisture at various temperatures (Fig. 1), the values of $(n_m - 1)/\rho_m$ and κ_m/ρ_m were retrieved. The latter are equal to the values of $(n_s - 1)/\rho_d$ and κ_s/ρ_d in the fitted models (Eqs. (2), (3)) at $m_g = 0$, respectively. Thus, the obtained temperature dependences for the values of the reduced RI, $(n_m - 1)/\rho_m$, and the NAC, κ_m/ρ_m, that characterize the soil solids were fitted with polynomials and yielded the following equations:

$$
\begin{aligned}
(n_m - 1)/\rho_m &= 0.581 - 0.001T, \quad -30°C \le T \le 25°C; \\
\kappa_m/\rho_m &= 0.011, \quad -30°C \le T \le 25°C.
\end{aligned}
\tag{9}
$$

Now that the dependence of the CRI on moisture has been established and the division between the frozen and thawed states has been clarified, we turn our attention to the methodology for retrieving the spectral parameters in Eqs. (5) and (6). The spectra for the DC and LF of moist soil samples measured at different moistures and temperatures will be used to determine these parameters. Fig. 2 shows several patterns of these spectra for the thawed and frozen soil samples. As can be deduced from the equations in Eq. (8), the samples shown in Fig. 2 contain only bound water ($m_g = 0.086$ g/g $< m_{g1}(20°C) = 0.165$ g/g for plot 1), bound water and transient bound water (m_{g1} (-20°C) $= 0.164$ g/ g $< m_g = 0.299$ g/g $< m_{g2}$ (-20°C) $= 0.343$ g/g for plot 2 and m_{g1} (20°C) $=$

TABLE 1 Dependence of the Dry Soil Density, ρ_d, and the Volumetric Moisture, m_v, on the Soil Gravimetric Moisture m_g

m_g (g/g)	0.01	0.086	0.126	0.144	0.176	0.202	0.237	0.299	0.339	0.377	0.385	0.441	0.563	0.602	0.763	0.942
ρ_d (g/cm^3)	0.666	0.632	0.625	0.591	0.604	0.568	0.564	0.538	0.581	0.564	0.574	0.601	0.596	0.609	0.603	0.608
m_v (cm^3/cm^3)	0.007	0.054	0.079	0.085	0.106	0.115	0.134	0.161	0.197	0.213	0.221	0.265	0.335	0.366	0.46	0.573

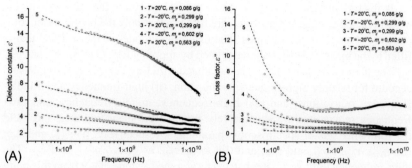

FIG. 2 Spectra of (A) dielectric constant, ε', and (B) loss factor, ε''. The measured data are shown by symbols. The fitted spectra are shown with lines and were calculated with (i) Eqs. (2)–(6), (8), and (9); and (ii) Eqs. (2)–(5) and (7)–(9) in the cases of dielectric constant and loss factor, respectively. The data represent both thawed ($T=20°C$) and frozen ($T=-20°C$) soils that contain only bound water ($m_g=0.086$), bound water and transient bound water ($m_g=0.299$), and bound water, transient bound water, and unbound-liquid water at $T=20°C$ or wet ice at $T=-20°C$ ($m_g=0.563$ and $m_g=0.602$).

0.165 g/g $< m_g = 0.299$ g/g $< m_{g2}$ ($20°C$) = 0.351 g/g for plot 3), all three soil water components ($m_g=0.602$ g/g $> m_{g2}$ ($-20°C$) = 0.343 g/g for plot 4 and $m_g=0.563$ g/g $> m_{g2}$ ($20°C$) = 0.351 g/g for plot 5).

The spectrum examples in Fig. 2 were especially selected to demonstrate all the cases observed in terms of (i) presence of different types of soil water and (ii) different thawed/frozen states of the measured soil samples. In fact, all the spectra corresponding to the moistures and temperatures shown in Table 1 and in Fig. 3, respectively, have been fitted to derive the spectroscopic parameters.

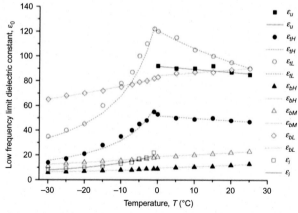

FIG. 3 Low frequency limit dielectric constants as a function of temperature. The measured data and their fits are shown with symbols and lines, respectively, for the bound water, ε_{0bQ}, transient bound water, ε_{0tQ}, unbound-liquid water, ε_{0uQ}, and wet ice, ε_{0iQ}, at high-frequency ($Q=H$), middle-frequency ($Q=M$), and low-frequency ($Q=L$) relaxations.

In the first phase of fitting, the measured DC spectra of the samples with the moistures in the range $0 < m_g < m_{g1}$ are fitted so that only the bound water component is present in the samples. The theoretical model of the soil DC that is fitted to the measured spectra is calculated by applying Eqs. (2)–(6), (8), and (9).

The theoretical spectra that are calculated with Eqs. (2)–(6), (8), and (9) and account for (i) only high-frequency relaxation; (ii) high-frequency and middle-frequency relaxations; and (iii) high-frequency, middle-frequency, and low-frequency relaxations are successively fitted to the measured data and yield the following sets of spectroscopic parameters, respectively: (i) ε_{0bH} and τ_{bH}; (ii) ε_{0bH}, τ_{bH}, ε_{0bM}, and τ_{bM}; and (iii) ε_{0bH}, τ_{bH}, ε_{0bM}, τ_{bM}, ε_{0bL}, and τ_{bL}. The high-frequency limit $\varepsilon_{\infty pH}$, $p = b, t, u, i$, is assigned a value of 4.9 from this point on. Figs. 3 and 4 show the spectroscopic parameters that are obtained for the bound water component as a function of temperature. The examples of theoretical spectra for the soil DC calculated at the derived values of the parameters ε_{0bH}, τ_{bH}, ε_{0bM}, τ_{bM}, ε_{0bL}, τ_{bL} are also shown in Fig. 2A and demonstrate good correlations with the measured data.

After all of the spectroscopic parameters for the bound soil water (ε_{0bH}, τ_{bH}, ε_{0bM}, τ_{bM}, ε_{0bL}, and τ_{bL}) were derived using only the DC spectra that were measured for moist soil samples, we can perform further fitting to obtain the specific conductivity of the bound soil water from the LF spectra measured for gravimetric soil moistures in the range $0 < m_g < m_{g1}$. In the course of fitting, to calculate the theoretical values of the soil LF, we apply Eqs. (2)–(5) and (7)–(9) using the values of the spectroscopic parameters that were previously derived and optimizing only the value of conductivity, σ_b. The examples of fitted theoretical model for the LF calculated using the derived values of ε_{0bH}, τ_{bH}, ε_{0bM}, τ_{bM}, ε_{0bL}, τ_{bL}, and σ_b are also given in Fig. 2B, exhibiting a good correlation with the measured data. As a result of this fitting, the value of the specific conductivity for the bound soil water, σ_b, was obtained as a function of temperature, and in the whole range of temperatures it was found to be approximately equal to zero.

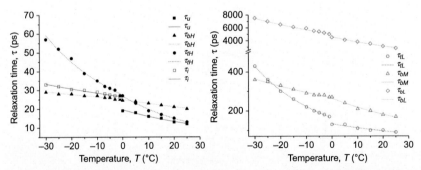

FIG. 4 Relaxation time as a function of temperature. The measured data and their fits are shown with symbols and lines, respectively, for the bound water, τ_{bQ}, transient bound water, τ_{tQ}, unbound-liquid water, and wet ice, τ_{iQ}, at high-frequency ($Q=H$), middle-frequency ($Q=M$), and low-frequency ($Q=L$) relaxations.

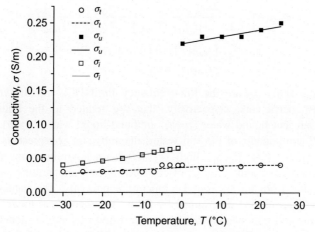

FIG. 5 Conductivity as a function of temperature. The measured data and their fits are shown with symbols and lines, respectively, for the transient bound water, σ_t, unbound-liquid water, σ_u, and wet ice, σ_i.

The spectra measured for the samples that contain both bound water and transient bound water components were used in the second phase of fitting, and the same fitting approach was applied, with the values of the parameters ε_{0bH}, τ_{bH}, ε_{0bM}, τ_{bM}, ε_{0bL}, τ_{bL}, and σ_b pertinent to the bound water being equal to those obtained in the first phase of fitting. The results of the second phase of fitting for the derived values ε_{0tH}, τ_{tH}, ε_{0tL}, τ_{tL}, and σ_t pertinent to the transient bound water are shown as a function of temperature in Figs. 3–5.

In the third phase of fitting, the spectra measured for the samples that contain all three components of soil water were used, with the values of the parameters ε_{0bH}, τ_{bH}, ε_{0bM}, τ_{bM}, ε_{0bL}, τ_{bL}, σ_b, ε_{0tH}, τ_{tH}, ε_{0tL}, τ_{tL}, and σ_t pertinent to the bound and transient bound water, as these were obtained in the first and second phases of fitting, being applied. The results of the third phase of fitting for the derived values of $\varepsilon_{0u,iH}$, $\tau_{u,iH}$, and $\sigma_{u,i}$ pertinent to the unbound water and moist ice are shown in Figs. 3–5. Now that the temperature dependences of all of the spectroscopic parameters and specific conductivities have been obtained, we will consider the TD MRSDM for moist soil following the methodology of Mironov et al. (2010), which outlined only a single relaxation case.

5 THE TEMPERATURE-DEPENDENT MULTIRELAXATION SPECTROSCOPIC DIELECTRIC MODEL

We suggest that the temperature dependences for the low-frequency limit DCs observed in Fig. 3 follow the equation that was obtained in Mironov et al. (2010) with the use of the Clausius-Mossotti law (Dorf, 1997):

$$\varepsilon_{0pQ}(T) = \frac{1 + 2\exp\left[F_{pQ}(T_{sepQ}) - \beta_{vpQ}(T - T_{sepQ})\right]}{1 - \exp\left[F_{pQ}(T_{sepQ}) - \beta_{vpQ}(T - T_{sepQ})\right]},$$

$$F_{pQ}(T) = \ln\left[\frac{\varepsilon_{0pQ}(T) - 1}{\varepsilon_{0qQ}(T) + 2}\right] \tag{10}$$

where ε_{0pQ} and β_{vpQ} are the low-frequency limit DCs and the volumetric expansion coefficients, respectively, that are related to the bound water ($p = b$), transient bound water ($p = t$), unbound-liquid water ($p = u$), and wet ice ($p = i$) components of the soil water; the subscript Q represents the low- ($Q = L$), middle- ($Q = M$), and high- ($Q = H$) frequency relaxations of the soil water components; and T_{sepQ} represents the starting temperature, which can be any value from the measured temperature intervals. The values of β_{vpQ} and $\varepsilon_{0pQ}(T_{sepQ})$ can be determined by fitting the theoretical model (Eq. 10) to the measured data shown in Fig. 3. The fitted theoretical models for the low-frequency limit DCs that were calculated with the values of β_{vpQ} and ε_{0pQ} (T_{sepQ}) obtained from the fitting are also shown in Fig. 3 and agree well with the measured data. This fitting is performed separately for the thawed and frozen soil samples. The values of β_{vpQ}, T_{sepQ}, and $\varepsilon_{0pQ}(T_{sepQ})$ derived from the fitting are given in Table 2A and B.

The measured relaxation time, which is shown as a function of temperature in Fig. 4, can be described by the Eyring equation (Dorf, 1997):

$$\ln\left(\frac{kT_K}{h}\tau_{pQ}\right) = \frac{\Delta H_{pQ}}{R}\frac{1}{T_K} - \frac{\Delta S_{pQ}}{R} \tag{11}$$

where h is the Planck's constant (6.624×10^{-34} Js), k is the Boltzmann constant (1.38×10^{-23} J/K), ΔH_{pQ} is the activation energy of the relaxation process, R is the universal gas constant (8.314×10^3 J/K/kmol), ΔS_{pQ} is the entropy of activation, and T_K is the temperature in Kelvin. The ratios $\Delta H_{pQ}/R$ and $\Delta S_{pQ}/R$, which are proportional to the activation energy and the entropy of activation, respectively, can be determined by linear fitting the measured value of $\ln$ ($kT_K\tau_{pQ}/h$), as shown in Fig. 4. The calculated values of $\Delta H_{pQ}/R$ and $\Delta S_{pQ}/R$ are given in Table 2A and B and can be used to calculate the relaxation time using Eq. (11). The theoretical models for the relaxation times that were calculated with the values of $\Delta H_{pQ}/R$ and $\Delta S_{pQ}/R$ obtained from the fitting are also shown in Fig. 4 and demonstrate good agreement with the measured data.

Finally, we suggest that the conductivity, σ_p, has a linear dependence with temperature as is characteristic of ionic solutions:

$$\sigma_p(T) = \sigma_p(T_{s\sigma p}) + \beta_{\sigma p}(T - T_{s\sigma p}) \tag{12}$$

where $\beta_{\sigma p}$ is the derivative of conductivity with respect to temperature, which is also called the conductivity temperature coefficient, and $\sigma_p(T_{s\sigma p})$ is the value of conductivity at an arbitrary starting temperature, $T_{s\sigma p}$, that is taken from the measured range.

TABLE 2 TD MRSDM Parameters for All Forms of Soil Water in the Temperature Range −30°C ≤ T ≤ +25°C

(A)

Soil Water Component

Relaxation						Bound Soil Water, p=b	
		High Frequency		Middle Frequency		Low Frequency	
Temperature range		$T \leq -1°C$	$T \geq 0°C$	$T \leq -1°C$	$T \geq 0°C$	$T \leq -1°C$	$T \geq 0°C$
Parameter	Units						
$\varepsilon_{0p}(T_{se0p})$	–	7	12	14	22	70	89
β_{v0p}	1/K	3.46×10^{-3}	2.74×10^{-3}	2.36×10^{-3}	1.09×10^{-3}	3.13×10^{-4}	9.6×10^{-5}
T_{se0p}	°C	−20	20	−20	20	−20	20
$\Delta H_p/R$	K	67	392	432	1218	632	1243
$\Delta S_p/R$	–	−4.72	−3.52	−5.74	−2.85	−7.94	−5.61
Temperature range		$T \leq -1°C$			$T \geq 0°C$		
$\sigma_p(T_{srp})$	S/m	0			0		
$\beta_{\sigma p}$	(S/m)/K	0			0		
T_{srp}	°C	−20			20		

Continued

TABLE 2 TD MRSDM Parameters for All Forms of Soil Water in the Temperature Range $-30°C \leq T \leq +25°C$—cont'd

(B)

Soil Water Component		Transient Bound Water (p=t)				Wet Ice Water (p=i)	Unbound-Liquid Water (p=u)
Relaxation		High Frequency		Low Frequency		High Frequency	High Frequency
Temperature range		$T \leq -1°C$	$T \geq 0°C$	$T \leq -1°C$	$T \geq 0°C$	$T \leq -1°C$	$T \geq 0°C$
Parameter	Units						
$\varepsilon_{0p}(T_{se0p})$	–	21	48	45	95	10	87
$\beta_{\upsilon 0p}$	1/K	4.63×10^{-3}	-2.23×10^{-4}	2.2×10^{-3}	-3.27×10^{-5}	7.32×10^{-3}	8.82×10^{-5}
T_{se0p}	°C	-20	20	-20	20	-20	20
$\Delta H_p/R$	K	1444	2065	1859	981	404	3071
$\Delta S_p/R$	–	0.24	2.54	-0.03	-3.03	-3.27	6.72
Temperature range		$T \leq -1°C$		$T \geq 0°C$		$T \leq -1°C$	$T \geq 0°C$
$\sigma_p(T_{se\sigma p})$	S/m	0.03		0.04		0.046	0.24
$\beta_{\sigma p}$	(S/m)/K	3.53×10^{-4}		1.14×10^{-4}		8.98×10^{-4}	1.03×10^{-3}
$T_{se\sigma p}$	°C	-20		20		-20	20

The values of σ_p and $\beta_{\sigma p}$ can be determined by fitting the theoretical model (Eq. 12) to the measured data shown in Fig. 5. This fitting was performed separately for the thawed and frozen soil samples. The theoretical models for the conductivity that were calculated with the values of σ_p and $\beta_{\sigma p}$ obtained from the fitting are also shown in Fig. 5 and agree well with the measured data. The values of σ_p and $\beta_{\sigma p}$ that are derived from the fitting are given in Table 2A and B.

As a result of the analyses conducted in this section, the TD MRSDM can be defined by the following steps, which form the algorithm procedure.

1. The temperature, T, must be assigned in cases of thawed or frozen soil.
2. The values of the Debye parameters, including the low frequency limit, relaxation time, and conductivity, for all types of soil water are calculated with Eqs. (10)–(12) and the data in Table 2A and B.
3. Once the values of $\varepsilon_{0pQ}(T)$, $\tau_{pQ}(T)$, and $\sigma_p(T)$ are known, the values of the DC, $\varepsilon_p'(f,T)$, and the LF $\varepsilon_p''(f,T)$, accounting for only bias electric currents, for all components of the soil water can be calculated as a function of frequency at a given temperature using the equations in Eq. (5).
4. The values of $\varepsilon_p'(f,T)$ and $\varepsilon_p''(f,T)$ are translated to the RI, n_p, and the NAC, κ_p, accounting for bias currents, for all of the components of the soil water using the equations in Eq. (4).
5. The gravimetric soil moisture, m_g, and the dry soil bulk density, ρ_d, must be assigned, and Eqs. (2), (3), (8), and (9) are applied to calculate the soil RI, $n_s(\rho_d, m_g, f, T)$, and the NAC, $\kappa_s(\rho_d, m_g, f, T)$ accounting for bias currents.
6. Finally, the values of $n_s(\rho_d, m_g, f, T)$ and $\kappa_s(\rho_d, m_g, f, T)$ are translated to the soil DC, $\varepsilon_s'(\rho_d, m_g, f, T)$ with Eq. (6). While the LF, $\varepsilon_s''(\rho_d, m_g, f, T)$, accounting for both bias and conductivity currents, is calculated with Eqs. (7), (8), (12), using the values of necessary parameters in Eq. (12) from Table 2A and B.

In the following section, the TD MRSDM that was developed for an organic soil will be evaluated in terms of the prediction error by correlating the predicted moist soil dielectric data with measured data in the multidimensional domain of temperature, frequency, and moisture.

6   EVALUATION OF THE TD MRSDM

Fig. 6 shows the values predicted with the TD MRSDM and the values measured at a frequency of 1.4 GHz as a function of temperature for samples with varying moistures and dry soil densities. The predicted and measured values agree well with each other. To estimate the deviations of the predicted CDC values from the measured values across the entire range of measured dry densities, moistures, temperatures, and frequencies, we calculated the coefficient of determination for the DC, $R_{\varepsilon'}^2$, and the LF, $R_{\varepsilon''}^2$. Their values were found to be $R_{\varepsilon'}^2 = 0.997$ and $R_{\varepsilon''}^2 = 0.993$. The root mean square errors (RMSEs) of the

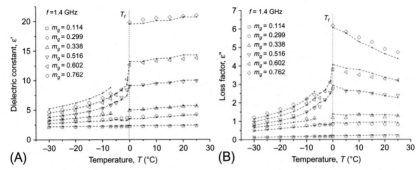

FIG. 6 Behavior of the dielectric constant (A) and loss factor (B) versus temperature at different values of gravimetric moisture and at a frequency of 1.4 GHz. The measured data and the data predicted with the TD MRSDM are shown with symbols and lines, respectively.

predicted values relative to the measured values were found to be of $\text{RMSE}_{\varepsilon'} = 0.266$ and $\text{RMSE}_{\varepsilon''} = 0.214$ for the DC and LF, respectively.

7 CONCLUSIONS

A TD MRSDM was developed for an organic-rich soil that was collected from a shrub tundra site located on the North Slope, Alaska. The model is presented as an ensemble of analytical expressions and gives the CDCs of the thawed and frozen soil using the dry soil density, gravimetric moisture, temperature, and wave frequency as input variables. The validation of this model demonstrates good agreement with the measured data for frequencies from 0.05 to 16 GHz, gravimetric moistures from 0.01 to 0.94 g/g, and temperatures from $-30°C$ to $+25°C$ with dry soil densities of 0.56 to 0.67 g/cm^3. With such wide variations of all of the input variables, the coefficients of determination for the DC, $R_{\varepsilon'}^2$, and the LF, $R_{\varepsilon''}^2$ were found to be $R_{\varepsilon'}^2 = 0.997$ and $R_{\varepsilon''}^2 = 0.993$. The RMSEs of the predicted values of the DC and the LF relative to the measured values were estimated to be $\text{RMSE}_{\varepsilon'} = 0.266$ and $\text{RMSE}_{\varepsilon''} = 0.214$, respectively.

A major advantage of the proposed dielectric model compared to the previous single-relaxation model of Mironov et al. (2010) is that it covers a wider frequency band of 0.05–16 GHz instead of 1–16 GHz, which was made possible by taking into account multiple relaxations of the different soil water components. The extension of the frequency band over the megahertz range addresses a distinct tendency to use the megahertz frequency range in the radar remote sensing that was confirmed by the European Space Agency, planning to launch the space mission with the P-band SAR on board (Arcioni et al., 2014). In addition, in the case of frozen soil, the temperature range from $-1°C$ to $-7°C$, missing in the model of Mironov et al. (2010), was included into consideration.

The prediction error estimates $\text{RMSE}_{\varepsilon'} = 0.266$ and $\text{RMSE}_{\varepsilon''} = 0.214$, for the developed dielectric model are less than those in the model of Mironov and

Fomin (2009), which is successfully used in the operational algorithm of SMOS to retrieve soil moisture (Mialon et al., 2015). From this point of view, the proposed model is acceptable for practical use to develop remote sensing algorithms provided the topsoil of the area being analyzed is predominantly composed of organic soil. Such a rigorous restriction may be realized in the zones of boreal forests and arctic tundra.

In general, the developed dielectric model forms the basis for data processing algorithms for modern remote sensing missions, such as AQUA, GCOM-W, SMOS, SMAP, RADARSAT, and ALOS PALSAR, as well as for perspective P-band sensors, for which the depth of sensing is expected to increase. In addition, it facilitates the application of GPR and TDR instruments, which operate in the megahertz band, to interpret the results of in situ measurements of the active permafrost layer, including studies of freeze/thaw processes.

REFERENCES

Arcioni, M., Bensi, P., Fehringer, M., Fois, F., Heliere, F., Lin, C.-C., et al., 2014. The biomass mission, status of the satellite system. In: 2014 IEEE International Geoscience and Remote Sensing Symposium (IGARSS), pp. 1413–1416. http://dx.doi.org/10.1109/IGARSS.2014.6946700.

Bircher, S., Balling, J.E., Skou, N., Kerr, Y.H., 2012. Validation of SMOS brightness temperatures during the HOBE airborne campaign, western Denmark. IEEE Trans. Geosci. Remote Sens. 50, 1468–1482.

De Roo, R.D., England, A.W., Gu, H., Pham, H., Elsaadi, H., 2006. Ground based radiobrightness observations of the active layer growth on the north slope near Toolik Lake, Alaska. In: Proc. IEEE IGARSS, Denver, CO, 31 July–4 August, pp. 2708–2711.

Dorf, R.C. (Ed.), 1997. Electrical Engineering Handbook. second ed. CRC Press, Boca Raton, FL.

Jones, L.A., Kimball, J.S., McDonald, K.C., Chan, S.T.K., Njoku, E.G., Oechel, W.C., 2007. Satellite microwave remote sensing of Boreal and Arctic soil temperatures from AMSR-E. IEEE Trans. Geosci. Remote Sens. 45, 2004–2018.

Kremer, F., Schonhals, A., Luck, W., 2002. Broadband Dielectric Spectroscopy. Springer Science & Business Media, New York.

Mialon, A., Richaume, P., Leroux, D., Bircher, S., Al Bitar, A., Pellarin, T., et al., 2015. Comparison of Dobson and Mironov dielectric models in the SMOS soil moisture retrieval algorithm. IEEE Trans. Geosci. Remote Sens. 53, 3084–3094.

Mironov, V.L., Fomin, S.V., 2009. Temperature and mineralogy dependable model for microwave dielectric spectra of moist soils. PIERS Online 5, 411–415.

Mironov, V.L., Savin, I.V., 2015. A temperature-dependent multi-relaxation spectroscopic dielectric model for thawed and frozen organic soil at 0.05–15 GHz. Phys. Chem. Earth A/B/C, 83–84, 57–64. http://dx.doi.org/10.1016/j.pce.2015.02.011.

Mironov, V.L., De Roo, R.D., Savin, I.V., 2010. Temperature-dependable microwave dielectric model for an Arctic soil. IEEE Trans. Geosci. Remote Sens. 48, 2544–2556.

Mironov, V.L., Muzalevskiy, K.V., Savin, I.V., 2013a. Retrieving temperature gradient in frozen active layer of Arctic Tundra soils from radiothermal observations in L-band—theoretical modeling. IEEE J. Sel. Topics Appl. Earth Observ. Remote Sens. 6, 1781–1785.

Mironov, V.L., Bobrov, P.P., Fomin, S.V., Karavaiskii, A.Yu., 2013b. Generalized refractive mixing dielectric model of moist soils considering ionic relaxation of soil water. Russ. Phys. J. 56, 319–324.

Mironov, V.L., Molostov, I.P., Lukin, Yu.I., Karavaisky, A.Yu., 2013c. Method of retrieving permittivity from S12 element of the waveguide scattering matrix. In: International Siberian Conference on Control and Communications (SIBCON), Krasnoyarsk. 978-1-4799-1062-5/13 ©2013 IEEE. http://dx.doi.org/10.1109/SIBCON.2013.6693609.

Rautiainen, K., Lemmetyinen, J., Pulliainen, J., Vehvilainen, J., Drusch, M., Kontu, A., et al., 2012. L-band radiometer observations of soil processes in Boreal and subarctic environments. IEEE Trans. Geosci. Remote Sens. 50, 1483–1497.

Watanabe, M., Kadosaki, G., Kim, Y., Ishikawa, M., Kushida, K., Sawada, Y., et al., 2012. Analysis of the sources of variation in L-band backscatter from terrains with permafrost. IEEE Trans. Geosci. Remote Sens. 50, 44–54.

Chapter 10

Active and Passive Microwave Remote Sensing Synergy for Soil Moisture Estimation

R. Akbar*, N. Das†, D. Entekhabi‡ and M. Moghaddam*

Ming Hsieh Department of Electrical Engineering, University of Southern California, Los Angeles, CA, †NASA/Jet Propulsion Laboratory (JPL), Pasadena, CA, ‡Department of Civil and Environmental Engineering, Massachusetts Institute of Technology, Cambridge, MA

1 INTRODUCTION

Simultaneous usage of active radar, especially synthetic aperture radar (SAR), and passive radiometer (RAD) microwave remote sensing observations to estimate surface soil moisture has gained significant interest in recent years and advancements have been made to develop so-called combined active-passive (CAP) methodologies to retrieve soil moisture under various moisture and vegetation regimes. Historically, and with a very long heritage, almost all existing soil moisture estimation schemes use a single data type; that is, as schematically shown in Fig. 1A, soil moisture estimation is performed using radar-only or radiometer-only microwave remote sensing measurements and not in a combined fashion.[1]

Significant developments of CAP have only a very recent history, having gained momentum with the formulation (2008) and launch (2015) of the NASA Soil Moisture Active and Passive (SMAP) mission (Entekhabi et al., 2010) and prior to that the Hydrosphere State (Entekhabi et al., 2004) mission. Conceptually, the idea behind CAP can be seen in Fig. 1B where within a single estimation and retrieval framework, to be discussed, both radar and radiometer measurements are used together.

The key driving force behind combined usage of SAR and RAD is the fact that over the past few decades science application requirements, especially in hydroclimatology, hydrometeorology, and carbon sciences, have converged toward three criteria for acceptable soil moisture estimates: (a) higher spatial

---

1. The reader is referred to other chapters within this book for more detailed discussion on active-only or passive-only methods.

Satellite Soil Moisture Retrieval. http://dx.doi.org/10.1016/B978-0-12-803388-3.00010-3

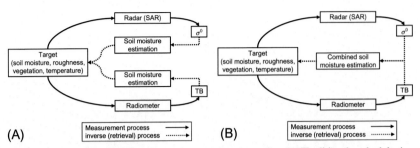

FIG. 1 Active and passive microwave remote sensing approaches. (A) Traditional methodologies. (B) Combined active-passive concept.

(<40 km) and temporal resolution (~3 days), (b) global coverage, and (c) volumetric soil moisture prediction accuracy of $0.04 \, \text{cm}^3/\text{cm}^3$ or less (Soil Moisture Active/Passive Mission, 2007). Soil moisture predictions that meet these requirements will assist in resolving water, energy, and carbon fluxes as well as improving short- and long-term weather and climate forecast skills.

However, with current airborne and spaceborne technologies and measurement approaches, addressing all these requirements at once is not immediately possible. To briefly highlight this fact, in Table 1 a qualitative comparison between different existing soil moisture observation schemes across multiple platforms is shown. Tower or truck-mounted and airborne observatories yield higher resolution predictions, but their geographical coverage is very limited. For example, airborne systems such as the passive active L- and S-band Sensor (PALS), uninhabited aerial vehicle synthetic aperture radar (UAVSAR), or airborne microwave observatory of subcanopy and subsurface (AirMOSS) typically cover an approximate area of $20 \times 100 \, \text{km}$ or less. Although on a regional scale the coverage is significant, frequent global mapping becomes

TABLE 1 Soil Moisture Remote Sensing Measurement Schemes

Platform	Instrument	Spatial Resolution	Temporal Resolution	Coverage	Accuracy
Probe	In situ sensor	Point	Frequent	Localized	High
Tower	Radar	~m	Frequent	Localized	High
	Radiometer				
Airborne	Radar (SAR)	~m–km	Daily–weekly	Regional	Medium
	Radiometer				
Spaceborne	Radar (SAR)	<10 km	Daily	Global	Medium
	Radiometer	>10 km			High

impractical. On the other hand, the near-daily global soil moisture coverage provided by spacecraft instruments are highly desirable. SMAP and the European Space Agency (ESA) Soil Moisture and Ocean Salinity (SMOS) mission (Kerr et al., 2001, 2010), for example, are capable of mapping the entire globe approximately every 2–3 days. From an instruments perspective, however, as well as soil moisture estimation abilities, neither SAR nor RAD alone are sufficient enough to meet all the science application requirements and standards at once.

Nonetheless, the drawbacks in estimating soil moisture via active-only or passive-only remote sensing approaches can be circumvented when the complementary set of strengths between the two techniques are merged and viewed collectivity. This feature can be considered a primary motivation behind the SMAP mission, where onboard a single spacecraft an L-band SAR and radiometer, sharing a single 6 m reflector mesh antenna, coexist. The mission's primary scientific goal is estimation of global surface soil moisture (top 5 cm) with a volumetric accuracy of ± 0.04 cm^3/cm^3 at $\sim$10 km spatial resolution or less. This can only be achieved through combined active and passive remote sensing.

The complementarity of SAR and RAD strengths can be evaluated as follows.

1.1 Measurement Spatial Resolution

On almost all moving platforms, especially spacecraft, radars operate in a synthetic aperture mode resulting in subkilometer scale backscatter cross-section measurements[2] at the cost of smaller swath widths. On the other hand, real aperture radiometers yield kilometer scale, or larger, brightness temperature (TB) data products covering a very large swath. To compare and contrast this spatial resolution disparity a nonexhaustive list of select SAR and RAD spacecraft and satellite missions as well as their approximate spatial resolutions is given in Table 2. A schematic representation of this resolution difference is also shown in Fig. 2B such that within a typical radiometer pixel, multiple SAR measurements can exist. For example, in the case of SMAP, Earth-gridded L-band radiometer TB measurements are reported at a 36 km resolution whereas SAR acquisitions are obtained at 250 m scale, reprocessed, and then reported at 1–3 km resolution.

Clearly, high-resolution soil moisture estimates obtained from radar-only measurements are preferred over the coarser resolution radiometer-only estimates and are potentially more suitable for high-level science applications.

---

2. It is important to note that soil moisture products derived from spaceborne SAR are typically reported at a coarser scale than the native radar resolution. This is due to SAR processing and downscaling to overcome speckle noise, etc.

TABLE 2 SAR and Radiometer Satellite Missions

Satellite or Instrument	Years of Operation	Instrument	Frequency	Nominal Resolution
SMOS	2009–present	Radiometer	L-band	35 km
Aquarius	2011–2015	Radiometer	L-band	40 km
AMSR-E	2002–2011	Radiometer	Multiband	5–50 km
SMAP	2015–present	Radiometer	L-band	36 km
SMAP	2015–present	SAR	L-band	300 m
Radarsat-1/2	1995–2013	SAR	C-band	3–8 m
	2007–present			
Sentinal-1	2014–present	SAR	C-band	5 m
PALSAR	2006–present	SAR	L-band	7–100 m
ERS-1/2	1991–2011	SAR	C-band	30 m

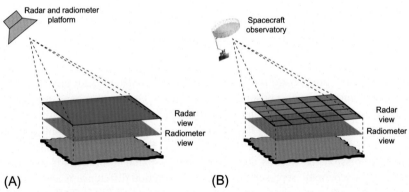

(A) (B)

FIG. 2 (A) Single-resolution radar-radiometer measurement schematic. (B) Multiresolution radar-radiometer measurement schematic from a spacecraft. It is assumed that Earth-gridded radar measurements are nested within a single radiometer measurement. Schematic (B) also represents a SMAP-like scenario.

However, in terms of estimation accuracy, the performance of radar-based and radiometer-based retrievals must be examined as well.

1.2 Soil Moisture Sensitivity and Estimation Accuracy

To estimate geophysical parameters of interest, using microwave remote sensing measurements that have the most sensitivity and least ambiguity across a wide range of soil moisture and vegetations conditions, will yield the most accurate estimate of that parameters. Extensive prior studies (Choudhury et al., 1995) have examined and evaluated σ^0 and TB sensitivity to changes in land surface parameters (LSP), land cover types, frequency bands (L-, C-, and X-band), as well as for different incidence angles. In short, for L-band soil moisture remote sensing, (a) for low to moderate amounts of vegetation water content (VWC) (<5 kg/m^2), L-band radiometry shows significant response to changes in soil moisture, at times up to $\sim$90 K from dry to wet conditions (Njoku and Entekhabi, 1996); (b) radar backscatter is highly sensitive to surface roughness as seen in Fig. 3; and (c) both TB and σ^0 observations are affected with increasing VWC, however vegetation scattering is more significant. The latter has also been a motivation to use SAR for land cover, crop, and wetlands classification applications. A thorough analysis and assessment on soil moisture products derived from advanced microwave scanning radiometer for earth observing (AMSR-E) (passive) and advanced scatterometer (ASCAT) (active) by Brocca et al. (2011) indicated a noticeable degradation in soil moisture estimation with increasing vegetation optical depth (i.e., more vegetation), which affirm the above emission and scattering sensitivity responses.

Although the physical processes of emission and scattering are interrelated, their individual responses and sensitivities to variations in LSP, especially soil moisture, are different. Scatter plots (see Fig. 4) of actual measured TB and σ^0 - (Njoku, 2003) across a range of soil moisture conditions clearly show this difference. As the surface becomes wetter, backscatter increases and typically tends to saturate. For the case of Fig. 4A (top), σ^0_{vv} varies from -25 to -10 dB due to a 0.3 cm^3/cm^3 increase in soil moisture. On the other hand, TB across the geophysical soil moisture ranges of interest, generally exhibits

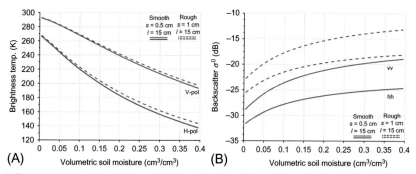

FIG. 3 (A) Variation in TB vs. soil moisture as the surface roughness root mean square (RMS) height(s) doubles at 300 K physical temperature. (B) Variation in radar backscatter vs. soil moisture for the same amount of roughness change. Radar backscatter shows significant sensitivity to roughness whereas emission is slightly affected. Approximately, backscatter changes 5 dB compared to $\sim$3 K at 0.02 cm^3/cm^3. *Solid lines* represent smooth surface ($s=0.5$ cm and $l=15$ cm); *dashed lines* are for rougher surface ($s=1$ cm and $l=15$ cm.); H-pol are in *blue* and V-pol are in *red*. The surface roughness profile is Gaussian with a correlation length (l).

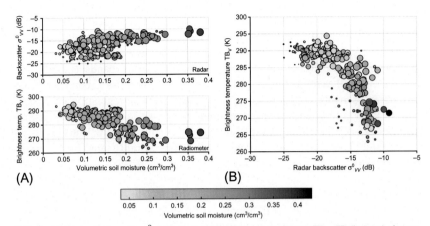

FIG. 4 (A) Radar backscatter σ_{vv}^0 (dB) (top) and brightness temperature TB_v (K) (bottom) plots vs. soil moisture (cm³/cm³). (B) Radar vs. radiometer general relationship indicates a negative correlation. The data is a subset of the PALS-SMEX02 campaign collocated with in situ sampling locations. H-polarized measurements have similar behaviors. Sizes of circles indicates the amount of vegetation water content (VWC) with min of 0 kg/m² and max of 6 kg/m²; that is, large VWC is *large circle* and vice versa.

a linear decrease and as seen in Fig. 4A (bottom) exhibits ~30 K change in TB at a rate of ~10 K/(cm³/cm³).

Simultaneously examining variations in both backscatter and emission, as shown in Fig. 4B, highlights the negative correlation between the two phenomena, such that with increasing radar backscatter (i.e., increasing soil moisture), the amount of emission, for the case presented here, decreases with an approximate slope of 1.6 K/dB and a correlation coefficient of −0.76.

Earliest known attempts at examining synergistic use of both active and passive soil moisture observations are from Ulaby et al. (1983) where reportedly an absolute estimation error accuracy of ±30% in soil moisture over corn and bare soil was determined when considering both TB and σ^0. This outcome stemmed from observing the different sensitivities of TB and σ^0 to soil moisture and then balancing and averaging independent active and passive overestimated and underestimated predictions.

Simple numerical simulations highlight how differing TB and σ^0 sensitivities affect the soil moisture retrieval process. Fig. 5 shows plots of normalized model predictions and data differences $|(\text{model-data})/\text{data}|^2$; that is, L_2 norm, for combinations of extremes of soil moisture (dry and wet) and VWC (low and high vegetation). This normalized L_2 norm is calculated for active-only, passive-only, and CAP scenarios. For very wet and moist soil where radar backscatter tends to saturate, the active-only norm exhibits very little change with respect to changes in soil moisture. Similarly, surface emission is masked by increased vegetation emission; therefore, calculation of the passive-only norm becomes invariant to soil moisture.

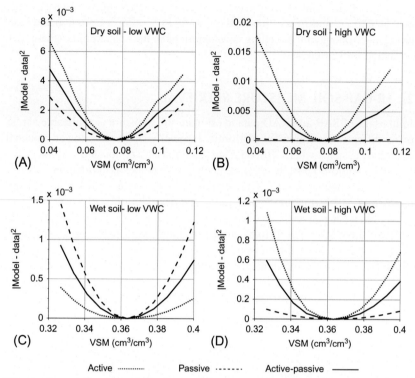

FIG. 5 Plots of $|(\text{model-data})/\text{data}|^2$ vs. volumetric soil moisture for (A) dry soil and low vegetation water content (VWC), (B) dry soil and high VWC, (C) wet soil and low VWC, and (D) wet soil and high VWC. For wetter soils, the response of the active scenario shows lack of sensitivity consistent with radar σ^0 saturation for wetter soil (Fig. 3). For high VWC (4.5 kg/m^2), passive comparisons lose sensitivity and are masked by vegetation emission effects. For all cases, the active-passive approach overcomes the lack of sensitivity of either instrument. Copyright © 2015 IEEE. *(Reprinted, with permission, from Akbar, R., Moghaddam, M., 2015. A combined active-passive soil moisture estimation algorithm with adaptive regularization in support of SMAP. IEEE Trans. Geosci. Remote Sens. 53, 3312–3324.)*

The CAP norm, which is the average of passive-only and active-only norms, shows detectable response for all cases and overcomes an individual measurement's lack of sensitivity by capturing the complementary behavior of the paring observation. More specifically, for example, in Fig. 5B, CAP is influenced more by radar data, as passive TB shows almost no sensitivity to soil moisture.

Therefore, it can be seen that by merging the benefits of SAR and radiometer remote sensing observation as well as contrasting their mutual sensitivities, predicting soil moisture estimates that meet science requirements is possible and the development of methods and algorithms to do so are of keen interest.

The perspective of this chapter henceforth is to present a general discussion on existing and potential CAP methods applicable to various platforms. In

Section 2 CAP from a single-resolution (SR) viewpoint is discussed and extended to the multiresolution (MR) scenario in Section 3. Emission and scattering electromagnetic modeling considerations in the context of CAP are then presented in Section 4.

2 SR CAP SOIL MOISTURE RETRIEVAL

Joint usage of SAR and RAD data to estimate soil moisture is applicable across many different measurement platforms, especially small-scale tower mounted or aircraft systems where TB and σ^0 observations are collocated with the same spatial resolution. Thus, the discussion on CAP methods is divided into SR and MR measurement scenarios. This section will focus on a SR CAP method and the case of MR estimation is deferred to the next section.

Within a SR active-passive framework (Fig. 2A) both σ^0 and TB are influenced by the same target features, therefore simultaneously taking advantage of both to estimate soil moisture proves to be a powerful technique. Akbar and Moghaddam (2015) previously developed a robust optimization-based SR CAP approach where, within a bicriteria objective function[3] $L(\bar{x})$, collocated and coresolution radar and radiometer contributions are constraint together. Explicit definition of their objective function is

$$L(\bar{x}) = \frac{1}{2} \left[\sum_{pq=\text{hh,vv}} \left| \frac{\sigma_{pq}^0 - \sigma_{pq}^0(\bar{x})}{\sigma_{pq}^0} \right|^2 + \gamma \cdot \sum_{p=\text{H,V}} \left| \frac{\text{TB}_p - \text{TB}_p(\bar{x})}{\text{TB}_p} \right|^2 \right] \qquad (1)$$

Here, σ_{pq}^0 and TB_p refer to the collocated and same resolution polarized active and passive measurements (hh and vv for σ^0; V-pol and H-pol for TB). $\sigma_{pq}^0(\bar{x})$ and $\text{TB}_p(\bar{x})$ are the corresponding electromagnetic scattering and emission models of choice sharing the same parameter kernel $\bar{x}$. In application, neither radar nor radiometer observations are perfect and are prone to noise. This imperfection also applies to scattering and emission models as well, since model predictions of observations are hampered by model accuracies and parameterization uncertainties. In SR CAP, to overcome these issues, a regularization or tunning term γ is included to balance active and passive contributions. When γ is small, radiometer measurements are discounted and radar-only soil moisture estimation is performed, corresponding to the first term on the right-hand side of Eq. (1). As the value of γ becomes larger, contributions from passive radiometer observations increase. By setting γ proportional to the ratio of measurements' noise standard deviations or prior model-data biases, or even sweeping across a range of γ, obtaining optimum model parameters $\bar{x}_{\text{opt}}$, which minimize $L(\bar{x})$ is possible.

Fig. 6 shows example soil moisture retrieval root mean squared (RMS) errors obtained using this method when applied to data gathered during the Soil

---

3. Also known as cost function or risk function.

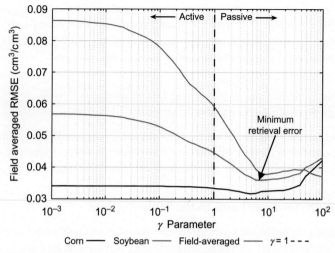

FIG. 6 CAP soil moisture estimation applied to PALS-SMEX02 data. When γ is small, radar-only estimation is performed and similarly, when γ is large, radiometer-only. Minimum field-averaged retrieval error of 0.035 cm³/cm³ is possible for $\gamma \sim 6$. High radar-only soybean soil moisture estimates (0.085 cm³/cm³) is indicative of data-model mismatches. In situ soil moisture varied 0.02–0.36 cm³/cm³ and VWC 0–4.5 kg/m².

Moisture Experiment 2002 (SMEX02) field campaign (Njoku, 2003; Wilson et al., 2001). The plot shows RMS errors as a function of the regularization parameter and clearly indicate that with the selection an appropriate balance between σ^0 and TB contributions, minimum field-averaged errors are possible. Observe that when $\gamma \sim 6$ field-averaged RMS errors outperform both radar-only and radiometer-only predictions, with a minimum error of 0.036 cm³/cm³ compared to 0.056 cm³/cm³ for radar and 0.044 cm³/cm³ for radiometer. Note that the volumetric moisture accuracy requirement for soil moisture is 0.04 cm³/cm³.

Such an adaptive SR CAP approach enables obtaining best retrievals based on proper regularization and weighting between σ^0 and TB contributions. By doing so, this algorithm is able to overcome data quality and model accuracy issues and is resilient against measurement noise. More importantly, "both" active and passive observation are simultaneously used to retrieve soil moisture and all the inherent sensitivities and complementary nature of the two phenomena are effectively captured within a single optimization framework.

3   MR CAP SOIL MOISTURE RETRIEVAL

For spacecraft missions, active and passive observations are often at different spatial resolutions. As was shown in Fig. 2B, MR CAP assumes Earth-gridded nested SAR measurements within a single coarse-resolution TB pixel. This is primarily the case for SMAP where level-1 Earth-gridded TB measurements (L1C_TB) are reported at a 36 km spatial resolution and corresponding SAR observations nested within the TB pixel at 1–3 km (L1C_S0_HiRes). This structured setup supports a variety of possible CAP solutions:

1. Physics-based estimation.
2. Probabilistic and machine learning (ML) techniques.
3. Image processing approaches.

As of yet, operational and practical CAP retrieval schemes are very limited in number and efforts by the active and passive remote sensing community are still ongoing. Thus, only a select few methods are presented here.

3.1 Multi-Temporal and Multi-Scale Method

Three main geophysical features affect both the radar and radiometer signals: soil dielectric constant (synonyms to soil moisture), surface roughness, and vegetation attenuation and scattering. Even with multiple polarized measurements, the number of parameters exceeds the number of observations. Correlation between polarizations further reduces the information content of measurements and prohibits robust retrieval of all the parameters of interest that affect the observations. There is, however, a separation of scales in the variability of these parameters that opens the opportunity to use multitemporal information to estimate the parameters. Surface soil moisture (and hence soil dielectric constant) varies at daily and subdaily time scales in response to precipitation, evaporation, and drainage. Vegetation structure and water content as well as soil roughness, however, vary on time scales of vegetation growth, which is typically weeks to months.

Therefore, over short periods of time, such that vegetation growth and surface roughness conditions remain relatively stable, TB and σ^0 are correlated with respect to variations in soil moisture, and as seen in Fig. 4B are in fact negatively correlated. Thus, a regression-based linear mapping between TB and σ^0 can be established, $TB \propto \beta \cdot \sigma + \alpha$. Das et al. (2011, 2014) extended this understanding to the SMAP MR scenario such that, prior to actual soil moisture retrieval, high-resolution co- and cross-pol SMAP active data are used to predict an intermediate TB product at medium resolution (9 km). This step is typically known as "brightness temperature disaggregation." To retrieve soil moisture, after the disaggregation step, a well-known radiometer-only retrieval algorithm, the single-channel algorithm (Jackson, 1993), is then used. The medium-resolution choice is to conform to SMAP hydrometeorological science requirements calling for ± 0.04 cm^3/cm^3 volumetric accuracy at ~ 10 km or less.

This two-step disaggregation-estimation approach is formally the SMAP baseline CAP algorithm (Entekhabi et al., 2012). Furthermore, the process is specific for every individual TB pixel such that time-series linear mapping is applied to obtain unique regression parameters. The TB disaggregation step is formulated as

$$TB_v(M) = TB_v(C) + \beta(C) \cdot \left\{ \left[\sigma_{vv}^0(M) - \sigma_{vv}^0(C) \right] + \Gamma \cdot \left[\sigma_{hv}^0(C) - \sigma_{hv}^0(M) \right] \right\} \quad (2)$$

where $TB_v(C)$ is the SMAP coarse-resolution V-pol passive measurement. $\beta(C)$, in units of (K/dB), is the data-driven time-series regression slope between TB and σ^0 (i.e., $TB_v(C) = \beta(C) \cdot \sigma_{vv}^0(C) + \alpha(C)$). The parameter β projects variations in the soil moisture contributions to the co-pol backscatter onto the TB space. The choice of V-pol observations stems from the observed higher linear correlation between $TB_V - \sigma_{vv}^0$ compared to $TB_H - \sigma_{hh}^0$.

The third term in Eq. (2), $\left[\sigma_{hv}^0(C) - \sigma_{hv}^0(M) \right]$, is a strong indicator of heterogeneity within a given pixel (C), which is then projected to the co-pol radar space through the factor Γ. Based on co- and cross-pol σ^0 statistical regression, specific to a given grid cell, Γ can be predicted. In other words, this third term is the soil contribution to the co-pol backscatter such that volume scattering component is isolated in the cross-pol measurement. Γ then represents a proportionality factor that depends on the architecture (distribution of the shape and orientation of the lossy dielectric elements in the vegetation canopy) within the pixel under consideration.

Disaggregation is performed for every medium-resolution pixel within $TB(C)$ such that grid averaged disaggregated $TB(M)$ should be close to $TB(C)$ $\left(TB(C) \sim \frac{1}{N} \sum_{i=1}^{N} TB(M_i) \right)$. A simple schematic overview of this algorithm is shown in Fig. 7.

A 3-day global composite of SMAP CAP surface soil moisture retrievals can be seen in Fig. 8. The figure shows unprecedented high-resolution

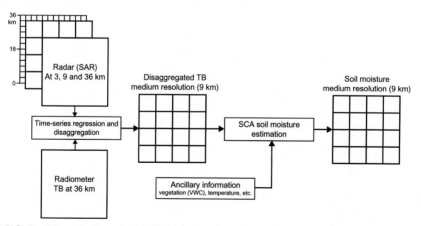

FIG. 7 Schematic diagram of SMAP baseline CAP soil moisture estimation approach. High-resolution radar and coarse-resolution radiometer are combined through regression analysis to derive medium-resolution TB (Eq. 2). Final soil moisture estimates are based on these TB predictions via the single-channel algorithm (SCA). This process, and derived regression parameters β and Γ, are specific and unique to each pixel. [Detailed discussion for disaggregation is in Das et al. (2014); for SCA see Jackson (1993).]

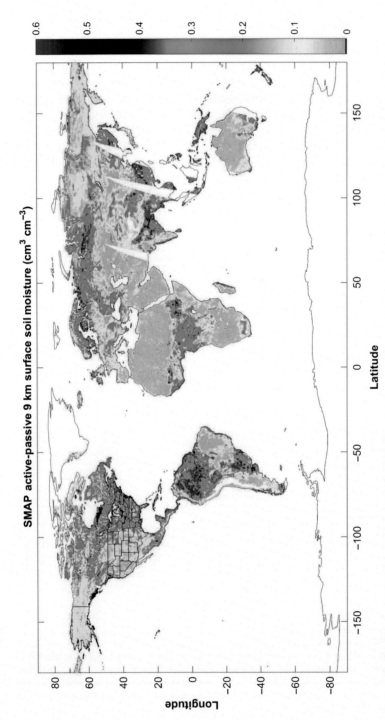

FIG. 8 SMAP L2 combined radar-radiometer global soil moisture retrieval at a 9 km spatial resolution. Image is a 3-day composite of soil moisture estimates covering Jun. 28–30, 2015. *(Image courtesy of Narenrda Das (NASA-JPL), and Dara Entekhabi (MIT).)*

(9 km) surface soil moisture spanning Jun. 28–30, 2015, obtained using the SMAP observatory.

In this approach, although radar observations are not directly used within the soil moisture estimation step, the derived linear relationship between TB and σ^0 based on their fundamental physical interrelatedness is utilized to relate one to the other. The availability of disaggregated TB also has additional science benefits.

Rather than explicit disaggregation-estimation, an alternative is estimation with implicit disaggregation. That is, as an extension to Eq. (1), radar-based soil moisture retrieval can be performed with the added constraint of minimizing the grid difference between measured coarse-resolution TB and mean disaggregated TB.

3.2   Probabilistic and Machine Learning Techniques

In a dynamic hydrological system with temporal evolutions, Kalman filtering (Kalman, 1960) can be applied to merge multiple observations in a single framework to update and even predict underlying state parameters. Decades of hydrological assimilation and forecasts studies such as Reichle et al. (2002), Entekhabi et al. (1994), and many more, have demonstrated the feasibility and practicality of this method. Within a Bayesian merging (BM) scheme, Zhan et al. (2006) have applied the Kalman filter state update equations $\mathbf{K}$ to a priori background soil moisture estimates $\mathbf{SM}_{ap}$ to directly estimate high-resolution soil moisture $\mathbf{SM}_f$. Notationally, such a system is written as

$$\mathbf{SM}_f = \mathbf{SM}_{ap} + \mathbf{K} \cdot \left[\mathbf{D} - \mathbf{FM}\left(\mathbf{SM}_{ap} \right) \right] \tag{3a}$$

$$\mathbf{K} = \frac{\mathbf{C}_{sm} \cdot \mathbf{J}_{FM}^t}{\mathbf{J}_{FM} \cdot \mathbf{C}_D \cdot \mathbf{J}_{FM}^t + \mathbf{C}_n} \tag{3b}$$

$\mathbf{D}$ is the vector of all coarse and fine scale observations, $\mathbf{C}_{sm}$ and $\mathbf{C}_n$ are state parameter (soil moisture) and measurement error covariance matrices, and $\mathbf{FM(SM)}$ is the system of scattering and emission electromagnetic models with the Jacobian matrix $\mathbf{J}_{FM}$. By temporally updating Eqs. (3a and 3b) with up-to-date observations and new covariances, predicting soil moisture then becomes possible. This scheme is also scalable to higher or lower resolutions dependent on the initial condition and driving parameters' spatial resolution. The confluence of active and passive observations is captured in the Kalman filter gain $\mathbf{K}$, which requires proper estimation of the covariance matrices $\mathbf{C}_{sm}$ and $\mathbf{C}_n$.

Within a SMAP-like observation system simulation experiment (Crow et al., 2005), mimicking hydrological processes over the Red-Arkansas River Basin using topographically-based land-atmosphere transfer scheme (Peters-Lidard et al., 1997), the performance of BM was tested and noted to be comparable and even superior to radar-only or radiometer-only soil moisture estimation outcomes. Figs. 9 and 10 show time series and spatial soil moisture RMS errors

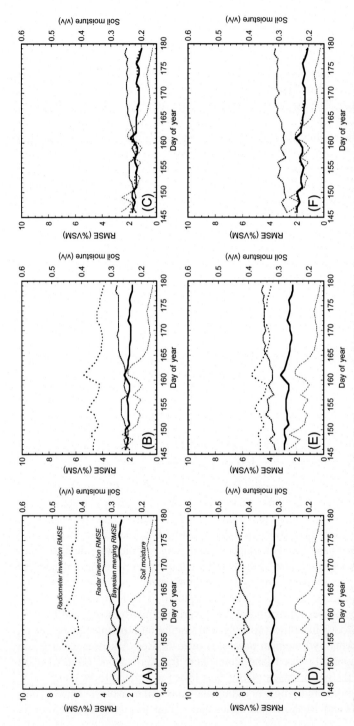

FIG. 9   Time-series RMS error of soil moisture retrievals from the Bayesian method (*thick solid line*), direct radar inversion method (*dotted line*), and direct radiometer inversion method (*dashed line*) against 3 km (left panels), 9 km (middle panels), and 36 km (right panels) soil moisture "truth" for the low (upper panels) and high (lower panels) noise data sets. The *thin dashed line* at the bottom of each panel is basin average of the "truth" surface soil moisture using the right-side scale. Copyright © 2015 IEEE. (Reprinted, with permission, from Zhan, X., Houser, P., Walker, J., Crow, W., 2006. A method for retrieving high-resolution surface soil moisture from hydros L-band radiometer and radar observations. IEEE Trans. Geosci. Remote Sens. 44, 1534–1544.)

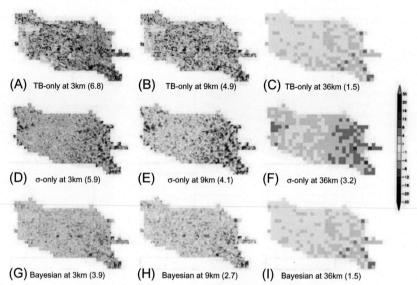

FIG. 10   Spatial distribution of the 3 km (left column), 9 km (middle), and 36 km (right) soil moisture errors from the radiometer-only (TB) inversion method (top row), 3 km radar backscatter (σ^0) inversion method (middle row), and the Bayesian merging method (bottom row) for the high noise data set (1–2 dB and 1.5 K Gaussian noise). The numbers in the parentheses are the RMS errors over the whole basin. Copyright © 2015 IEEE. *(Reprinted, with permission, from Zhan, X., Houser, P., Walker, J., Crow, W., 2006. A method for retrieving high-resolution surface soil moisture from hydros L-band radiometer and radar observations. IEEE Trans. Geosci. Remote Sens. 44, 1534–1544.)*

over the simulation area. For the numerical simulations presented here, over almost the entire time series and at various spatial resolutions, BM CAP outperforms all other methods, even in the presence of high measurement noise.

In a well-formulated system such as Eqs. (3a and 3b) two data types (σ^0 and TB) have been merged together for soil moisture retrieval. An extension of this approach is to merge even more data layers and observations of different characteristics. In addition to microwave remote sensing, knowledge of soil texture, aspect, topography, precipitation, and many more, directly or indirectly relate to soil moisture. Existing in situ moisture sensing networks also provide valuable near real-time information, although at a single point scale. Proper assimilation and convergence of these data layers can effectively increase the amount of available information to obtain better estimates.

ML techniques and algorithms (Bishop, 2006) are very powerful tools for such investigations. Generally, multiple data layers are used to train, validate, and test an ML algorithm, say for soil moisture prediction, as seen in Fig. 11. Although obtaining closed-form expressions are not always possible, ML techniques are capable of extracting abstract high-order relationships between inputs and outputs. Once trained, ML-based estimation of soil moisture with new data and observations becomes possible. In particular, neural networks

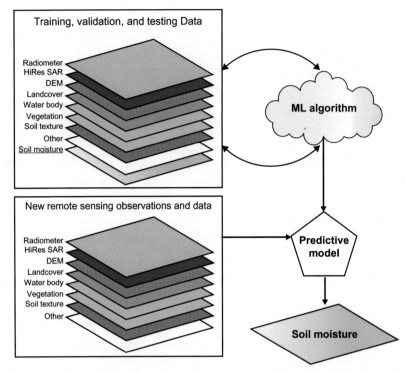

FIG. 11 Multiple geophysical and remote sensing data layers are input to machine learning algorithms for training, validation, and testing. Once complete, with new observations, the "predictive model," is capable of estimating soil moisture.

(NN) have widespread applications in developing global retrieval schemes. Prior radar-only (Paloscia et al., 2013; Notarnicola et al., 2008) or radiometer-only (Del Frate et al., 2003; Liu et al., 2002) attempts have shown reasonable soil moisture estimation using NN under various conditions capturing both spatial and temporal variations. Particularly, Rodriguez-Fernandez et al. (2015) applied feedforward NNs to SMOS TB observations and reported acceptable soil moisture predictions when using both TB_V and TB_H measurements between 26 degree and 60 degree. A further reduction in estimation error (10–15%) was also discovered when additional information such as ASCAT backscatter coefficient, soil temperature, and soil texture were included in the NN training phase.

A common theme and consideration among feedforward NN and ML methods is the identification and selection of appropriate training sets to prevent underfitting or overfitting of derived relationships. For CAP, obvious inputs are high-resolution radar, coarse-resolution TB and ancillary data such as the moderate resolution imaging spectroradiometer and normalized difference vegetation index (NDVI) (Justice et al., 1998) as an indicator on the amount of vegetation, digital elevation model, and soil texture, aspect, precipitation information, and many other static or dynamic ancillary information. An even further extension would be the concept of "hybrid training"; that is, along with informative and relevant data layers, electromagnetic scattering and emission models can also be incorporated in the ML training phase. The algorithm not only can benefit from physics-based modeling, remote sensing observations can supplement model deficiencies and shortcomings.

From an operational standpoint, especially for spacecraft missions, a major advantage of ML or NN soil moisture estimation is the possible reduction in product delivery latency times. Once training and cross validation is complete, upon delivery of new satellite acquisitions and ancillary data, near real-time science processing is possible.

4 FORWARD ELECTROMAGNETIC SCATTERING AND EMISSION MODEL CONSIDERATIONS

From a fundamental physical point of view and in a resolution-independent scenario such that both radar and radiometer observe the same scene, the amount of measured emission and backscatter are influenced by a set of core and shared geophysical properties. Here denoted by $\bar{x}$, these properties, or parameter kernel, typically consist of soil moisture, surface roughness, and information on the overlaying vegetation cover. Emission is also influenced by the target's physical temperature; however, in most remote sensing applications it is assumed known via ancillary data sources or models. Furthermore, theoretically Peake (1959) showed that under Kirchhoff's Law of Thermodynamics in equilibrium and while taking into account electromagnetic reciprocity, the emissivity $e_p(\theta_i, \phi_i)$ of an object can be related to its scattering properties as

$$r_p(\theta_i, \phi_i) = \frac{1}{4\pi} \sum_{q=h,v} \int_0^{\pi/2} d\theta_s \sin\theta_s \int_0^{2\pi} d\phi_s \gamma_{pq}(\theta_i, \phi_i; \theta_s, \phi_s) \qquad (4a)$$

$$e_p(\theta_i, \phi_i) = 1 - r_p(\theta_i, \phi_i) \qquad (4b)$$

where γ_{pq} is the bistatic scattering coefficient of the object in the direction (θ_s, ϕ_s) from (θ_i, ϕ_i) and is a function of the target's electromagnetic properties (i.e., $\bar{x}$). Therefore, it can been seen that both TB and σ^0 are closely interrelated.

In model-based CAP methods, as σ^0 and TB are simultaneously used to estimate soil moisture, it becomes imperative to ensure that model development and responses with respect to changes in model parameters $\bar{x}$ are consistent with each other. For example, in most radar scattering models, surface roughness effects are analytically derived in the form of the surface's Gaussian or exponential spectral density functions (Tsang et al., 2004). On the other hand, in TB models such as the widely used zeroth order radiative transfer solution (Jackson and Schmugge, 1991), commonly known as the $\tau - \omega$ model, surface roughness effects are modeled as exponential modification to the Fresnel equations r_{0p}

(i.e., $r_p = r_{0p} \cdot e^{\left(-h^2/\cos\theta_i\right)}$), where h is an empirically determined roughness parameter, usually set to 0.1. From a CAP perspective, this is considered inconsistent because the effects of roughness on either process are modeled from two very dissimilar approaches. Not only may the range of validity of either approach be different, but to initiate the retrieval process, two different roughness parameters, each suiting their own model, must be chosen.

Similarly, modeling of vegetation between operational emission and scattering models is also different. In the $\tau - \omega$ model for example, equivalent VWC along with an empirically determined parameter b are used to predict the vegetation opacity and emission attenuation through the vegetation layer, $e^{(-b \cdot \text{VWC})}$. On the other hand, radar scattering models (Burgin et al., 2011; Huang and Tsang, 2012; Huang et al., 2010) use prior knowledge of vegetation density and geometry to determine the individual contributing scattering mechanisms.

Thus, it can be seen that inconsistent forward modeling complicates usages of proper parameterization, even though physically both TB and σ^0 are influenced by the same target features. Application of Peake's formulation (Eqs. 4a and 4b), to rough surfaces (Tsang et al., 2013) and grasslands (Saatchi et al., 1994) have shown acceptable and consistent radar and radiometer comparisons with tower-based observations while using the same parameter kernel. However, CAP inverse processes using joint and self-consistent models have not been fully examined nor demonstrated.

5 FURTHER DISCUSSIONS

This chapter has highlighted the case for combined usage of active radar and passive radiometer remote sensing observations to obtain surface soil moisture retrievals. As was mentioned in Section 1, this field is fairly young, and significant opportunities to develop new and novel methodologies exist. Additional practical considerations also exist, which are briefly mentioned in the following section:

1. Discussion on assessment and validation of retrieved soil moisture using CAP, or other, is well beyond the scope of this chapter. However, over a given region or designated validation site, the predicted soil moisture trends should be spatial and temporally compared to radar-only and radiometer-only estimates to ensure consistency, as a relative measure of accuracy (see Figs. 9 and 10).
2. Using the information theoretic metrics of entropy and mutual information Konings et al. (2015) showed that, in general, for N active and passive observations, the degrees of information is at least $N - 1$. In the case of CAP, with the availability of multiple radar and radiometer observations at different polarizations, it then becomes possible to assume parameters, other than soil moisture, to be unknown. This is especially applicable to surface roughness as

on a global scale knowledge of roughness is limited and difficult to measure. CAP, thus, allows both surface soil moisture and roughness to be unknowns.

3. When merging observations from multiple different platforms, care must be taken to ensure that active and passive measurements are temporally and spatially as close as possible and are representative of the same scene under observation. This is to minimize incidence and azimuthal angle effects as well as topographical changes. For SMAP, since SAR and RAD acquisitions are concurrent, with an almost constant viewing angle, these effects are negligible.

REFERENCES

Akbar, R., Moghaddam, M., 2015. A combined active-passive soil moisture estimation algorithm with adaptive regularization in support of SMAP. IEEE Trans. Geosci. Remote Sens. 53, 3312–3324.

Bishop, C.M., 2006. Pattern Recognition and Machine Learning. Springer, New York.

Brocca, L., Hasenauer, S., Lacava, T., Melone, F., Moramarco, T., Wagner, W., Dorigo, W., Matgen, P., Martínez-Fernández, J., Llorens, P., et al., 2011. Soil moisture estimation through ASCAT and AMSR-E sensors: an intercomparison and validation study across Europe. Remote Sens. Environ. 115 (12), 3390–3408.

Burgin, M., Clewley, D., Lucas, R., Moghaddam, M., 2011. A generalized radar backscattering model based on wave theory for multilayer multispecies vegetation. IEEE Trans. Geosci. Remote Sens. 49, 4832–4845.

Choudhury, B.J., Kerr, Y.H., Njoku, E.G., Pampaloni, P., 1995. Passive Microwave Remote Sensing of Land-Atmosphere Interactions. Taylor & Francis, London.

Crow, W., Chan, S., Entekhabi, D., Houser, P., Hsu, A., Jackson, T., Njoku, E., O'Neill, P., Shi, J., Zhan, X., 2005. An observing system simulation experiment for HYDROS radiometer-only soil moisture products. IEEE Trans. Geosci. Remote Sens. 43, 1289–1303.

Das, N., Entekhabi, D., Njoku, E., 2011. An algorithm for merging SMAP radiometer and radar data for high-resolution soil-moisture retrieval. IEEE Trans. Geosci. Remote Sens. 49, 1504–1512.

Das, N., Entekhabi, D., Njoku, E., Shi, J., Johnson, J., Colliander, A., 2014. Tests of the SMAP combined radar and radiometer algorithm using airborne field campaign observations and simulated data. IEEE Trans. Geosci. Remote Sens. 52, 2018–2028.

Del Frate, F., Ferrazzoli, P., Schiavon, G., 2003. Retrieving soil moisture and agricultural variables by microwave radiometry using neural networks. Remote Sens. Environ. 84 (2), 174–183.

Entekhabi, D., Nakamura, H., Njoku, E., 1994. Solving the inverse problem for soil moisture and temperature profiles by sequential assimilation of multifrequency remotely sensed observations. IEEE Trans. Geosci. Remote Sens. 32, 438–448.

Entekhabi, D., Njoku, E., Houser, P., Spencer, M., Doiron, T., Kim, Y., Smith, J., Girard, R., Belair, S., Crow, W., Jackson, T., Kerr, Y., Kimball, J., Koster, R., McDonald, K., O'Neill, P., Pultz, T., Running, S., Shi, J., Wood, E., Van Zyl, J., 2004. The Hydrosphere State (HYDROS) satellite mission: an earth system pathfinder for global mapping of soil moisture and land freeze/thaw. IEEE Trans. Geosci. Remote Sens. 42, 2184–2195.

Entekhabi, D., Njoku, E., O'Neill, P., Kellogg, K., Crow, W., Edelstein, W., Entin, J., Goodman, S., Jackson, T., Johnson, J., Kimball, J., Piepmeier, J., Koster, R., Martin, N., McDonald, K., Moghaddam, M., Moran, S., Reichle, R., Shi, J.-C., Spencer, M., Thurman, S., Tsang, L., Van Zyl, J., 2010. The soil moisture active passive (SMAP) mission. Proc. IEEE 98, 704–716.

Entekhabi, D., et al., 2012. SMAP Algorithm Theoretical Basis Document (ATBD) L2 and L3 Radar/Radiometer Soil Moisture (Active/Passive) Data Products @Online. USDA, Washington, DC.

Huang, S., Tsang, L., 2012. Electromagnetic scattering of randomly rough soil surfaces based on numerical solutions of Maxwell equations in three-dimensional simulations using a hybrid UV/PBTG/SMCG method. IEEE Trans. Geosci. Remote Sens. 50, 4025–4035.

Huang, S., Tsang, L., Njoku, E., Chan, K.S., 2010. Backscattering coefficients, coherent reflectivities, and emissivities of randomly rough soil surfaces at L-band for smap applications based on numerical solutions of maxwell equations in three-dimensional simulations. IEEE Trans. Geosci. Remote Sens. 48, 2557–2568.

Jackson, T.J., 1993. III: measuring surface soil moisture using passive microwave remote sensing. Hydrol. Process. 7 (2), 139–152.

Jackson, T., Schmugge, T., 1991. Vegetation effects on the microwave emission of soils. Remote Sens. Environ. 36 (3), 203–212.

Justice, C.O., Vermote, E., Townshend, J.R., Defries, R., Roy, D.P., Hall, D.K., Salomonson, V.V., Privette, J.L., Riggs, G., Strahler, A., et al., 1998. The moderate resolution imaging spectroradiometer (MODIS): land remote sensing for global change research. IEEE Trans. Geosci. Remote Sens. 36 (4), 1228–1249.

Kalman, R.E., 1960. A new approach to linear filtering and prediction problems. J. Fluids Eng. 82 (1), 35–45.

Kerr, Y., Waldteufel, P., Wigneron, J.-P., Martinuzzi, J., Font, J., Berger, M., 2001. Soil moisture retrieval from space: the soil moisture and ocean salinity (SMOS) mission. IEEE Trans. Geosci. Remote Sens. 39, 1729–1735.

Kerr, Y., Waldteufel, P., Wigneron, J.-P., Delwart, S., Cabot, F., Boutin, J., Escorihuela, M.-J., Font, J., Reul, N., Gruhier, C., Juglea, S., Drinkwater, M., Hahne, A., Martin-Neira, M., Mecklenburg, S., 2010. The SMOS mission: new tool for monitoring key elements of the global water cycle. Proc. IEEE 98, 666–687.

Konings, A., McColl, K., Piles, M., Entekhabi, D., 2015. How many parameters can be maximally estimated from a set of measurements? IEEE Geosci. Remote Sens. Lett. 12, 1081–1085.

Liu, S.-F., Liou, Y.-A., Wang, W.-J., Wigneron, J.-P., Lee, J.-B., 2002. Retrieval of crop biomass and soil moisture from measured 1.4 and 10.65 GHz brightness temperatures. IEEE Trans. Geosci. Remote Sens. 40 (6), 1260–1268.

Njoku, E., 2003. Smex02 Passive and Active L and S Band System (PALS) Data. NASA National Snow and Ice Data Center, Boulder, CO.

Njoku, E.G., Entekhabi, D., 1996. Passive microwave remote sensing of soil moisture. J. Hydrol. 184 (1), 101–129.

Notarnicola, C., Angiulli, M., Posa, F., 2008. Soil moisture retrieval from remotely sensed data: neural network approach versus Bayesian method. IEEE Trans. Geosci. Remote Sens. 46 (2), 547–557.

Paloscia, S., Pettinato, S., Santi, E., Notarnicola, C., Pasolli, L., Reppucci, A., 2013. Soil moisture mapping using sentinel-1 images: algorithm and preliminary validation. Remote Sens. Environ. 134, 234–248.

Peake, W., 1959. Interaction of electromagnetic waves with some natural surfaces. IRE Trans. Antennas Propag. 7, 324–329.

Peters-Lidard, C.D., Zion, M.S., Wood, E.F., 1997. A soil-vegetation-atmosphere transfer scheme for modeling spatially variable water and energy balance processes. J. Geophys. Res. Atmos. 102 (D4), 4303–4324.

Reichle, R.H., McLaughlin, D.B., Entekhabi, D., 2002. Hydrologic data assimilation with the ensemble Lalman filter. Mon. Weather Rev. 130 (1), 103–114.

Rodriguez-Fernandez, N., Aires, F., Richaume, P., Kerr, Y., Prigent, C., Kolassa, J., Cabot, F., Jimenez, C., Mahmoodi, A., Drusch, M., 2015. Soil moisture retrieval using neural networks: application to SMOS. IEEE Trans. Geosci. Remote Sens. 99, 1–17.

Saatchi, S., Le Vine, D., Lang, R., 1994. Microwave backscattering and emission model for grass canopies. IEEE Trans. Geosci. Remote Sens. 32, 177–186.

Soil Moisture Active/Passive Mission, 2007. NASA workshop report, soil moisture active/passive mission. In: Report of NASA HQ Community Workshop.

Tsang, L., Kong, J.A., Ding, K.-H., 2004. In: Scattering of Electromagnetic Waves, Theories and Applications, vol. 27. Wiley, New York.

Tsang, L., Suek Koh, I., Liao, T.-H., Huang, S., Xu, X., Njoku, E., Kerr, Y., 2013. Active and passive vegetated surface models with rough surface boundary conditions from nmm3d. IEEE J. Select. Top. Appl. Earth Observ. Remote Sens. 6, 1698–1709.

Ulaby, F., Dobson, M., Brunfeldt, D., 1983. Improvement of moisture estimation accuracy of vegetation-covered soil by combined active/passive microwave remote sensing. IEEE Trans. Geosci. Remote Sens. GE-21, 300–307.

Wilson, W., Yueh, S., Dinardo, S., Chazanoff, S., Kitiyakara, A., Li, F.K., Rahmat-Samii, Y., 2001. Passive active L- and S-band (PALS) microwave sensor for ocean salinity and soil moisture measurements. IEEE Trans. Geosci. Remote Sens. 39, 1039–1048.

Zhan, X., Houser, P., Walker, J., Crow, W., 2006. A method for retrieving high-resolution surface soil moisture from hydros L-band radiometer and radar observations. IEEE Trans. Geosci. Remote Sens. 44, 1534–1544.

Chapter 11

Intercomparison of Soil Moisture Retrievals From In Situ, ASAR, and ECV SM Data Sets Over Different European Sites

B. Barrett[*,a], C. Pratola[*], A. Gruber[†] and E. Dwyer[‡]

University College Cork (UCC), Cork, Ireland, †Vienna University of Technology, Vienna, Austria, ‡EurOcean—European Centre for Information on Marine Science and Technology, Lisbon, Portugal

1 INTRODUCTION

The amount of water stored in the soil is a key parameter for the energy and mass fluxes at the land surface-atmosphere boundary and is of fundamental importance to many agricultural, meteorological, biological, and biogeochemical processes (Koster et al., 2004; Seneviratne et al., 2010; Bolten and Crow, 2012). Soil moisture dynamics are dependent on both meteorological conditions and soil physical characteristics and, as a result, exhibit large spatial and temporal variations between different areas, seasons, and years (Western and Blöschl, 1999; Schulte et al., 2005). The spatial and temporal coverage attainable by spaceborne remote sensing has demonstrated the capability to monitor soil moisture over large areas at regular time intervals, and several approaches for soil moisture retrieval have been developed using optical, thermal infrared, and microwave (MW) sensors over the last three decades (Barrett and Petropoulos, 2013; Petropoulos et al., 2015). Since the late 1970s, coarse resolution (25–50 km) soil moisture products derived from past and present microwave radiometers (Advanced Microwave Scanning Radiometer for Earth Observing (AMSR-E) (Njoku et al., 2003) and WindSat (Li et al., 2010)) and scatterometers (European Remote Sensing Satellite Scatterometer (ERS SCAT) (Wagner et al., 2003), and meteorological operational satellite (MetOp) Advanced Scatterometer (ASCAT) (Bartalis et al., 2007; Naeimi et al., 2009) have been available on

---

a. Present address: University of Glasgow, Glasgow, United Kingdom.

Satellite Soil Moisture Retrieval. http://dx.doi.org/10.1016/B978-0-12-803388-3.00011-5

an operational basis. Data from the European Space Agency (ESA) Soil Moisture and Ocean Salinity (SMOS) (Kerr et al., 2012; Mecklenburg et al., 2012) and the National Aeronautics and Space Administration (NASA) Soil Moisture Active and Passive (SMAP) (Entekhabi et al., 2010) dedicated soil moisture missions are strengthening this record of observations and further facilitating the study of long-term soil moisture behavior (please see Petropoulos et al. (2015) and Zeng et al. (2015) for further details of available satellite-derived soil moisture products). With the availability of these products, it is necessary to validate them using independently derived soil moisture observations obtained through in situ monitoring, models, or with different satellite sensors (van Doninck et al., 2012; Ochsner et al., 2013; Al-Yaari et al., 2014). In situ validation has generally been achieved over small temporal and spatial scales but has been significantly advanced since the establishment of the Global Soil Moisture Data Bank (Robock et al., 2000) and the International Soil Moisture Network (ISMN) (Dorigo et al., 2011). For example, Albergel et al. (2013c) validated three global soil moisture products using a combination of 196 in situ stations taken from five different soil moisture networks across the world. Similarly, Paulik et al. (2014) and Dorigo et al. (2015) used over 600 in situ stations for validating ASCAT and essential climate variable soil moisture (ECV SM) products, respectively. All these studies generally found good agreement between the satellite-derived and in situ observations.

The comparison of time series of soil moisture data sets acquired by different sources and representing different spatial scales is challenging, however, due to the scale differences between products and/or observations (Crow et al., 2012). In situ networks represent single point locations and usually cover only limited observation periods. Gruber et al. (2013) investigated the quality of over 1400 in situ stations of the ISMN for representing soil moisture at satellite footprint scales (~25 km) on a global basis using triple collocation and highlighted the need for a comprehensive characterization of in situ representativeness errors in addition to measurement errors when considering satellite-derived soil moisture—in situ soil moisture intercomparisons. Consequently, the spatial and temporal resolution provided by synthetic aperture radar (SAR) data make them a promising additional data source for measuring seasonal and long-term variations in surface soil moisture content and characterizing the errors of coarse-scale soil moisture products. The advanced synthetic aperture radar (ASAR) instrument onboard the Envisat satellite was capable of providing global measurements at 1 km and 150 m spatial resolution every 4–7 days, depending on the acquisition plan (Desnos et al., 2000). Although there are certain technical limitations in retrieving surface soil moisture from SAR data, significant progress has been made in recent years to the point that SAR data could be used not only as another validation source for coarse-scale soil moisture products but also for applications that require finer spatial resolution soil moisture data such as hydrological or runoff modeling (Dostálová et al., 2014; Pratola et al., 2014). Furthermore, the regular temporal coverage and higher spatial resolution of current C-band sensors such as

Sentinel-1 will provide greater opportunities to characterize surface soil moisture within the large areas covered by coarse-scale product cells and help strengthen the understanding of such products. In this study, the capability of the coarse-scale ECV SM product in representing the temporal variations in surface soil moisture content at finer scales is evaluated using both in situ and ASAR-derived soil moisture observations in three different European countries, characterized by contrasting climate and vegetation conditions.

2 MATERIALS AND METHODS

2.1 In Situ Soil Moisture Observations

The Irish in situ soil moisture measurements were collected at two grassland sites: Kilworth and Solohead, located in southern Ireland (see Fig. 1) using

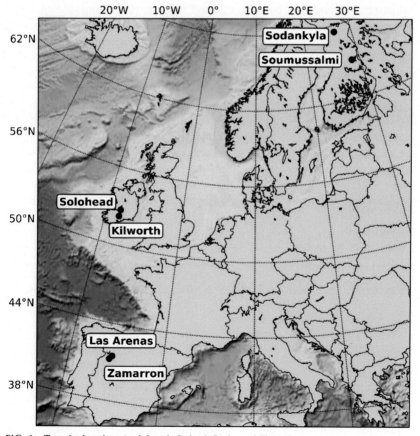

FIG. 1   Test site locations *(red dots)* in Ireland, Spain, and Finland.

Campbell Scientific CS616 water content reflectometers, installed horizontally at a depth of 5 cm below the surface under the framework of the Aeon project (http://aeon.ucc.ie). Measurements were recorded continuously at 30-min intervals between 2007 and 2010 and are expressed as the soil water-filled pore space. These values were subsequently converted to volumetric units ($m^3 m^{-3}$) by multiplying by the associated soil porosity values. The network also measured precipitation and soil temperature and has been used predominantly for modeling N_2O fluxes from agricultural grasslands, but has also been used for the validation of soil moisture products (e.g., Pratola et al., 2014). The sites have a temperate maritime climate with annual precipitation of 900–1200 mm and an annual average temperature of 10°C.

The Red de Medición de la Humedad del Suelo (REMEDHUS) soil moisture network (Martínez-Fernández and Ceballos, 2005) is located in the semiarid region of the Duero basin (Zamora) in Spain. It has a continental Mediterranean climate with a mean annual temperature of 12°C and mean annual precipitation of 400 mm. The land use is predominantly agricultural, with small areas of forest and pasture. The network comprises 20 soil moisture monitoring stations, each using a Stevens Hydra probe sensor integrated over a depth of 0–5 cm below the surface. The network has been used for several purposes, including calibration and validation campaigns in support of the SMOS mission (Sanchez et al., 2012), and the evaluation of different satellite-derived soil moisture products (Ceballos et al., 2005; Wagner et al., 2008; Albergel et al., 2012).

The Sodankyla (Finnish Meteorological Institute (FMI)) and Suomussalmi (Geological Survey of Finland (GTK)) sites are located in northern Finland and have a boreal climate, with average summer temperatures of 12–13°C and average winter temperatures of −12°C. Annual rainfall is 600–650 mm and snow generally covers the soil surface from the beginning of Nov. to the end of Apr. The predominant land cover type is forest. Table 1 details the main characteristics for each of the study sites. For the Spanish and Finnish sites, daily

TABLE 1 Study Site Characteristics

Site	Latitude	Longitude	Elevation (masl)	Land Cover	Soil Type
Kilworth	52°10′ N	−8°14′ E	51	Grassland	Sandy loam
Solohead	52°30′ N	−8°12′ E	98	Grassland	Loam
Zamarron	41°14′ N	−5°32′ E	855	Cropland	Sandy loam
Las Arenas	41°22′ N	−5°32′ E	745	Cropland	Sandy loam
Sodankyla	67°22′ N	26°37′ E	179	Forest	Sandy loam
Suomussalmi	64°54′ N	28°41′ E	220	Forest	Sandy loam

precipitation data from nearby meteorological stations were acquired through the European Climate Assessment and Data Set (Klein Tank et al., 2002) (data available at http://www.ecad.eu). In situ soil moisture data sets for Spain and Finland were retrieved from the ISMN (Dorigo et al., 2011, 2013) at https://ismn.geo.tuwien.ac.at/. At all sites, soil moisture values have been evaluated on a daily basis by averaging the measurements recorded at 30-min intervals from 00:00 to 23:30 of the same day as the ASAR and ECV SM acquisition.

2.2   ECV Soil Moisture Observations

The ECV SM product (Liu et al., 2011, 2012; Wagner et al., 2012) was initiated within the water cycle multimission observation strategy project (see http://wacmos.itc.nl/) and is being continued and refined in the context of the ESA climate change initiative (CCI) program (http://www.esa-soilmoisture-cci.org/). It currently provides daily global surface soil moisture values ($m^3 \, m^{-3}$) covering the period 1978–2014 at a 0.25 degree spatial resolution. To achieve this, the ECV SM product combines observations from multiple C-band scatterometers (ERS AMI and MetOp ASCAT) and multifrequency radiometers AMSR, scanning multichannel microwave radiometer (SMMR), special sensor microwave/imager (SSM/I), tropical rainfall measuring mission microwave imager (TRMM-TMI), and WindSat. The Water Retrieval Package developed by TU Wien (Wagner et al., 1999a,b; Naeimi et al., 2009) is used to convert backscatter measurements to soil moisture values and the Land Parameter Retrieval Model (LPRM) developed jointly by VU University Amsterdam and the NASA Goddard Space Flight Center (Owe et al., 2001, 2008; De Jeu and Owe, 2003) is used to convert brightness temperatures to soil moisture, respectively. The ECV SM product has been validated across different regions using in situ, model, and SAR-derived soil moisture data sets in previous studies (e.g., Albergel et al., 2013b; Loew et al., 2013; Dorigo et al., 2015; Pratola et al., 2014; Zeng et al., 2015) where good agreement between the data sets was generally found. For example, Zeng et al. (2015) found the ECV SM product to be highly related to in situ data from two different soil moisture networks at the Tibetan Plateau, with the highest R values (0.70–0.85) and smallest unbiased root mean square difference (*ubRMSD*) values (0.034–0.042 $m^3 \, m^{-3}$) compared to six other satellite-derived soil moisture products (AMSR-E (NASA product), AMSR-E (JAXA product), AMSR-E (LPRM product), AMSR-2, ASCAT, and SMOS). Similarly, Pratola et al. (2014) found strong correlations ($R = 0.72$–0.88) and associated low *ubRMSD* values (0.0–0.06) between ECV SM and SAR-derived soil moisture values across three grassland sites in Ireland.

2.3   ASAR Soil Moisture Observations

The Envisat satellite was launched on Mar. 1, 2002, by ESA and operated until Apr. 8, 2012. The ASAR instrument onboard the satellite operated at C-band (5.3 GHz) and was capable of acquiring data in multiple modes (image, alternating polarization, wave, ScanSAR (wide swath), and ScanSAR (global monitoring)) at various incidence angles and in several polarizations. This study focused on the use of wide swath (WS) mode data rather than global monitoring (GM) mode data due to its higher radiometric accuracy (0.6 dB compared to 1.2 dB) and also due to its higher native spatial resolution (150 m compared to 1 km). WS data have a 405 km swath width and could potentially acquire three to five images a month. Acquisitions between 2005 and 2010 (see Fig. 2) from both ascending and descending orbits were considered in VV polarization. As WS mode data use ScanSAR technology to cover a much larger swath width, effects on the backscatter due to varying incidence angle and distance from the sensor are usually present in the scene. To limit the influence of the large incidence angle range (17–42 degrees) and to ensure intercomparability between the different data scenes, an angular normalization to an incidence angle of 30 degrees was applied to all scenes. The WS data were geometrically and radiometrically calibrated and resampled to a regular grid with a 15 arc-second sampling interval. Consequently, the ASAR WS data were aggregated to 1 km spatial resolution, supporting the comparison with the ECV product and also improving the radiometric accuracy of the satellite data.

There are different approaches to soil moisture estimation using SAR data (Barrett et al., 2009; Petropoulos et al., 2015) and in this study, soil moisture values were retrieved from the ASAR WS acquisitions by applying the TU Wien change detection algorithm (Wagner et al., 1999a,b). This technique was originally developed for ERS SCAT and ASCAT data but has been successfully adapted to both ASAR WS and GM data (e.g., Pathe et al., 2009; Mladenova et al., 2010; Doubková et al., 2012; Peters et al., 2012; Dostálová et al., 2014; Zribi et al., 2014). The TU Wien change detection approach derives relative changes in surface soil moisture and indirectly accounts for surface roughness and vegetation by assuming changes in these parameters will generally occur at longer temporal scales than soil moisture changes. It is based on the assumption of a linear relationship between the surface soil moisture content and the backscatter coefficient and provides soil moisture values expressed in terms of the degree of saturation, whereby variations in a time series of soil moisture values are adjusted between the historically lowest (0% for dry soil) and highest (100% for saturated soil) values.

The retrieved soil moisture values were masked using the Corine Land Cover Map 2006 (EEA, 1995) to exclude pixels representing areas where the soil moisture values were unreliable (e.g., urban, water bodies, and snow and ice). Furthermore, and as demonstrated in Wagner et al. (2008), the temporal stability of soil moisture fields gives rise to an associated temporal stability in the backscatter signal. Strong correlations between local and regional backscatter is usually a good indicator of high sensitivity to soil moisture dynamics at the local scale (similarly, if the signal observed at the local scale correlates with the coarse-scale measurement, then the local-scale measurement is sensitive to the dynamics at the coarse scale). Areas where there are weak

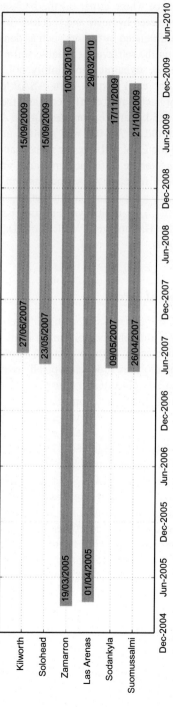

FIG. 2 Temporal interval of study period for each of the study sites used in the data set comparison.

correlations are indicative of where either (a) the backscatter response to soil moisture dynamics is dominated by noise and speckle; (b) the backscatter characteristics are adversely influenced by factors such as dense vegetation or complex topography, inhibiting the retrieval of reliable soil moisture values; or (c) the local-scale soil moisture dynamics are simply not representative of the coarse-scale dynamics. For each 1 km × 1 km ASAR pixel, the correlation between the time series of backscatter coefficients and the average of the backscattering over a 25 km × 25 km area encompassing the ASAR pixel was evaluated. A minimum threshold of $R = 0.3$ was applied and only those ASAR acquisitions that covered the ECV cell size for more than 50% of the available pixels were selected. The average of the soil moisture values retrieved within the corresponding ECV cell for each ASAR acquisition is considered for the data set intercomparison. In a final step, soil moisture values were converted to volumetric units ($m^3 \, m^{-3}$) by multiplying by the associated soil porosity values, obtained from the Harmonized World Soil Database (FAO/IIASA/ ISRIC/ISS-CAS/JRC, 2009).

2.4 Characterization of Errors

In addition to large-scale differences, systematic differences between satellite-derived and in situ soil moisture observations make it difficult to have absolute agreement between the time series of these data sets (Brocca et al., 2013). As a result, satellite-derived soil moisture products typically require scaling and/or filtering before being compared to in situ soil moisture measurements. The cumulative distribution function matching approach has been demonstrated with various satellite-derived soil moisture data sets (e.g., SMMR (Reichle and Koster, 2004), TRMM-TMI (Drusch et al., 2005)), and combined SMMR, SSM/I, TMI, and AMSR-E data sets (Liu et al., 2009) and was used in this study to adjust the satellite-derived soil moisture values to the same range and distribution as the in situ measurements. Only ECV SM and in situ soil moisture data corresponding to the ASAR WS acquisition dates have been considered in this study.

Different metrics are commonly used for the validation of soil moisture products. In this study, the correlation coefficient (R) was calculated to provide details of the temporal agreement between the different soil moisture data sets. In addition, the *ubRMSD* was calculated instead of the conventional *RMSD* in order to correct for biases in the mean of the satellite-derived data sets (Albergel et al., 2013a), and is given by

$$ubRMSD = \sqrt{\frac{1}{N} \sum_{N=1}^{N} \left\{ \left[(sat_n - \overline{sat_n}) - \left(insitu_n - \overline{insitu_n} \right) \right]^2 \right\}} \qquad (1)$$

where sat_n and $insitu_n$ represent the satellite-derived and in situ soil moisture measurements, respectively, and the overbars represent averaged quantities.

In addition to the whole time series analysis, a seasonal comparison was also carried out to help evaluate the performance of ASAR and ECV SM soil moisture products in capturing the soil moisture dynamics. The time series data were analyzed by season: winter (Dec., Jan., Feb.), spring (Mar., Apr., May), summer (Jun., Jul., Aug.), and autumn (Sep., Oct., Nov.) and analyzed with respect to the in situ measurements, which were taken as a reference.

3   RESULTS AND DISCUSSION

3.1   Time Series Temporal Analysis

The temporal evolution of the ECV SM-scaled surface soil moisture estimates compared to the ASAR-scaled estimates and in situ observations from the Irish, Spanish, and Finnish sites are displayed in Figs. 3–5. In general, the satellite-derived data sets and the in situ observations were in good agreement. The Irish sites, to a large extent, displayed the same temporal pattern with the highest soil moisture levels observed between Dec. and Mar., although the dynamic ranges differ between the sites. Soils are generally dry between Mar. and Jul. as a result of increased surface temperature, evapotranspiration, and decreasing precipitation. The characteristically high soil moisture after snowmelt can be observed at both the Finnish sites (Fig. 5) and is more pronounced for the Sodankyla site. The Suomussalmi site generally displayed higher soil moisture variability during the summer months, compared to Sodankyla.

In Fig. 3, the ability of the ECV SM and ASAR data to represent the soil moisture variability at Kilworth and Solohead is well represented. From

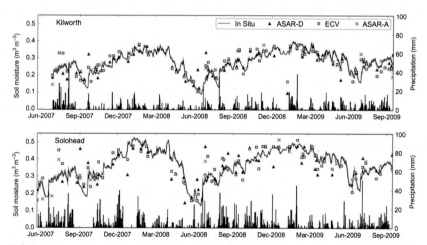

FIG. 3   Time series of in situ *(continuous black line)*, advanced synthetic aperture radar (ASAR) (ascending *(green cross)* and descending *(blue triangle)* passes), and essential climate variable (ECV) *(red square)* soil moisture values for Kilworth and Solohead sites in Ireland. The daily accumulated precipitation is represented by vertical bars along the *x*-axis.

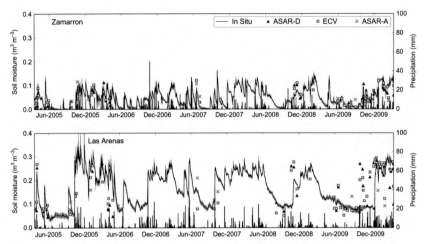

FIG. 4   Time series of in situ *(continuous black line)*, advanced synthetic aperture radar (ASAR) (ascending *(green cross)* and descending *(blue triangle)* passes), and essential climate variable (ECV) *(red square)* soil moisture values for Zamarron and Las Arenas sites in Spain. The daily accumulated precipitation is represented by vertical bars along the x-axis.

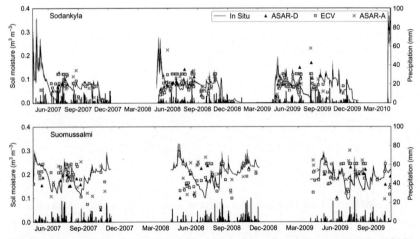

FIG. 5   Time series of in situ *(continuous black line)*, advanced synthetic aperture radar (ASAR) (ascending *(green cross)* and descending *(blue triangle)* passes), and essential climate variable (ECV) *(red square)* soil moisture values for Sodankyla and Suomussalmi sites in Finland. The daily accumulated precipitation is represented by vertical bars along the x-axis. There are missing values between the months of Dec. and Apr. as the soil is frozen during this time period.

Table 2, it can be observed that there is essentially no difference between the correlation coefficients for the ascending orbit ($R = 0.71$), descending orbit ($R = 0.71$) or combined ascending and descending orbit ($R = 0.70$) ASAR acquisitions when compared to the in situ measurements at the Kilworth site. When ascending and descending data were available for the same day, the average soil

TABLE 2 Comparison (Correlation Coefficient—*R*, and Unbiased Root Mean Square Difference—*ubRMSD*) Between ASAR-Derived Soil Moisture and In Situ Soil Moisture, ECV and In Situ, and ASAR and ECV at the Six Study Sites

		n	ASAR vs. In Situ			ECV vs. In Situ			ASAR vs. ECV		
			R	p	ubRMSD	R	p	ubRMSD	R	p	ubRMSD
Kilworth	AD	85	0.70	<0.001	0.05	0.77	<0.001	0.04	0.81	<0.001	0.04
	A	17	0.71	0.001	0.05	0.63	0.007	0.05	0.73	<0.001	0.05
	D	68	0.71	<0.001	0.05	0.79	<0.001	0.04	0.83	<0.001	0.04
Solohead	AD	76	0.71	<0.001	0.07	0.85	<0.001	0.05	0.77	<0.001	0.06
	A	16	0.87	<0.001	0.04	0.84	<0.001	0.05	0.92	<0.001	0.03
	D	60	0.71	<0.001	0.07	0.85	<0.001	0.05	0.77	<0.001	0.06
Zamarron	AD	51	0.71	<0.001	0.03	0.86	<0.001	0.02	0.77	<0.001	0.02
	A	16	0.67	0.004	0.03	0.90	<0.001	0.02	0.79	<0.001	0.02
	D	32	0.63	<0.001	0.03	0.87	<0.001	0.02	0.71	<0.001	0.02
Las Arenas	AD	60	0.56	<0.001	0.07	0.64	<0.001	0.07	0.80	<0.001	0.05
	A	19	0.63	0.004	0.06	0.72	<0.001	0.05	0.69	0.001	0.05
	D	41	0.53	<0.001	0.08	0.66	<0.001	0.07	0.82	<0.001	0.05

Continued

TABLE 2 Comparison (Correlation Coefficient—*R*, and Unbiased Root Mean Square Difference—*ubRMSD*) Between ASAR-Derived Soil Moisture and In Situ Soil Moisture, ECV and In Situ, and ASAR and ECV at the Six Study Sites—cont'd

		ASAR vs. In Situ			ECV vs. In Situ			ASAR vs. ECV		
	n	*R*	*p*	*ubRMSD*	*R*	*p*	*ubRMSD*	*R*	*p*	*ubRMSD*
Sodankyla										
AD	121	−0.18	0.048	0.05	−0.24	0.008	0.05	0.52	<0.001	0.03
A	73	−0.17	0.150	0.05	−0.24	0.041	0.05	0.55	<0.001	0.03
D	48	−0.19	0.195	0.05	−0.25	0.086	0.05	0.47	<0.001	0.03
Suomussalmi										
AD	101	−0.09	0.370	0.07	0.14	0.162	0.06	0.41	<0.001	0.05
A	62	−0.15	0.244	0.07	0.30	0.018	0.05	0.28	0.027	0.05
D	39	0.05	0.762	0.07	−0.11	0.505	0.07	0.95	<0.001	0.04

"A" represents ASAR ascending acquisitions, "D" represents ASAR descending acquisitions, and "AD" represents combined ASAR ascending and descending acquisitions (daily averages were calculated if there were ascending and descending passes on the same day).

moisture value was taken. Although the number of ascending acquisitions is far fewer than descending, the results were still statistically significant ($p < 0.001$) and the *ubRMSD* values remained low (0.05). The correlations between the ECV SM and in situ measurements were also high (R between 0.63 and 0.79) with strong correlations between the ASAR and ECV SM soil moisture values (R between 0.73 and 0.83). There is a larger difference at the Solohead site between the correlations for ascending ($R = 0.87$) and descending ASAR acquisitions ($R = 0.71$) with the in situ measurements. Solohead is a very wet site and a possible explanation for the difference between both passes could be due to the effects of diurnal solar heating cycles (van der Velde et al., 2012). Additionally, the ECV SM data displayed higher correlations (R between 0.84 and 0.85) with the in situ measurements in comparison to the Kilworth site. As identified in Dorigo et al. (2015), sites that exhibit a pronounced annual soil moisture cycle and where this seasonality is suitably detected by the ECV SM product, usually result in high correlations between the ECV SM and in situ data such as those obtained at the Irish sites.

At the Spanish sites, the soil moisture is a lot less variable at Zamarron, compared to Las Arenas (see Fig. 4). Interestingly, the ASAR and ECV SM data sets displayed a better overall agreement with the in situ measurements at Zamarron, compared to Las Arenas, given the lower soil moisture variability ($R = 0.71$ compared to 0.56 for the ASAR-in situ comparison, and $R = 0.86$ compared to 0.64 for the EC-in situ comparison). As expected, the Zamarron site had a much lower *ubRMSD* (0.02–0.03) compared to the Las Arenas site (0.05–0.08), as the area is much drier and the soil moisture less variable in this region. At Las Arenas, the highest soil moisture values were observed between Dec. and May for most years, with the lowest values occurring between Jun. and Nov. The trend is similar for Zamarron, although the overall range of soil moisture values were much lower. Similar to the Irish sites, the correlation between the ASAR and ECV SM soil moisture values was high for the Zamarron and Las Arenas sites, with $R = 0.77$ and 0.80, respectively. However, the correlations for ASAR against the in situ measurements were generally weaker than for the ECV SM data set, despite the scale gap between measurements being less pronounced. Similar findings were observed by Pathe et al. (2009) when comparing ASAR-GM derived soil moisture to ERS scatterometer and in situ soil moisture measurements (Oklahoma MESONET). Zribi et al. (2014) also found similar results for ASAR WS derived soil moisture (resampled to 1 km resolution) and ERS/ASCAT products covering a semiarid region in central Tunisia. Moreover, both of these studies demonstrated a strong correlation between coarse-scale soil moisture products and the finer resolution ASAR-derived soil moisture products ($R > 0.8$). This is an indication of the representativeness errors in the in situ sites, which were very likely to be larger than those of the ASAR data and supports the argument of using finer spatial resolution SAR data as another source for intercomparison.

The weakest correlations across all data set comparisons occurred at the Finnish sites (Fig. 5). There are several possible reasons for this poor performance. First, the area contained within these ECV cells is dominated by forests. Dense vegetation cover attenuates the backscattered signal and decreases the sensitivity of the radar backscatter to soil moisture (Ulaby et al., 1986). Second, the GTK Suomussalmi in situ soil moisture sensor is buried at a depth of 0.1 m, which may be considered beyond the depth at which the satellite is sensitive to surface soil moisture (generally only sensitive to the top few centimeters of the soil surface at C-band). Despite this, the correlations were no better for the FMI Sodankyla in situ sensor, which is buried at a depth of 0.05 m. Similar low correlations have been observed by Paulik et al. (2014) using ASCAT data and by Al-Yaari et al. (2014) for the northern latitudes. Conversely, Griesfeller et al. (2016) found relatively high correlations at stations (with soil sensors buried at a depth of 0.1 m) located in Norway (ranging in latitude from 58°45′ to 69°01′) for both ASCAT (R ranging from 0.68 to 0.72) and AMSR-E (R ranging from 0.52 to 0.64) data. Despite this, the ASAR-ECV SM comparison at both Suomussalmi and Sodankyla displayed moderate agreement with R values of 0.41 and 0.51, respectively, indicating the two independently satellite-derived data sets were in reasonable agreement with one another. A further possible explanation for the lower correlation values, in comparison to the Irish and Spanish sites, may be as a result of the increased presence of surface water bodies. Most soil moisture validation studies occur around the central latitudes and there is a need for further studies concentrating on the northern high latitudes, as indicated by our results and those of Al-Yaari et al. (2014) and Griesfeller et al. (2016), and in light of the rapid warming occurring in these regions in recent decades (Xu et al., 2013).

3.2 Seasonal Analysis

Fig. 6 displays four Taylor diagrams, illustrating the statistics from the comparison between ASAR and in situ, and ECV and in situ soil moisture measurements for the six different study sites on a seasonal basis. There was less agreement during winter acquisitions, where negative correlations were not displayed, and hence there were fewer symbols. Additionally, as the soil at the Finnish sites was frozen during the winter months, no values were included in the analysis. The ECV (red) and ASAR (blue) soil moisture values were less variable in spring, as demonstrated by their proximity to the dashed arc, representing a normalized standard deviation value (SDV) of one. The ECV values for Kilworth, Zamarron, and Las Arenas exhibited less variability than the in situ measurements (SDV < 1). However, there was generally no tendency for the stations located in Ireland, Spain, and Finland, as symbols were present on either side of the dashed line for all seasons. Overall, the highest correlation values and lowest SDV values occurred in spring for the Irish and Spanish sites. This would suggest that the ECV SM and ASAR product have a reduced capability for capturing the driest and wettest soil conditions at these sites, occurring during summer and winter, respectively. The grassland vegetation also reaches

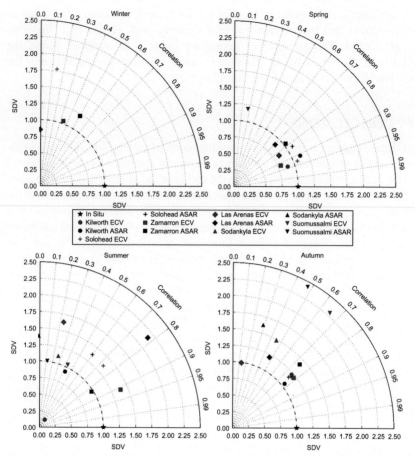

FIG. 6 Taylor diagram illustrating the seasonal comparison metrics between advanced synthetic aperture radar (ASAR) and in situ soil moisture *(blue)*, and essential climate variable (ECV) and in situ soil moisture *(red)* for the six study sites. The standard deviation value (SDV) represents the normalized standard deviation and is calculated as the ratio between satellite and in situ standard deviations.

maximum growth during the summer months which may reduce the quality of the soil moisture retrievals. In contrast, the Zamarron ECV SM and ASAR values displayed high correlations during summer and autumn, while the results for Las Arenas were more variable.

The poorest agreement generally occurred at the Finnish sites across spring and summer but was much improved during autumn. Nonetheless, they still represented weak to moderate correlations, where the ECV SM product outperformed the ASAR product. These poor results were likely due to the dense forest cover in these areas and also the reduced capability of the ASAR soil moisture retrieval algorithm for data acquired above 60 degree latitude. The challenges for deriving soil moisture in high latitudes has been documented by Bartsch et al. (2011) and further explored by Gouttevin et al. (2013) and Högström et al. (2014). The presence of permanent open water surfaces within the ECV cell size may be an additional source of disagreement between the satellite-derived and in situ soil moisture values.

4 CONCLUSIONS

An intercomparison study of two satellite-derived surface soil moisture products with in situ measurements in three European countries has been presented in this chapter. The study focused on the ability of the satellite soil moisture data sets to capture the same relative temporal behavior and this was compared to the in situ soil moisture observations as a reference. The study demonstrated that the coarse-scale ECV SM product was representative of the temporal soil moisture variations observed through finer-scale ASAR-derived and in situ soil moisture observations at the selected study sites. Strong correlations were observed over humid (Irish) and semiarid (Spanish) sites, while weak correlations were observed over the boreal (Finnish) sites. Given the current large data volumes being generated by Sentinel-1A, and the significant increase when Sentinel-1B is launched in 2016, soil moisture change detection techniques such as the TU Wien approach adapted for ASAR data are likely to be applicable and improved for Sentinel-1 data (Hornacek et al., 2012). The benefits of using SAR data with a higher radiometric resolution has been demonstrated by Dostálová et al. (2014) and this new data set could be used not only as another validation source for coarse-scale soil moisture products but also for applications that require finer spatial resolution soil moisture data such as hydrological or runoff modeling.

The NASA SMAP mission (Entekhabi et al. 2010), launched on January 31st 2015 and a successor to the canceled Hydrosphere State mission (Entekhabi et al., 2004), was intended to integrate L-band radar and radiometer measurements from a single platform to provide high-resolution and high-accuracy global maps of surface soil moisture every 2–3 days and thereby overcome some of the limitations of using individual active or passive MW observations to determine soil moisture content. However the failure of the radar in July 2015 represents a significant setback and the mission may now look towards incorporating C-band Sentinel-1a radar data as a replacement. Further improvements in the ECV SM product such as eliminating the use of any ancillary data (eg, soil texture maps (de Jeu et al., 2014) or rescaling to GLDAS Noah land surface model) within the retrieval routine are highly desired by the scientific community (Loew et al., 2013). Incorporating additional sensors and improving their intercalibration will also lead to an improved product that will further strengthen our knowledge on observed soil moisture dynamics.

ACKNOWLEDGMENTS

The authors gratefully acknowledge the ESA CCI Soil Moisture project for supporting this work (ESRIN Contract No. 4000104814/11/I-NB) and Wolfgang Wagner and Wouter Dorigo for their guidance. We would also like to thank Ger Kiely for providing access to the Irish in situ soil moisture data sets. The authors would also like to thank the anonymous reviewer and George Petropoulos for their helpful suggestions and comments.

REFERENCES

Albergel, C., et al., 2012. Evaluation of remotely sensed and modelled soil moisture products using global ground-based in situ observations. Remote Sens. Environ. 118, 215–226.

Albergel, C., Brocca, L., Wagner, W., de Rosnay, P., Calvet, J.-C., 2013a. Selection of performance metrics for global soil moisture products: the case of ASCAT product. In: Petropoulos, G.P. (Ed.), Remote Sensing of Energy Fluxes and Soil Moisture Content. CRC Press, Boca Raton, FL, pp. 427–444.

Albergel, C., et al., 2013b. Monitoring multi-decadal satellite earth observation of soil moisture products through land surface reanalyses. Remote Sens. Environ. 138, 77–89.

Albergel, C., et al., 2013c. Skill and global trend analysis of soil moisture from reanalyses and microwave remote sensing. J. Hydrometeorol. 14 (4), 1259–1277.

Al-Yaari, A., et al., 2014. Global-scale comparison of passive (SMOS) and active (ASCAT) satellite based microwave soil moisture retrievals with soil moisture simulations (MERRA-Land). Remote Sens. Environ. 152, 614–626.

Barrett, B., Petropoulos, G.P., 2013. Satellite remote sensing of surface soil moisture. In: Petropoulos, G.P. (Ed.), Remote Sensing of Energy Fluxes and Soil Moisture Content. CRC Press, Boca Raton, FL, pp. 85–120.

Barrett, B., Dwyer, E., Whelan, P., 2009. Soil moisture retrieval from active spaceborne microwave observations: an evaluation of current techniques. Remote Sens. 1 (3), 210–242.

Bartalis, Z., et al., 2007. Initial soil moisture retrievals from the METOP-A Advanced Scatterometer (ASCAT). Geophys. Res. Lett. 34 (20), L20401.

Bartsch, A., Sabel, D., Wagner, W., Park, S.E., 2011. Considerations for derivation and use of soil moisture data from active microwave satellites at high latitudes. In: Proceedings of IEEE International Geoscience and Remote Sensing Symposium (IGARSS 2011), Vancouver, British Columbia, Canada, 24–29 July.

Bolten, J., Crow, W., 2012. Improved prediction of quasi-global vegetation conditions using remotely-sensed surface soil moisture. Geophys. Res. Lett. 39(19).

Brocca, L., Melone, F., Moramarco, T., Wagner, W., Albergel, C., 2013. Scaling and filtering approaches for the use of satellite soil moisture observations. In: Petropoulos, G.P. (Ed.), Remote Sensing of Energy Fluxes and Soil Moisture Content. CRC Press, Boca Raton, FL, pp. 411–442.

Ceballos, A., Scipal, K., Wagner, W., Martínez-Fernández, J., 2005. Validation of ERS scatterometer-derived soil moisture data in the central part of the Duero Basin, Spain. Hydrol. Process. 19 (8), 1549–1566.

Crow, W.T., et al., 2012. Upscaling sparse ground-based soil moisture observations for the validation of coarse-resolution satellite soil moisture products. Rev. Geophys. 50 (2), RG2002.

De Jeu, R., Owe, M., 2003. Further validation of a new methodology for surface moisture and vegetation optical depth retrieval. Int. J. Remote Sens. 24 (22), 4559–4578.

De Jeu, R.A., Holmes, T.R., Parinussa, R.M., Owe, M., 2014. A spatially coherent global soil moisture product with improved temporal resolution. J. Hydrol. 516, 284–296.

Desnos, Y., et al., 2000. ASAR—Envisat's advanced synthetic aperture radar. ESA Bull. 102, 91–100.

Dorigo, W., et al., 2011. The international soil moisture network: a data hosting facility for global in situ soil moisture measurements. Hydrol. Earth Syst. Sci. 15 (5), 1675–1698.

Dorigo, W., et al., 2013. Global automated quality control of in situ soil moisture data from the International Soil Moisture Network. Vadose Zone J. 12(3).

Dorigo, W., et al., 2015. Evaluation of the ESA CCI soil moisture product using ground-based observations. Remote Sens. Environ. 162, 380–395.

Dostálová, A., Doubková, M., Sabel, D., Bauer-Marschallinger, B., Wagner, W., 2014. Seven years of advanced synthetic aperture radar (ASAR) global monitoring (GM) of surface soil moisture over Africa. Remote Sens. 6 (8), 7683–7707.

Doubková, M., Van Dijk, A.I., Sabel, D., Wagner, W., Blöschl, G., 2012. Evaluation of the predicted error of the soil moisture retrieval from C-band SAR by comparison against modelled soil moisture estimates over Australia. Remote Sens. Environ. 120, 188–196.

Drusch, M., Wood, E., Gao, H., 2005. Observation operators for the direct assimilation of TRMM microwave imager retrieved soil moisture. Geophys. Res. Lett. 32 (15), L15403.

Entekhabi, D., et al., 2004. The hydrosphere State (hydros) Satellite mission: an Earth system pathfinder for global mapping of soil moisture and land freeze/thaw. IEEE Trans. Geosci. Remote Sens. 42 (10), 2184–2195.

Entekhabi, D., et al., 2010. The soil moisture active passive (SMAP) mission. Proc. IEEE 98 (5), 704–716.

European Environment Agency (EEA), 1995. CORINE Land Cover Project. Commission of the European Communities.

FAO/IIASA/ISRIC/ISS-CAS/JRC, 2009. Harmonized World Soil Database (version 1.1). FAO, Rome, Italy and IIASA, Laxenburg, Austria. http://www.fao.org/docrep/018/aq361e/aq361e.pdf (accessed 01.07.15).

Gouttevin, I., Bartsch, A., Krinner, G., Naeimi, V., 2013. A comparison between remotely-sensed and modelled surface soil moisture (and frozen status) at high latitudes. Hydrol. Earth Syst. Sci. Discuss. 10, 11241–11291.

Griesfeller, A., et al., 2016. Evaluation of satellite soil moisture products over Norway using ground-based observations. Int. J. Appl. Earth Obs. Geoinf. 45, 155–164.

Gruber, A., Dorigo, W., Zwieback, S., Xaver, A., Wagner, W., 2013. Characterizing coarse-scale representativeness of in situ soil moisture measurements from the international soil moisture network. Vadose Zone J. 12(2).

Högström, E., Trofaier, A.M., Gouttevin, I., Bartsch, A., 2014. Assessing seasonal backscatter variations with respect to uncertainties in soil moisture retrieval in Siberian Tundra regions. Remote Sens. 6 (9), 8718–8738.

Hornacek, M., et al., 2012. Potential for high resolution systematic global surface soil moisture retrieval via change detection using Sentinel-1. IEEE J. Sel. Top. Appl. Earth Obs. Remote Sens. 5 (4), 1303–1311.

Kerr, Y.H., et al., 2012. The SMOS soil moisture retrieval algorithm. IEEE Trans. Geosci. Remote Sens. 50 (5), 1384–1403.

Klein Tank, A., et al., 2002. Daily dataset of 20th century surface air temperature and precipitation series for the European Climate Assessment. Int. J. Climatol. 22 (12), 1441–1453.

Koster, R.D., et al., 2004. Regions of strong coupling between soil moisture and precipitation. Science 305 (5687), 1138–1140.

Li, L., et al., 2010. WindSat global soil moisture retrieval and validation. IEEE Trans. Geosci. Remote Sens. 48 (5), 2224–2241.

Liu, Y.Y., van Dijk, A.I., de Jeu, R.A., Holmes, T.R., 2009. An analysis of spatiotemporal variations of soil and vegetation moisture from a 29-year satellite-derived data set over mainland Australia. Water Resour. Res. 45(7).

Liu, Y., et al., 2011. Developing an improved soil moisture dataset by blending passive and active microwave satellite-based retrievals. Hydrol. Earth Syst. Sci. 15 (2), 425–436.

Liu, Y., et al., 2012. Trend-preserving blending of passive and active microwave soil moisture retrievals. Remote Sens. Environ. 123, 280–297.

Loew, A., Stacke, T., Dorigo, W., Jeu, R.d., Hagemann, S., 2013. Potential and limitations of multidecadal satellite soil moisture observations for selected climate model evaluation studies. Hydrol. Earth Syst. Sci. 17 (9), 3523–3542.

Martínez-Fernández, J., Ceballos, A., 2005. Mean soil moisture estimation using temporal stability analysis. J. Hydrol. 312 (1), 28–38.

Mecklenburg, S., et al., 2012. ESA's soil moisture and ocean salinity mission: mission performance and operations. IEEE Trans. Geosci. Remote Sens. 50 (5), 1354–1366.

Mladenova, I., et al., 2010. Validation of the ASAR global monitoring mode soil moisture product using the NAFE'05 data set. IEEE Trans. Geosci. Remote Sens. 48 (6), 2498–2508.

Naeimi, V., Scipal, K., Bartalis, Z., Hasenauer, S., Wagner, W., 2009. An improved soil moisture retrieval algorithm for ERS and METOP scatterometer observations. IEEE Trans. Geosci. Remote Sens. 47 (7), 1999–2013.

Njoku, E.G., Jackson, T.J., Lakshmi, V., Chan, T.K., Nghiem, S.V., 2003. Soil moisture retrieval from AMSR-E. IEEE Trans. Geosci. Remote Sens. 41 (2), 215–229.

Ochsner, T.E., et al., 2013. State of the art in large-scale soil moisture monitoring. Soil Sci. Soc. Am. J. 77 (6), 1888–1919.

Owe, M., de Jeu, R., Walker, J., 2001. A methodology for surface soil moisture and vegetation optical depth retrieval using the microwave polarization difference index. IEEE Trans. Geosci. Remote Sens. 39 (8), 1643–1654.

Owe, M., de Jeu, R., Holmes, T., 2008. Multisensor historical climatology of satellite-derived global land surface moisture. J. Geophys. Res. 113 (F1), F01002.

Pathe, C., Wagner, W., Sabel, D., Doubkova, M., Basara, J.B., 2009. Using ENVISAT ASAR global mode data for surface soil moisture retrieval over Oklahoma, USA. IEEE Trans. Geosci. Remote Sens. 47 (2), 468–480.

Paulik, C., Dorigo, W., Wagner, W., Kidd, R., 2014. Validation of the ASCAT Soil Water Index using in situ data from the International Soil Moisture Network. Int. J. Appl. Earth Obs. Geoinf. 30, 1–8.

Peters, J., Lievens, H., Baets, B.D., Verhoest, N., 2012. Accounting for seasonality in a soil moisture change detection algorithm for ASAR Wide Swath time series. Hydrol. Earth Syst. Sci. 16 (3), 773–786.

Petropoulos, G.P., Ireland, G., Barrett, B., 2015. Surface soil moisture retrievals from remote sensing: current status, products & future trends. Phys. Chem. Earth A/B/C 83–84, 36–56.

Pratola, P., Barrett, B., Gruber, A., Kiely, G., Dwyer, N., 2014. Evaluation of a global soil moisture product from finer spatial resolution SAR data and ground measurements at Irish sites. Remote Sens. 6, 8190–8219.

Reichle, R.H., Koster, R.D., 2004. Bias reduction in short records of satellite soil moisture. Geophys. Res. Lett. 31 (19), L19501.

Robock, A., et al., 2000. The global soil moisture data bank. Bull. Am. Meteorol. Soc. 81 (6), 1281–1299.

Sanchez, N., Martinez-Fernandez, J., Scaini, A., Perez-Gutierrez, C., 2012. Validation of the SMOS L2 soil moisture data in the REMEDHUS network (Spain). IEEE Trans. Geosci. Remote Sens. 50 (5), 1602–1611.

Schulte, R., Diamond, J., Finkele, K., Holden, N., Brereton, A., 2005. Predicting the soil moisture conditions of Irish grasslands. Irish J. Agric. Food Res. 44 (1), 95–110.

Seneviratne, S.I., et al., 2010. Investigating soil moisture-climate interactions in a changing climate: a review. Earth Sci. Rev. 99 (3), 125–161.

Ulaby, F.T., Moore, R.K., Fung, A.K., 1986. Microwave Remote Sensing: Active and Passive. Volume Scattering and Emission Theory, Advanced Systems and Applications, vol.III. Artech House, Dedham, MA.

van der Velde, R., et al., 2012. Soil moisture mapping over the central part of the Tibetan Plateau using a series of ASAR WS images. Remote Sens. Environ. 120, 175–187.

van Doninck, J., Peters, J., Lievens, H., de Baets, B., Verhoest, N., 2012. Accounting for seasonality in a soil moisture change detection algorithm for ASAR Wide Swath time series. Hydrol. Earth Syst. Sci. 16, 773–786.

Wagner, W., Lemoine, G., Borgeaud, M., Rott, H., 1999a. A study of vegetation cover effects on ERS scatterometer data. IEEE Trans. Geosci. Remote Sens. 37 (2), 938–948.

Wagner, W., Lemoine, G., Rott, H., 1999b. A method for estimating soil moisture from ERS scatterometer and soil data. Remote Sens. Environ. 70 (2), 191–207.

Wagner, W., et al., 2003. Evaluation of the agreement between the first global remotely sensed soil moisture data with model and precipitation data. J. Geophys. Res. 108 (D19), 4611.

Wagner, W., et al., 2008. Temporal stability of soil moisture and radar backscatter observed by the advanced Synthetic aperture radar (ASAR). Sensors 8 (2), 1174–1197.

Wagner, W., et al., 2012. Fusion of active and passive microwave observations to create an essential climate variable data record on soil moisture. In: XXII ISPRS Congress, Melbourne, Australia, pp. 315–321.

Western, A.W., Blöschl, G., 1999. On the spatial scaling of soil moisture. J. Hydrol. 217 (3), 203–224.

Xu, L., et al., 2013. Temperature and vegetation seasonality diminishment over northern lands. Nat. Clim. Chang. 3 (6), 581–586.

Zeng, J., et al., 2015. Evaluation of remotely sensed and reanalysis soil moisture products over the Tibetan Plateau using in-situ observations. Remote Sens. Environ. 163, 91–110.

Zribi, M., et al., 2014. Soil moisture mapping in a semiarid region, based on ASAR/Wide Swath satellite data. Water Resour. Res. 50 (2), 823–835.

Section IV

Advanced Applications of Soil Moisture

Chapter 12

Use of Satellite Soil Moisture Products for the Operational Mitigation of Landslides Risk in Central Italy

L. Brocca*, L. Ciabatta*, T. Moramarco*, F. Ponziani[†], N. Berni[†] and W. Wagner[‡]
*National Research Council, Perugia, Italy, †Civil Protection Centre, Foligno, Italy
‡Vienna University of Technology, Vienna, Austria

1 INTRODUCTION

Shallow landslides are one of the most common and dangerous natural hazards, causing fatalities and high economic damage worldwide, and Italy is one of the earth locations most prone to landslide risk (Guzzetti et al., 2005). Additionally, climate change is expected to exacerbate the impact of landslides on global environmental trends, mainly due to the expected increase in heavy rainfall (Fischer and Knutti, 2015; Trenberth et al., 2015; Ciabatta et al., in press).

To prevent and mitigate landslide risk, several countries have developed early warning systems (Baum and Godt, 2010), mainly based on the use of intensity-duration rainfall thresholds (e.g., Brunetti et al., 2010; Gariano et al., 2015). Basically, these systems rely on the definition of empirical rainfall amounts needed for triggering landslides and have the great advantage of being easy to implement (requiring only rainfall data) and easy to use for stakeholders. However, rainfall thresholds alone might be not reliable as they do not take into account soil moisture conditions. Indeed, thanks to the recent increased availability of soil moisture measurements from in situ observations, several studies showed that soil moisture plays a key role in landslide triggering (e.g., Godt et al., 2006; Hawke and McConchie, 2011; Bittelli et al., 2012; Lepore et al., 2013; Bordoni et al., 2015). For instance, Hawke and McConchie (2011), by analyzing the relationship between in situ soil moisture, piezometric head, rainfall, and landslide occurrence in New Zealand, observed that the slope failure occurred when maximum soil moisture conditions, though not maximum

Satellite Soil Moisture Retrieval. http://dx.doi.org/10.1016/B978-0-12-803388-3.00012-7

pore-water pressures, were recorded. Bittelli et al. (2012), for a site in Northern Italy, recognized the importance of soil moisture and soil matric suction monitoring for obtaining significant information on the parameterization of slope stability models and data interpretation. Specifically, Bittelli et al. (2012) concluded that oscillations in soil matric suction were the dominant variables determining soil failure. From these studies, it is evident that the in situ monitoring of soil wetness conditions has the potential to largely improve landslide prediction.

In the last decade, the use of remote sensing for soil moisture retrieval has largely advanced, as demonstrated from the two recent missions: Soil Moisture and Ocean Salinity (SMOS) by the European Space Agency (ESA) and Soil Moisture Active and Passive (SMAP) by the National Aeronautics and Space Administration (NASA), specifically dedicated to the measurement of soil moisture from space (Das et al., 2016; Kerr et al., 2016). Moreover, other microwave sensors were found to provide accurate soil moisture estimates (e.g., Brocca et al., 2011) including the Advanced Scatterometer (ASCAT) on board MetOp-A and MetOp-B satellites (Bartalis et al., 2007; Wagner et al., 2013), and the Advanced Microwave Scanning Radiometer (AMSR) for Earth Observing (AMSR-E) and 2 (AMSR2) onboard Aqua and Global Change Observation Mission-Water1 (GCOM-W1) satellites, respectively (De Jeu et al., 2014; Parinussa et al., 2015). Despite these satellite soil moisture products having been freely available on a global scale since 2002 (from AMSR-E), the studies attempting to use them for landslide-related applications is nearly absent. Indeed, to our knowledge, only three pioneering studies by Ray et al. (2010, 2011), with AMSR-E, and Brocca et al. (2012a), with ASCAT, have demonstrated that satellite soil moisture data, notwithstanding their coarse resolution ($\sim$20 km), can provide useful information for the detection and prediction of landslide events. The large contrast between the quality and maturity of satellite soil moisture products and the scarcity of landslide studies clearly suggests that further investigations are necessary.

Since 2012, an early warning system for landslide risk assessment, named PRESSCA, is operating at the Umbria Region Civil Protection Centre (CPC) (http://www.cfumbria.it/) in central Italy. PRESSCA is based on rainfall-soil moisture thresholds (Ponziani et al., 2012, 2013) in which soil moisture conditions are estimated from a soil water balance model (SWBM) calibrated and tested with local soil moisture observations (e.g., Brocca et al., 2008, 2013a, 2014). The system is found to be a useful support for day-by-day decision making of authorities and stakeholders involved in landslide risk management. Moreover, in a very recent assessment of the PRESSCA system by using more than 200 landslide events, Ciabatta et al. (in press) obtained that 86% of landslide events are correctly identified by the operational system.

In this chapter, the recent improvements of the PRESSCA system related to the incorporation of in situ and satellite measurements for enhancing the estimation of soil moisture over the Umbria territory are described. Specifically, the intercomparison of modeled, in situ and satellite soil moisture data is carried out in the period 2007–13 to investigate the spatial-temporal agreement between the different data sets. Moreover, the impact of soil moisture conditions on landslide triggering is highlighted by considering rainfall events characterized by different initial soil moisture conditions. Finally, the future steps to be taken for further improving the reliability of the PRESSCA system through the assimilation of satellite soil moisture data are outlined.

2 PRESSCA EARLY WARNING SYSTEM

PRESSCA (Ponziani et al., 2012, 2013) is an operational early warning system for landslide risk prediction and mitigation (preparedness "nonstructural" measure for risk reduction). Specifically, landslide risk is evaluated by coupling a first dynamic component that provides landslide hazard maps for current and near-future conditions and a second static component that identifies the susceptibility and exposure of the territory to landslides. By coupling the two components, dynamic landslide risk scenarios are built for day-by-day operations of the Umbria Region CPC both for ordinary activities (alert emissions) both for "real-time" monitoring phases, if necessary.

In this study, only the first "hazard component" is considered (see Fig. 1). It is composed of four parts: (1) a meteorological module providing rainfall and air temperature data for current and two-day ahead conditions from real-time hydrometeorological observations and numerical weather prediction modeling; (2) a soil moisture module in which soil moisture is estimated through a SWBM (Brocca et al., 2013a, 2014; the code is freely available at http://dx.doi.org/10.13140/2.1.1460.8323), and also compared with in situ and satellite data; (3) a set of rainfall-soil moisture thresholds evaluated for 24, 36, 48, and 72 h of cumulated rainfall; and (4) semaphoric landslide hazard maps identifying the level of warning spatially distributed over the territory.

More explicitly, observed and forecasted rainfall and air temperature data (input data sets) are used to force SWBM that simulates soil moisture conditions. Successively, rainfall and simulated soil moisture data for each grid point in the Umbria region are computed and compared with the historical rainfall-soil moisture thresholds to predict the occurrence of landslide events. Based on the proximity, or exceeding, to the rainfall thresholds, early warning indicator maps (see Fig. 1) are finally obtained (output data sets). The reader is referred to Ponziani et al. (2012) for details on the procedure for developing the historical rainfall-soil moisture thresholds, to Ponziani et al. (2012) for a detailed description of the PRESSCA system, and to Brocca et al. (2013a, 2014) for the assessment of SWBM in central Italy and Europe (Spain, France, and Luxembourg). As mentioned earlier, Ponziani et al. (2012) demonstrated the significance of soil moisture conditions in the triggering landslides in the study region. Based on this theoretical background, the PRESSCA system was developed, considering rainfall-soil moisture thresholds (instead of rainfall thresholds alone).

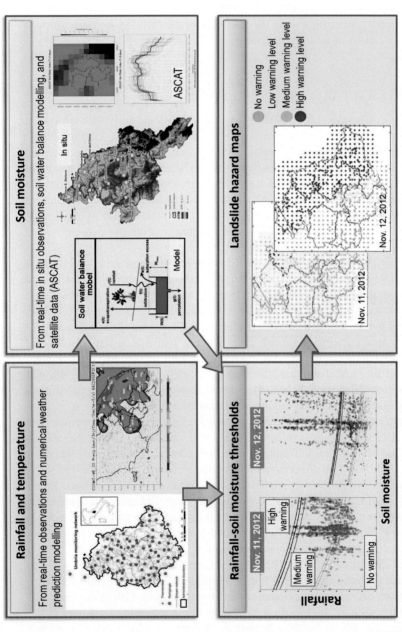

FIG. 1 Scheme of the landslide hazard component of the PRESSCA system. *Top left*: Hydrometeorological monitoring network with the location of the Umbria region and a rainfall map from numerical weather prediction modeling. *Top right*: Soil moisture monitoring system by the soil water balance model, the ground monitoring network, and satellite data (ASCAT). *Bottom left*: Rainfall-soil moisture threshold and landslide hazard assessment for Nov. 11 and 12, 2012. *Bottom right*: Landslide hazard maps with the identification of the different warning level for Nov. 11 and 12, 2012.

The intercomparison between modeled (via SWBM), satellite (from ASCAT, see in the following section), and in situ soil moisture data is here carried out by using as performance scores the Pearson correlation coefficient, R, and the root mean square error (RMSE).

For assigning the parameter values to SWBM, a soil texture map was built ad hoc for the Umbria region (Fig. 2A). The map is obtained by analyzing more than 1600 soil surveys and by coupling this information with the regional geologic settings. The model parameter values are derived from the tabulated values of Rawls et al. (1982) and they are shown in Table 1. The soil thickness is assumed equal to 150 mm and a constant value of 0.84 is considered for the correction coefficient of the potential evapotranspiration (see Brocca et al., 2013a for more details on the model parameterization). The model is applied to a regular grid with spacing of ∼5 km (611 grid points) that covers the whole study area. For each grid point, the corresponding soil textural class is obtained from Fig. 2A and, then, Table 1 is used to assign the corresponding parameter values.

3 CASE STUDY AND DATA SETS

3.1 Study Area and Ground Meteorological Observations

The Umbria region is located in central Italy (∼200 km from Rome) and covers an area of approximately 8450 km^2. The landscape is mainly hilly, with a mountainous area in the eastern sector and a flood plain that stretches along the north-south direction (Fig. 2B). Due to the conformation of the territory, the Umbria region is prone to landslides, and specifically to rainfall-induced landslides. Indeed, more than 500 landslides triggered by rainfall events were recorded and listed in a regional catalog by the Umbria region CPC during the period 1990–2013.

The Umbria region meteorological network currently consists of 90 rain gauges and 77 thermometers distributed throughout the territory (see the upper left box in Fig. 1). The monitoring network provides semihourly data for which a quality-check step is performed to remove anomalous values and to fill any temporal gaps. For this study, data covering the period 2004–13 are used to force SWBM and for the dynamic assessment of the rainfall-soil moisture conditions.

Since the end of 2009, a soil moisture monitoring network was installed by the Umbria Region CPC for floods and landslides monitoring and prediction throughout the territory. Currently, the network consists of 12 frequency domain reflectometry (FDR) stations and 4 time domain reflectometry (TDR) stations with sensors located at three different depths (10, 20, and 40 cm). All stations, except one, are operating in real-time and provide measurements every 30 min. We used here seven stations (listed in Table 2) for which data have shown to be of good quality, consistent over time, and available for multiple years.

3.2 Satellite Soil Moisture Observations

ASCAT is a real-aperture radar instrument successfully launched on board the satellites MetOp-A in 2006 and MetOp-B in 2012, which measures radar

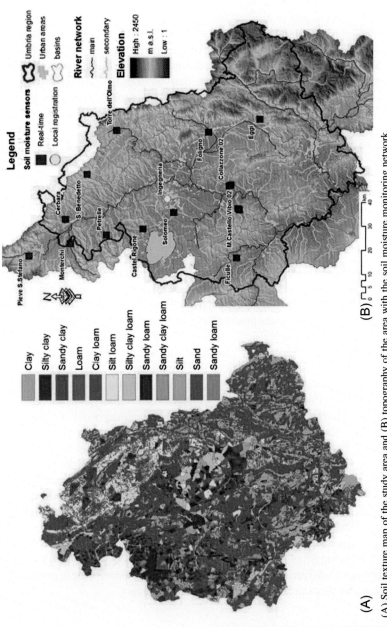

FIG. 2 (A) Soil texture map of the study area and (B) topography of the area with the soil moisture monitoring network.

TABLE 1 Relationship Between Soil Texture Classes and the Parameter
Values Used in the Soil Water Balance Model (SWBM)

Soil Texture	Residual Porosity, θ_r (m³/m³)	Effective Porosity, θ_s (m³/m³)	Bubbling Pressure, ψ (cm)	Pore Size Distribution, λ (−)	Saturated Hydraulic Conductivity, K_s (mm/h)
Sand	0.020	0.417	72.6	0.592	210.0
Loamy sand	0.035	0.401	86.9	0.474	61.1
Sandy loam	0.041	0.412	146.6	0.322	25.9
Loam	0.027	0.434	111.5	0.220	13.2
Silt loam	0.015	0.486	207.6	0.211	6.8
Sandy clay loam	0.068	0.330	280.8	0.250	4.3
Clay loam	0.075	0.390	258.9	0.194	2.3
Silty clay loam	0.040	0.432	325.6	0.151	1.5
Sandy clay	0.109	0.321	291.7	0.168	1.2
Silty clay	0.056	0.423	341.9	0.127	0.9
Clay	0.090	0.385	373.0	0.131	0.6

backscatter at C-band (5.255 GHz) in VV polarization. The spatial resolution of
ASCAT is 25 km (resampled at 12.5 km) and, in central Italy, measurements are
generally obtained at least once a day. The surface soil moisture product is
retrieved from the ASCAT backscatter measurements using a time series-based
change detection approach previously developed for the ERS-1/2 scatterometer
by Wagner et al. (1999). In this approach soil moisture is considered to have
a linear relationship to backscatter in the decibel space, the surface roughness
is assumed to have a constant contribution in time, and the typical yearly
vegetation cycle is modeled as a function of the backscatter-incidence angle

TABLE 2 Comparison of Modeled (SWBM) and Linearly Rescaled Satellite
Data (ASCAT SWI) With and In Situ Soil Moisture Observations

	ASCAT SWI			SWBM		
Station	R	RMSE (m^3/m^3)	N	R	RMSE (m^3/m^3)	N
Ficulle	0.83	0.05	583	0.90	0.05	35,088
Monterchi	0.70	0.02	642	0.93	0.04	35,088
San Benedetto Vecchio	0.79	0.06	644	0.94	0.04	35,088
Pieve Santo Stefano	0.71	0.09	636	0.92	0.06	35,088
Cerbara	0.67	0.07	726	0.88	0.09	35,075
Petrelle	0.69	0.07	646	0.80	0.07	31,088
Torre dell'Olmo	0.78	0.05	620	0.51	0.12	30,387

R, Pearson's correlation coefficient; RMSE, root mean square error; N, sample size.

relationship. The derived surface soil moisture product (corresponding to a depth of 2–3 cm) ranges between 0% (dry) and 100% (wet). Validation studies of the ASCAT soil moisture product assessed its reliability for estimating both in situ and modeled soil moisture observations across different regions in Europe (Brocca et al., 2011), and globally (Alyaari et al., 2014; Leroux et al., 2014; Paulik et al., 2014; Wagner et al., 2014), thus addressing their use for practical applications. In particular, it was found that the ASCAT soil moisture product has some skill even over mountainous and complex regions (Brocca et al., 2013b), thus making particularly feasible its use for landslides studies.

In many applications, including shallow landslide prediction, the knowledge of soil moisture for a very thin surface layer is not sufficient. In this study, the semiempirical approach (also known as exponential filter) proposed by Wagner et al. (1999) is adopted to obtain a root-zone soil moisture product (Soil Wetness Index (SWI)) that depends on a single parameter, T (characteristic time length), representing the time scale of soil moisture variation with depth. The reader is referred to Wagner et al. (1999) for a detailed description of the ASCAT retrieval algorithm and the exponential filter approach while the emerging applications using the ASCAT soil moisture product can be found in Wagner et al. (2013).

4 RESULTS AND DISCUSSION

4.1 Comparison Between Satellite, In Situ, and Modeled Soil Moisture Data

The preliminary step for using satellite soil moisture products in a landslide (or any other) application is to test their reliability and accuracy. For that, we first compare satellite and in situ observations at the seven stations listed in Table 2. For making satellite and in situ measurements comparable, the SWI is applied with the T-value that is obtained by maximizing the correlation between satellite and modeled data (see later). Specifically, T-value is found to be ranging between 3 and 8 days, with a median value equal to 4.8 days. Then, the SWI data, expressed in term of degree of saturation (between 0 and 1), are linearly rescaled to the range of variability of in situ observations, and expressed in volumetric term (m^3/m^3). As shown in Table 2, and accordingly to previous studies (e.g., Brocca et al., 2011), the ASCAT SWI is found to perform satisfactorily in reproducing in situ soil moisture temporal variability in central Italy notwithstanding the spatial scale differences between the measurements. Indeed, as found by many authors (starting from the original study by Vachaud et al. (1985)), soil moisture exhibits scaling properties in space allowing to upscale/downscale soil moisture data to different scales without losing significant information. In other words, coarse-scale satellite measurements can be used for small-scale applications (e.g., landslide studies) obtaining meaningful results. Specifically, the R-values range between 0.67 and 0.83 (RMSE $=0.02$–0.09 m^3/m^3) and the sample size exceeds 580, thus ensuring the robustness of the obtained scores (Wagner et al., 2014).

Successively, the performance of SWBM are analyzed by performing the comparison with the same seven stations as mentioned earlier. The agreement between modeled and in situ stations is highly satisfactory with R-values in the range of 0.51–0.92, slightly higher than those obtained with satellite data. Only for the Torre dell'Olmo station is the performance less good ($R = 0.51$), likely due to some malfunctioning in the in situ station. Fig. 3 shows the comparison between modeled and in situ soil moisture data for the Cerbara station in the period 2009–13. It is evident that SWBM is able to reproduce both the overall seasonal temporal variability of soil moisture, and also the short-time variability due to the frequent rainfall events occurring in the winter period. These results confirm previous studies (e.g., Brocca et al., 2013a,b, 2014) and underline the high reliability of SWBM that can be effectively used for simulating soil moisture in the study area.

However, it is expected that satellite soil moisture data can provide a further benefit as an independent source of soil moisture information that should be merged with that coming from modeling. For testing this hypothesis, the comparison between modeled and satellite observations was also carried out

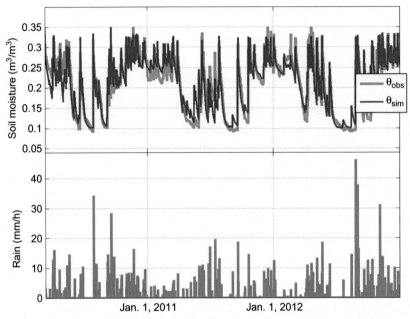

FIG. 3 Comparison (*upper panel*) between modeled, θ_{sim}, and in situ, θ_{obs}, soil moisture data for Cerbara station in the period 2009–13. In the *lower panel* the rainfall time series is shown.

in the 7-year period 2007–13. The map in the *left panel* of Fig. 4 shows the pixel-by-pixel R-value and it is obtained that R is always higher than 0.64 with maximum value of 0.86. The median R-value for the whole area is equal to 0.80. The time series for four randomly selected pixels confirm the good agreement between satellite and modeled data. Only during some winter periods (e.g., Jan.–Feb. 2012), the ASCAT SWI shows a sudden decrease due to soil freezing conditions. Indeed, in these periods the satellite data should be masked out as microwave sensors are only able to measure the liquid water in the soil. Based on these encouraging results, an advanced monitoring system was developed in which ASCAT SWI is compared with both in situ and modeled data in real-time (see Muñoz Sabater et al., 2015). The combined use of the three data sources allows a robust and reliable assessment of soil moisture conditions throughout the Umbria region territory that is fundamental for the prediction of possible occurrence of landslides (see the next paragraph).

4.2 Impact of Soil Moisture Condition on Landslide Hazard

The impact of soil moisture conditions in the prediction of landslide hazard was finally investigated by considering different significant rainfall events (rainfall amount >30 mm/24 h) that occurred in the period 2004–13. For that period meteorological data were used to force SWBM and the temporal evolution of modeled soil moisture data was computed for each grid point. By considering

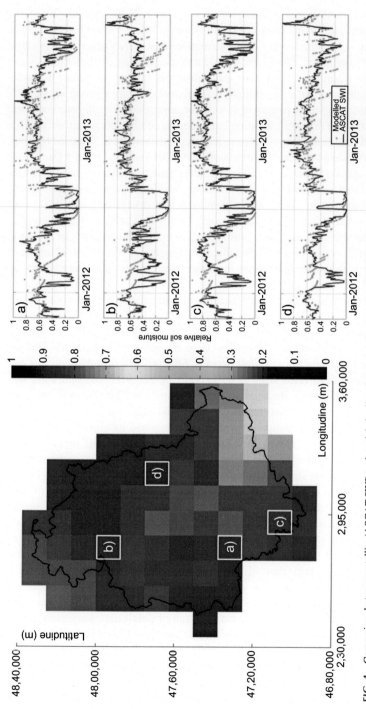

FIG. 4 Comparison between satellite (ASCAT SWI) and modeled soil moisture data for the whole Umbria region. On the *left*, the temporal correlation map between SWI and modeled data is shown. The highlighted pixels identify the locations for which the time series are shown on the *right panels* (*R*: Pearson's correlation coefficient).

the historical rainfall-soil moisture threshold for 24 h (Ponziani et al., 2012), the rainfall threshold for moderate criticality ("orange" alert, second last alert level) is equal to ~100 mm for dry (20% saturation) and to ~50 mm for wet (80% saturation) conditions. Therefore, it is clear that depending on the wetness state, a landslide warning is issued for quite different rainfall amounts. It should be also mentioned that even though soil moisture exhibits a strong seasonality with a distinctive yearly pattern in Mediterranean regions (i.e., dry summers and wet winters), the dry-wet transition occurring in autumn may start also with a 1–2 month time delay from year to year. A typical example is observed by comparing the modeled soil moisture data of 1 Dec, 2010 and 2011. On 1 Dec, 2010, the soil was very wet (saturation at 90%) and a moderate rainfall event (~50 mm/36 h) occurred after 2 days causing flooding on the Tiber River floodplain. Conversely, on 1 Dec, 2011, the soil was very dry (saturation at 10%) with the direct consequence of a severe drought for the first half of 2012. Therefore, on the same date, soil moisture was at the two extremes (very wet in 2010 and very dry in 2011) thus highlighting that assuming dry or wet soil moisture state, according with the season, might introduce large errors in the assessment of actual soil wetness conditions (at least as observed in the Umbria region).

To further underline the significance of soil moisture conditions, Fig. 5 shows the landslide hazard maps provided by the PRESSCA system for two rainfall events that occurred on Nov. 28, 2010, and Oct. 12, 2012. In the first event (Nov. 2010), several landslides were triggered throughout the Umbria region due to a significant rainfall event (~60 mm/24 h) occurring on wet soil moisture conditions (saturation at 85%). On Oct. 12, 2012, the saturation was lower (50% saturation), the total amount of rainfall in 24 h was ~50 mm, and landslide events were not recorded. The landslide warning maps of the PRESSCA system are compared with maps computed by considering a fixed rainfall threshold of 60 mm in 24 h; that is, not taking soil moisture into account. We underline that this is the case in the widely adopted early warning systems that employ intensity-duration rainfall thresholds. As shown in Fig. 5, on the Nov. 2010 event, the PRESSCA system correctly identified the critical conditions for landslide hazard with the exceedance of the threshold for nearly half of the regional territory. The constant rainfall approach also identified the critical conditions, even though probably underestimated due to the very wet conditions at the beginning of the rainfall event. On the Oct. 2012 event, only the southwestern part of the region exceeded the threshold for the PRESSCA system while most of the southern part of the region exceeded the fixed thresholds of 60 mm in 24 h. Therefore, a large overestimation of landslide hazard would have been identified if only rainfall data were considered. Moreover, comparing the two landslide hazard maps provided by PRESSCA (for the two rainfall events), the importance of soil moisture conditions is clearer. Indeed, for a similar amount of rainfall, the landslide hazard strongly increases, moving from intermediate to wet initial soil moisture values. In other words, for the same rainfall amount the initial soil moisture state discriminates

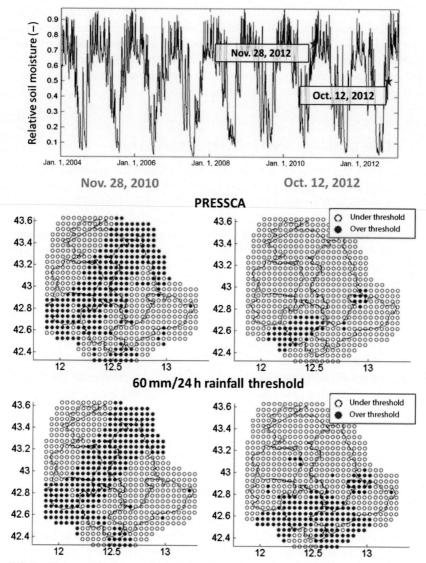

FIG. 5 (*Top panel*) areal mean modeled soil moisture for the Umbria region in the period 2004–13. The soil moisture conditions for the two rainfall events of Nov. 28, 2010, and Oct. 12, 2012, are highlighted. (*Middle panels*) Binary maps showing the locations for which the moderate criticality rainfall-soil moisture threshold of the PRESSCA system was exceeded for a 24-h duration during the two selected dates. (*Lower panels*) Binary maps showing the locations for which a threshold of 60 mm over 24 h was exceeded during the two selected dates.

between the occurrence, or not, of critical warning level to landslide hazard. Overall, in the analysis of different critical rainfall events (not shown for brevity), the PRESSCA system is found to outperform the approach based only on rainfall data for issuing landslide alerts, with a significant reduction of false alarms and, also, a small improvement in the correct identification of actual critical conditions (Ciabatta et al., in press).

5 CONCLUSIONS AND FUTURE PERSPECTIVES

In this study, the use of satellite soil moisture products for improving the prediction of landslide hazard is investigated by considering, as a case study, the early warning system named PRESSCA operating at the Umbria Region Civil Protection Centre (Fig. 1). Specifically, the intercomparison of satellite, modeled, and in situ soil moisture observations showed a quite good agreement between the data sets (Figs. 3 and 4). Modeled data provides slightly better performance with respect to satellite data when compared with in situ observation (Table 2). However, satellite soil moisture data could represent an additional and independent source of information that can improve the large-scale soil moisture estimation and, consequently, landslide prediction. Indeed, the results also highlight the significant impact of soil moisture conditions on the identification of landslide hazard conditions in the day-by-day operations of civil protection officers involved in landslide risk mitigation. Moreover, the identification of soil moisture state is obtained only by considering the season might introduce large errors in the evaluation of the real soil moisture conditions. In addition, for similar amounts of rainfall, the warning level provided by the PRESSCA system is significantly different even moving from intermediate to wet soil moisture conditions (Fig. 5). The PRESSCA system is also found to provide improved performance with respect to approaches only based on rainfall information with a large reduction of false alarms.

Based on these results, future steps will be addressed to the effective use of satellite data for improving the spatial-temporal representation of soil moisture conditions in the Umbria region. Specifically, it is foreseen to make different attempts for integrating satellite and modeled soil moisture data starting from simple nudging approaches to more complex data assimilation techniques that fully take into account the expected errors of the different data sources (similar to the works by Brocca et al. (2010, 2012b), which addressed the soil moisture data assimilation into flood modeling). Through the assimilation of satellite soil moisture data, it will be tested if the performance of the PRESSCA system will improve by a thorough comparison with the landslide events catalog that were collected for the Umbria region. Finally, it is also expected that different satellite soil moisture products coming from new (e.g., SMAP and Sentinel-1) and recent (AMSR2) satellite sensors will be used. Specifically, the SMAP and Sentinel-1 soil moisture products will be characterized by a much better spatial resolution (3–9 km for SMAP, 1 km for Sentinel-1) that might be highly beneficial for the detailed spatial identification of critical landslide conditions.

ACKNOWLEDGMENTS

The authors acknowledge the support of the European Organisation for the Exploitation of Meteorological Satellites (EUMETSAT) through the "Satellite Application Facility in Support to Operational Hydrology and Water Management (H-SAF)" project and are grateful to Umbria Region Civil Protection Centre for providing most of the analyzed data and for the implementation of the PRESSCA system.

REFERENCES

Alyaari, A., Wigneron, J.-P., Ducharne, A., Kerr, Y.H., Wagner, W., De Lannoy, G., Reichle, R.H., Al-Bitar, A., Dorigo, W., Richaume, P., Mialon, A., 2014. Global-scale comparison of passive (SMOS) and active (ASCAT) satellite based microwave soil moisture retrievals with soil moisture simulations (MERRA-Land). Remote Sens. Environ. 152, 614–626.

Bartalis, Z., Wagner, W., Naeimi, V., Hasenauer, S., Scipal, K., Bonekamp, H., Figa, J., Anderson, C., 2007. Initial soil moisture retrievals from the METOP-A Advanced Scatterometer (ASCAT). Geophys. Res. Lett. 34.

Baum, R.L., Godt, J.W., 2010. Early warning of rainfall-induced shallow landslides and debris flows in the USA. Landslides 7, 259–272.

Bittelli, M., Valentino, R., Salvatorelli, F., Rossi Pisa, P., 2012. Monitoring soil-water and displacement conditions leading to landslide occurrence in partially saturated clays. Geomorphology 173–174, 161–173.

Bordoni, M., Meisina, C., Valentino, R., Lu, N., Bitteli, M., Chersich, S., 2015. Hydrological factors affecting rainfall-induced shallow landslides: from the field monitoring to a simplified slope stability analysis. Eng. Geol. 193, 19–37.

Brocca, L., Melone, F., Moramarco, T., 2008. On the estimation of antecedent wetness conditions in rainfall-runoff modelling. Hydrol. Process. 22 (5), 629–642.

Brocca, L., Melone, F., Moramarco, T., Wagner, W., Naeimi, V., Bartalis, Z., Hasenauer, S., 2010. Improving runoff prediction through the assimilation of the ASCAT soil moisture product. Hydrol. Earth Syst. Sci. 14, 1881–1893.

Brocca, L., Hasenauer, S., Lacava, T., Melone, F., Moramarco, T., Wagner, W., Dorigo, W., Matgen, P., Martínez-Fernández, J., Llorens, P., Latron, J., Martin, C., Bittelli, M., 2011. Soil moisture estimation through ASCAT and AMSR-E sensors: an intercomparison and validation study across Europe. Remote Sens. Environ. 115, 3390–3408.

Brocca, L., Ponziani, F., Moramarco, T., Melone, F., Berni, N., Wagner, W., 2012a. Improving landslide forecasting using ASCAT-derived soil moisture data: a case study of the Torgiovannetto landslide in central Italy. Remote Sens. 4 (5), 1232–1244.

Brocca, L., Moramarco, T., Melone, F., Wagner, W., Hasenauer, S., Hahn, S., 2012b. Assimilation of surface and root-zone ASCAT soil moisture products into rainfall-runoff modelling. IEEE Trans. Geosci. Remote Sens. 50 (7), 2542–2555.

Brocca, L., Zucco, G., Moramarco, T., Morbidelli, R., 2013a. Developing and testing a long-term soil moisture dataset at the catchment scale. J. Hydrol. 490, 144–151.

Brocca, L., Tarpanelli, A., Melone, F., Moramarco, T., Caudaro, M., Ratto, S., Ferraris, S., Berni, N., Ponziani, F., Wagner, W., Melzer, T., 2013b. Soil moisture estimation in alpine catchments through modelling and satellite observations. Vadose Zone J. 12 (3), 10.

Brocca, L., Camici, S., Melone, F., Moramarco, T., Martinez-Fernandez, J., Didon-Lescot, J.-F., Morbidelli, R., 2014. Improving the representation of soil moisture by using a semi-analytical infiltration model. Hydrol. Process. 28 (4), 2103–2115.

Brunetti, M.T., Perruccacci, S., Rossi, M., Luciani, S., Valigi, D., Guzzetti, F., 2010. Rainfall thresholds for the possible occurrence of landslides in Italy. Nat. Hazards Earth Syst. Sci. 10, 447–458.

Ciabatta, L., Camici, S., Brocca, L., Ponziani, F., Stelluti, M., Berni, N., Moramarco, T., in press. Assessing the impact of climate-change scenarios on landslide occurrence in Umbria region, Italy. J. Hydrol. http://dx.doi.org/10.1016/j.jhydrol.2016.02.007

Das, N.N., Entekhabi, D., Dunbar, R.S., Njoku, E.G., Yueh, S.H., 2016. Uncertainty estimates in the SMAP combined active-passive downscaled brightness temperature. IEEE Trans. Geosci. Rem. Sens. 54 (2), 640–650.

de Jeu, R.A.M., Holmes, T.R., Parinussa, R.M., Owe, M., 2014. A spatially coherent global soil moisture product with improved temporal resolution. J. Hydrol. 516, 284–296.

Fischer, E.M., Knutti, R., 2015. Anthropogenic contribution to global occurrence of heavy-precipitation and high-temperature extremes. Nat. Clim. Change 5, 560–564.

Gariano, S.L., Brunetti, M.T., Iovine, G., Melillo, M., Perruccacci, S., Terranova, O., Vennari, C., Guzzetti, F., 2015. Calibration and validation of rainfall thresholds for shallow landslide forecasting in Sicily, southern Italy. Geomorphology 228, 653–665.

Godt, J.W., Baum, R.L., Chleborad, A.F., 2006. Rainfall characteristics for shallow landsliding in Seattle, Washington, USA. Earth Surf. Processes Landform 31, 97–110.

Guzzetti, F., Stark, C.P., Salvati, P., 2005. Evaluation of flood and landslide risk to the population of Italy. Environ. Manag. 36 (1), 15–36.

Hawke, R., McConchie, J., 2011. In situ measurement of soil moisture and pore-water pressures in an 'incipient' landslide: Lake Tutira, New Zealand. J. Environ. Manag. 92, 266–274.

Kerr, Y.H., Al-Yaari, A., Rodriguez-Fernandez, N., Parrens, M., Molero, B., Leroux, D., Bircher, S., Mahmoodi, A., Mialon, A., Richaume, P., Delwart, S., Al Bitar, A., Pellarin, T., Bindlish, R., Jackson, T.J., Rudiger, C., Waldteufel, P., Mecklenburg, S., Wigneron, J.-P., in press. Overview of SMOS performance in terms of global soil moisture monitoring after six years in operation. Rem. Sens. Environ.

Lepore, C., Arnone, E., Noto, L.V., Sivandran, G., Bras, R.L., 2013. Physically based modeling of rainfall triggered landslides: a case study in the Luquillo forest, Puerto Rico. Hydrol. Earth Syst. Sci. 17 (9), 3371–3387.

Leroux, D.J., Kerr, Y.H., Al Bitar, A., Bindlish, R., Jackson, T.J., Berthelot, B., Portet, G., 2014. Comparison between SMOS, VUA, ASCAT, and ECMWF soil moisture products over four watersheds in US. IEEE Trans. Geosci. Rem. Sens. 52 (3), 1562–1571.

Parinussa, R.M., Holmes, T.R., Wanders, N., Dorigo, W.A., de Jeu, R.A.M., 2015. A preliminary study toward consistent soil moisture from AMSR2. J. Hydrometeorol. 16 (2), 932–947.

Paulik, C., Dorigo, W., Wagner, W., Kidd, R., 2014. Validation of the ASCAT Soil Water Index using in situ data from the International Soil Moisture Network. Int. J. Appl. Earth Obs. Geoinf. 30, 1–8.

Ponziani, F., Pandolfo, C., Stelluti, M., Berni, N., Brocca, L., Moramarco, T., 2012. Assessment of rainfall thresholds and soil moisture modeling for operational hydrogeological risk prevention in the Umbria region (central Italy). Landslides 9 (2), 229–237.

Ponziani, F., Berni, N., Stelluti, M., Zauri, R., Brocca, L., Moramarco, T., Salciarini, D., Tamagnini, C., 2013. Landwarn: an operative early warning system for landslides forecasting based on rainfall thresholds and soil moisture. In: Margottini, C. et al., (Ed.), Landslide Science and Practice. vol. 2. Springer-Verlag, Berlin, pp. 627–634.

Ray, R.L., Jacobs, J.M., Cosh, M.H., 2010. Landslide susceptibility mapping using downscaled AMSR-E soil moisture: a case study from Cleveland Corral, California, US. Remote Sens. Environ. 114 (11), 2624–2632.

Ray, R.L., Jacobs, J.M., Ballestero, T.P., 2011. Regional landslide susceptibility: spatiotemporal variations under dynamic soil moisture conditions. Nat. Hazards 59, 1317–1337.

Rawls, W.J., Brakensiek, D.L., Saxton, K.E., 1982. Estimation of soil water properties. Trans. ASAE 25 (5), 1316–1320.

Muñoz Sabater, J., Al-Bitar, A., Brocca, L., 2015. State of the art soil moisture retrievals with active and passive microwave data. Operational applications. In: Remote Sensing of Land Surface Turbulent Fluxes and Soil Surface moisture Content: State of the Art. Elsevier, New York.

Trenberth, K.E., Fasullo, J.T., Shepherd, T.G., 2015. Attribution of climate extreme events. Nat. Clim. Change. http://dx.doi.org/10.1038/nclimate2657.

Vachaud, G., Passerat de Silans, A., Balabanis, P., Vauclin, M., 1985. Temporal stability of spatially measured soil water probability density function. Soil Sci. Soc. Am. J. 49, 822–828.

Wagner, W., Lemoine, G., Rott, H., 1999. A method for estimating soil moisture from ERS scatterometer and soil data. Remote Sens. Environ. 70, 191–207.

Wagner, W., Hahn, S., Kidd, R., Melzer, T., Bartalis, Z., Hasenauer, S., Figa, J., de Rosnay, P., Jann, A., Schneider, S., Komma, J., Kubu, G., Brugger, K., Aubrecht, C., Zuger, J., Gangkofner, U., Kienberger, S., Brocca, L., Wang, Y., Bloeschl, G., Eitzinger, J., Steinnocher, K., Zeil, P., Rubel, F., 2013. The ASCAT soil moisture product: specifications, validation results, and emerging applications. Meteorol. Z. 22 (1), 5–33.

Wagner, W., Brocca, L., Naeimi, V., Reichle, R., Draper, C., de Jeu, R., Ryu, D., Su, C.-H., Western, A., Calvet, J.-C., Kerr, Y., Leroux, D., Drusch, M., Jackson, T., Hahn, S., Dorigo, W., Paulik, C., 2014. Clarifications on the "Comparison between SMOS, VUA, ASCAT, and ECMWF soil moisture products over four watersheds in U.S." IEEE Trans. Geosci. Remote Sens. 52 (3), 1901–1906.

Chapter 13

Remotely Sensed Soil Moisture as a Key Variable in Wildfires Prevention Services: Towards New Prediction Tools Using SMOS and SMAP Data

D. Chaparro*,†, M. Piles*,† and M. Vall-llossera*,†

*Universitat Politècnica de Catalunya, IEEC/UPC, Barcelona, Spain, †Barcelona Expert Center, Institute of Marine Sciences, CSIC, Barcelona, Spain

1 INTRODUCTION

Wildfires are an increasingly concerning issue due to their impact on safety, environment, and the economy. Although natural fires are in part essential to maintain dynamics, biodiversity, and productivity in some ecosystems, their damages are huge in economic and environmental terms (FAO, 2006; Ireland and Petropoulos, 2015; Karamesouti et al., 2016). Usually, forest fires occur in patchy landscapes, affecting agricultural and urban areas, and in some cases, seriously threatening human lives.

The human factor plays a critical role on fire dynamics directly or indirectly, thus increasing the complexity of the phenomenon as well as its analysis and prevention. Indeed, direct effects of human activities such as land clearing, agriculture, resettlements, negligence, and arson are the main causes of fire ignition in most areas of the world. In the Mediterranean regions 95% of fires are due to these causes, and similar percentages are found in other areas (e.g., 90% in South Asia, 85% in South America, 80% in Northern Asia; FAO, 2006). The impact of CO_2 emissions on climate change is predicted to lead to an increase of temperatures and of the duration and intensity of droughts, which are already threatening woodlands with larger and more frequent forest fires due to facilitated drying and combustion of vegetation fuels (Oppenheimer et al., 2013).

The importance of weather on wildfires is notorious, and particularly its role determining soil moisture is a keystone of the climate-fire relation as it partly

Satellite Soil Moisture Retrieval. http://dx.doi.org/10.1016/B978-0-12-803388-3.00013-9

determines prefire conditions. Because litter and soil water content are very dependent on immediate weather changes (Chuvieco et al., 2004), the estimation of soil moisture through variables such as precipitation and temperature is usually considered in common fire risk indexes. Still, the relation between climate change, forest fires, and soil moisture is not yet well understood. In this regard, the use of measurements from the first missions specifically dedicated to measuring soil moisture, i.e. the ESA's Soil Moisture and Ocean Salinity (SMOS, 2009–2017) and the NASA's Soil Moisture Active Passive (SMAP, 2015–2018), opens a path forward to the study of fire ignition and propagation risks.

2 REMOTELY SENSED SOIL MOISTURE, CLIMATE CHANGE, AND FIRE RISK

2.1 Remote Sensing of the Earth's Soil Moisture

Soil moisture is a key element of the water cycle clearly influencing land-atmosphere interactions. Actually, the ESA's Climate Change Initiative Program recognized soil moisture as an essential climate variable (ECV) in 2010 (ESA, 2015).

In the field of remote sensing techniques for soil moisture estimation, satisfactory results were initially obtained with radiometers such as the Scanning Multichannel Microwave Radiometer (SMMR, measuring at 6.6 GHz) and the Advanced Microwave Scanning Radiometer for Earth Observing (AMSR-E, measuring at 6.8 GHz) among others. However, their measurements are limited by low penetration through vegetation and sensitivity only to the skin surface soil layer (Kerr et al., 2010). The ESA's SMOS mission was launched in Nov. 2009 to overcome these limitations: it works at the protected L-band frequency, providing information of the top 5 cm of the soil, including areas with vegetation up to high densities (<5 kg/m^2). L-band radiometry is an optimal frequency band for soil sensing due to the direct relationship of soil emissivity with its soil water content. Also, at these frequencies, the atmosphere is almost transparent and measurements can be obtained day and night. SMOS unique payload is the first 2D interferometric radiometer in space, the Microwave Imaging Radiometer with Aperture Synthesis. After more than 6 years in orbit, SMOS continues providing global views of the Earth's soil moisture every 3 days with an estimated accuracy of 0.04 m^3/m^3 (Kerr et al., 2012).

Following SMOS, NASA's SMAP mission was launched on Jan. 2015 and is composed of an L-band radiometer ($\sim$36 km resolution) and a synthetic aperture radar ($\sim$3 km) to improve the spatial resolution of the estimates to $\sim$9 km resolution. The SMAP instrument was designed to make coincident measurements of surface emission and backscatter, providing more insight into multiscale soil moisture estimates. However, the production of the higher resolution products ceased abruptly with the failure of the SMAP radar after about ten weeks of operation. As for SMOS, soil moisture under vegetation covers with less than 5 kg/m^2 water content will be measurable and global coverage will be achieved within 3 days at the equator (Entekhabi et al., 2010).

SMOS soil moisture estimates have been extensively validated and proved useful for global studies (Piles et al., 2011a; Sánchez et al., 2012; Piles et al., 2014; Petropoulos et al., 2015). However, its coarse resolution limits its applicability in local- or regional-scale studies, where soil moisture at higher spatial resolutions (1–10 km) is needed. Several downscaling approaches have been developed to bridge this gap, based on the synergy of SMOS with visible/infrared information at higher spatial resolution, indirectly related to soil moisture changes (Merlin et al., 2005, 2008; Piles et al., 2011a; Piles et al., in press). Since 2012, the methodology first presented in Piles et al. (2011a) and upgraded in Piles et al. (2014) is operational at SMOS-Barcelona Expert Centre (BEC) and is providing the first SMOS-derived soil moisture maps at fine scale (1 km) over the Iberian Peninsula (http://cp34-bec.cmima.csic.es/NRT). These maps are provided in near real-time and meet the spatial and temporal requirements of forest fire prevention studies (Piles et al., 2013) and are the maps used in this study (see Section 5).

2.2 Remotely Sensed Soil Moisture, Climate Change, and Fire Risk

The prediction of climate change effects is largely dependent on the amount of CO_2 released to the atmosphere. In that sense, the Intergovernmental Panel on Climate Change (IPCC) expects increasing CO_2 emissions at least until 2050 depending on the scenario (IPCC, 2000) and shows increasing evidence that changes in climate observed from the 1950s are attributable to human influences. These changes are evident in terms of increasing temperatures. Changes in soil moisture, in turn, are difficult to assess, because observational records available are either sparse and/or short (Bindoff et al., 2013). At the moment, recent remotely sensed measurements of precipitation from the Tropical Rainfall Measuring Mission provide evidence of significant trends towards drier soils in some regions (Jung et al., 2010). In this context, SMOS and SMAP satellites are expected to improve the knowledge of soil moisture as a key variable on climate change, which is needed to better understand the observed and predicted global warming and drying tendencies.

Concerning forest fires, Settele et al. (2014) explain that present drying trends as well as the rising number of extreme temperature events are probably leading to a significant increase in the number of forest fires from 2004 to 2100. The authors also analyze the McArthur Forest Fire Danger Index (FFDI) (McArthur, 1967; see Section 3.2) for the period 1970–99 and predict an important increase in fire risk for the decades 2070–99 (based on the scenario Representative Concentration Pathway, RCP 8.5). It has been shown that the increased burned area in Canada during the last four decades is the result of human-induced climate change (Gillett et al., 2004). Also, a raise in fire risk is expected for IPCC A2 and B2 scenarios in the EU Mediterranean countries, especially affecting areas with extensive forest cover (mainly Balkans, Pyrenees, and Alps; Moriondo et al., 2006).

The increase in fire occurrences is causing changes in the terrestrial ecosystem that, together with rising temperatures and droughts, could lead to forests' and soils' carbon stocks becoming a weaker sink, or a net carbon source, before the end of the century (Settele et al., 2014). Wildfires could be reinforcing climate change creating a positive fire-global warming feedback at least in the immediate years after burning (Amiro et al., 2001; Randerson et al., 2006). However, this feedback is possibly not occurring in the long term (80 years after fire; Randerson et al., 2006).

The direct impact of increasing fire activity is not only environmental but also economic. As examples, it was estimated that the average annual economic cost of fires in the United States was of $261 million (Dale et al., 2001), and that fire management agencies in Canada spent about $500 million yearly in fire suppression (Flannigan et al., 2009). It is suggested that the increase in the number of fires will raise these costs and also that the capacity of fire management agencies will be overtaken, reaching a tipping point in some regions in one or two decades. Hence, the expected increase of burned area could be produced not only due to climate trends but also due to the limited human capacity to cope with the new burning rates in some regions (Flannigan et al., 2009). Consequently, there is an important need to improve prediction tools to facilitate prevention and extinction tasks.

In this context, a deeper comprehension of the relationships among climate, fires, and drought is needed and may become easier with the launch of satellite missions dedicated to monitoring the Earth's global soil moisture. More than 6 years of soil moisture measurements from SMOS and complementary information from SMAP are already available, providing stable, long-term, soil moisture measurements at a global scale for climate modelers and researchers. Hence, these data will help to better understand and predict hazards influenced by climate change, such as fires, droughts, or floods. In the case of fire risk studies, the use of remotely sensed information on soil moisture rather than or complementarily with precipitation or weather-based indices is very promising.

3 THE ROLE OF SOIL MOISTURE IN FOREST FIRES

3.1 Droughts and High Temperatures Lead to Extreme Forest Fires Events

The literature reporting variables associated with burning risk is huge and diverse, but several climate conditions are generally common: heat and lack of precipitation (leading to low moisture in soil and fuels), low relative humidity, and high wind speed are some of the determinants of fire, but not the only ones (e.g. orography, soil use, and distances to human settlements or roads are also crucial, but are out of the scope of this chapter).

In Europe, Portugal and Spain are usually the countries with the highest burned area each year (European Commission, 2015) as dry and hot summers are frequent in the Iberian Peninsula with the consequent drying of soils and fuels. Verdú et al. (2012) explained that fires in the Spanish Mediterranean areas are few but frequently large due to limited moisture, while the Euro-Siberian (northwestern) region accounts for frequent but little fires due to a more humid climate limiting the flammability of fuels and thus the propagation of fires. Padilla and Vega-García (2011) showed that one of the most important variables in fire prediction in northwestern Spain was the Fine Fuel Moisture Code (FFMC), strongly related with soil moisture (see Section 3.2); FFMC is the limiting factor in fire propagation in this region, as it determines fuels' flammability. Only when unusual drought episodes dry the available fuel in NW Spain, large fires occur. This pattern is similarly followed in Portugal, where the first 2 weeks of Aug. 2003 were dramatic, with more than 400,000 ha burned. The situation was mainly caused by the heat wave produced during that summer in Europe, with maximum temperatures in Portugal reaching 47.3°C (Trigo et al., 2006). Fischer et al. (2007) studied the interaction between soil moisture and the atmosphere during and previously to this heat wave performing simulations using European Centre for Medium-Range Weather Forecasts (ECMWF) reanalysis and climate models. The authors explained that the lack of precipitation and the strong positive radiative anomalies that were registered in spring 2003 brought an important and rapid decrease of soil moisture. That year, the high temperatures from Jun. to Aug. were initiated by persistent anticyclonic anomalies, but this situation was reinforced by persistent dry soils, which strongly reduced latent cooling and consequently amplified the temperature anomalies. García-Herrera et al. (2010) also pointed to the lack of soil moisture as one of the influencing factors in the 2003 heat wave, and analyzed the record-breaking forest fires in Portugal in that summer due to this heat wave.

The abovementioned studies suggest that estimating soil moisture at different time scales can be crucial in fire prevention by providing early indication of the presence of severe drought situations. In that sense, several studies have been directed to the use of SMOS in monitoring drought conditions (Scaini et al., 2014; Martínez-Fernández et al., 2015; Portal et al., 2015). Scaini et al. (2014) calculated soil moisture anomalies for the SMOS L2 product (40 km resolution) as well as for in situ measurements in the REMEDHUS network (Red de Estaciones de Medición de la Humedad del Suelo de la Universidad de Salamanca, Duero basin, Spain; Sánchez et al., 2010), at different time scales (from 10 to 120 days). The anomalies were compared with the common drought indices Standard Precipitation Index (McKee et al., 1993) and Standard Evaporation Precipitation Index (Vicente-Serrano et al., 2010) also at time scales from 10 to 120 days. Results showed the best agreement between SMOS anomalies and drought indices at the 1-month time scale (Scaini et al., 2014). Another interesting example reported a drought episode in Castilla y León (Spain) during Feb. and Mar. 2012. In this case, Portal et al. (2015) found soil moisture during this period to be fairly below average in the region using different SMOS products and resolutions (L2: 40 km; L3: 25 km; L4: 1 km) as well as using soil moisture ECV, which estimates soil moisture (0.25° resolution) from different satellites' information (Liu et al., 2012; Wagner et al., 2012).

Drought indices using remotely sensed soil moisture can be used in different applications. In the field of agriculture, Martínez-Fernández et al. (2015) proposed a Soil Water Deficit Index (SWDI) using in situ soil moisture measurements from REMEDHUS for the period 2009–13. The authors observed how the annual number of drought weeks (SWDI < 0) was linked to the annual cereal production in the region, this being lower in the drier years. As the REMEDHUS network is used as a validation site for SMOS, it may be inferred that obtaining a similar index using SMOS data would be feasible. Also, for agricultural purposes, Stoyanova et al. (2014) proposed a drought index, using as a proxy of soil water availability the difference between land surface temperature (LST) (from Meteosat Second Generation satellites), and air temperature at 2 m high (obtained from meteorological stations). The index was used to prevent drought from affecting crop yields in Bulgaria. Otherwise, Ross et al. (2014) showed dieback in grasslands suffering soil moisture deficit in the United States. They also analyzed these patterns in Australia using SMOS data.

Despite these applications relating soil moisture remote sensing and vegetation, to the authors' knowledge no research has been conducted to estimate the link between fuel moisture content (either live or dead fuels) and satellite-derived soil moisture information. Estimation of live fuel moisture content in grasslands was obtained by Chuvieco et al. (2004) and García et al. (2008) with the Normalized Difference Vegetation Index (NDVI) and surface temperature from the Advanced Very High Resolution Radiometer (AVHRR). Nevertheless, dead fuels deposited on the forest floor are the most dangerous as they are drier than live fuels and depend on rapid atmospheric changes. Due to this reason, the most common fire risk indices estimate the moisture of organic layers in the soil surface.

3.2 Dead Fuels Moisture Is a Key Variable in Forest Fire Risk Indices

Moisture of dead fuels mainly depends on their thickness and on weather changes. Thinner fuels rapidly change their moisture as a function of weather conditions, while changes in thicker fuels are slower. Several fire risk indices take into account fuels and soil moisture to estimate the dryness of fuels as a crucial parameter on fire risk assessment. They are listed and detailed below.

- The Canadian Forest Fire Danger Rating System uses information on dead-fuel moisture as a main driver in its Fire Weather Index (FWI) system (Van Wagner, 1974, 1987). In particular, moisture information is used in three FWI components: (i) FFMC, which estimates moisture of the first centimeter of litter; (ii) Duff Moisture Code, which accounts for the organic material at 5–10 cm depth; and (iii) Drought Code (DC), estimating moisture of compact fuel at 10–20 cm depth. In Spain, the previously mentioned study of Padilla and Vega-García (2011) ranked FWI as the first variable and FFMC as the third variable in importance within their models of fire risk prediction in several regions.

- The European Forest Fires Information System (EFFIS) (European Comission, 2010) also uses the FWI. The EFFIS studies fires in a comprehensive manner, from risk assessment to fire detection and mapping. EFFIS has been used to check how extreme danger situations reported by FWI and, in particular, by the DC, were linked to a maximum number of occurrences of big fires (>500 ha) in Portugal in 2003 (San-Miguel-Ayanz et al., 2013).
- The McArthur, 1967 fuel moisture model (McArthur, 1967) is the basis for the McArthur FFDI, widely used in Australia. In FFDI, the use of a drought factor is included, and partly based on the soil moisture deficit.
- In the United States, the National Fire Danger Rating System (Deeming et al., 1977; Bradshaw et al., 1984) studies the moisture of fuels at different time lags (from 1 h, for the finest dead fuels, to 1000 h, for the largest ones). This is a good example of the importance of analyzing different time scales when studying the relation between fires and their predisposing moisture conditions.

The calculation of the above-mentioned indices is based on data from meteorological stations. In that sense, Shvetsov (2013) considers that the main drawback of doing so is an absence of a regular network of weather stations in some regions, as in Siberia. Bourgeau-Chávez et al. (2013) also report the lack of weather stations—in this case for Alaska—and they suggest that remotely sensed soil moisture information is needed to improve the index currently used in the region (FWI) not only by providing spatially explicit data of soil moisture, but also for accounting for hydrological processes not driven by weather (e.g., permafrost changes).

Despite the importance that remotely sensed soil moisture could have in forest fire assessment (e.g., improving forest fire prevention services), advances in this field are just beginning.

4   LINKING REMOTELY SENSED SOIL MOISTURE WITH FOREST FIRES IGNITION AND PROPAGATION

The use of satellite soil moisture data can provide new insight to the study of forest fires preconditions and help improving fire risk indices. The soil moisture conditions prior to fire occurrences have been studied by different authors.

Bartsch et al. (2009) analyzed prefire anomalies of soil moisture for the period 1992–2000 in central Siberia. They used satellite-derived soil moisture data from ERS-1 and ERS-2 scatterometers (Earth Resources Satellites, measuring at C-band), at 50 km spatial resolution. The authors calculated soil moisture anomalies at a 1-month time scale and compared this data set with the number of fires and burned area, which were higher during Jun. and Jul. Dry soils (negative anomalies of soil moisture) favored the occurrence of fires and their extension in continuous permafrost regions. In nonpermafrost sites, dry conditions were only associated with a higher occurrence of fires in comparison with wetter forests. Bartsch et al. concluded that wet soil surface limited the extent of burned area and that 80% of fires occurred under dry soil conditions in summer.

Further insight into the relation of soil moisture, permafrost, and occurred fires was presented by Forkel et al. (2012), who identified surface moisture as an important factor determining the evolution of extreme fire events in the region of Lake Baikal (Siberia). The authors studied fires for the period 2003–09 and used the surface moisture product derived from AMSR-E onboard the Aqua satellite as one of the explanatory variables in their study. Forkel et al. mentioned that 2003 fires occurred mainly in permafrost soils, which presented negative surface moisture anomalies. Interestingly, the authors concluded that lack of precipitation during the previous summer (2002) led to frozen soils storing an unusual low water content during winter 2002/2003. Consequently, the amount of water released by permafrost melting in spring 2003 was low, as detected by remotely sensed soil moisture products. Clearly, this situation favored the extreme fires that occurred in Jun. 2003.

Shvetsov (2013) also pointed to the lack of meteorological stations in Siberia as a problem for the interpolation of FWI, and suggested the use of remotely sensed soil moisture to solve this drawback. The author compared the FWI index with SMOS data in Siberia during 2011 and 2012 and found quite good correlations in specific areas, where increasing soil moisture (SMOS) and decreasing risk (FWI) were concordant. However, a good FWI-SMOS relationship was difficult to obtain for the whole region of Siberia. Thus, further research in this line is needed and is probably urgent in some regions due to the lack of in situ meteorological data.

At the Iberian Peninsula, several research studies have been recently conducted to analyze prefire conditions of soil moisture (from SMOS) and LST. Piles et al. (2011b) studied the 12 fires (>35 ha) registered in the Iberian Peninsula during 2010 using soil moisture data from SMOS (BEC L4 product, at 1 km resolution; Piles et al., 2011a) and LST data from the Moderate Resolution Imaging Spectroradiometer (MODIS) (1 km). Piles et al. (2011b) compared these data from burned areas with the same information obtained for randomly selected unburned zones. Soil moisture was found to be lower and LST to be higher in burned areas prior to fire occurrences. Moreover, using these differences between burned and unburned sites, it was concluded that high fire risk was found when SMOS-derived soil moisture was below $0.04 \text{ m}^3/\text{m}^3$ and MODIS LST was above 33°C. A validation test with 67 randomly selected fires was done achieving an accuracy of 58%.

Using a similar approach, D. Chaparro (2012, unpublished results) found little but significant differences when comparing soil moisture conditions between burned and unburned pixels, the first ones being drier. The study was performed in the province of Barcelona (NE Spain) for 2010 and 2011 summers, when 113 fires were registered, most of them (90%) being very small in size (<1 ha). Similar results were obtained when comparing seven fires >10 ha with seven random unburned zones (fires occurred in Catalonia during 2010 and 2011). The author concluded that broader differences could be achieved using a bigger sample of big fires. This conclusion led to a new study reported in Chaparro et al. (2014), which analyzed 63 fires occurring in Catalonia (NE Spain) for the period 2010–13. The methodology was the same as in the previous study, but the comparison between burned and unburned pixels ensured similar conditions between the two groups in two variables that can influence soil moisture (SM): solar radiation and aspect. Results (Fig. 1) found significant differences between burned areas (median $SM = 0.10 \, m^3/m^3$) and unburned ones (median $SM = 0.13 \, m^3/m^3$).

With 4 years of data (2010–14), forest fires that occurred in the Iberian Peninsula and Balearic Islands were further studied by Chaparro et al. (2015). The study used the EFFIS database (European Comission, 2010), which collected 2096 fires bigger than 10 ha in this period. The authors classified the fires in three burned-area categories: <500 ha, 500–3000 ha, and >3000 ha. The BEC Level 4 v.3. SM product was used, which provides soil moisture estimates at 1 km spatial resolution over the entire Iberian Peninsula at least every 3 days independently of weather conditions (ERA-interim LST are used when clouds are masking MODIS LST measurements, BEC, 2015). Surface temperatures (LSTs) from ERA-Interim reanalysis models at noon were also used in the analysis (ECMWF, 2015). They had an initial grid of 0.125 degree and were linearly interpolated to 1 km. A soil moisture climatology was built from monthly means for the period 2010–14. Anomalies at a 9-day time scale were computed for the whole study period with respect to this climatology. The same procedure

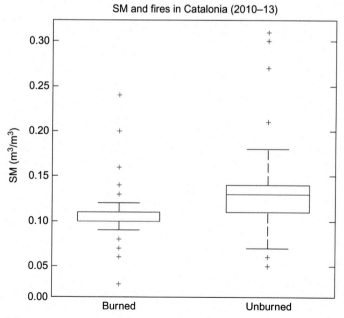

FIG. 1   Soil moisture for unburned and burned conditions prior to 63 fires in Catalonia (2010–13).

was applied for LST. As will be shown later in this section, only the combined use of absolute values and anomalies for SM and LST permits improving the detection of fires in all the seasons.

The four studied variables (SM, LST, and their anomalies An_SM and An_LST) were introduced in the fires database to study each prefire occurrence condition. Only one value per variable was assigned to each fire. This was the median of all burned pixels considering the first day with available data (from the fire occurrence day until 3 days before the burning date). Fires without data in one of the four variables were removed from the database, leading to a final data set of 2076 fires. Another database was compiled with daily values of SM and LST for unburned cells, yearly, to obtain the distribution of both variables in unburned conditions. Unburned cells (1 km^2) were introduced in this database if they accomplished the following criteria: (i) contain flammable land use and types based on the CORINE Land Cover Map (EEA, 2006), and (ii) not having suffered from burns after 2005, as previously burned cells are less likely to be affected in the same way by fires as other areas, until their vegetation recovers some years later. The fire database from EFFIS for the period 2006–09 was used to this end.

Results showed that the conditions observed before most fires occurrences were particularly hot and dry, with values of LST above 300 K and of SM below 0.10 m^3/m^3 (Table 1). Considering unburned regions, dry soils and high temperatures are usual in the Iberian Peninsula, with a broad SM-LST distribution (see Fig. 2 where year 2013 is used as an example). Concerning anomalies, most fires started on soils drier and hotter than usual (i.e., positive anomaly of LST and negative anomaly of SM). Also, the percentage of fires under dry and hot conditions (both in absolute values and anomalies) increases with burned area (Table 1).

Despite the previous results, 268 fires (most of them <500 ha) occurred on Feb. and Mar. 2012 in colder and wetter conditions than the other studied. However, they presented anomalous, hotter and drier conditions than usual, considering the average moisture and temperatures in the season (Fig. 3). These results outline the importance of SM anomalies as complementary information

TABLE 1 First row shows the percentage of fires burning in drier and hotter conditions than the median SM-LST conditions in the unburned pixels for the corresponding year. Second row shows the percentage of fires occurring under drier and hotter than usual conditions in the site

	Area (ha)		
Prefire conditions	*<500*	*500–3000*	*>3000*
SM < annual median and LST > annual median	70%	88%	90%
An SM < 0 and An LST > 0	74%	80%	94%

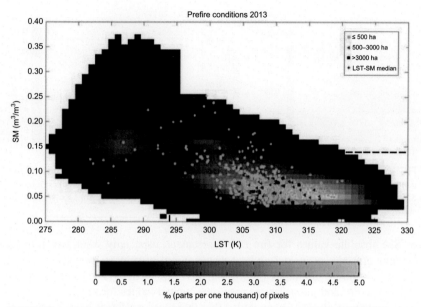

FIG. 2   Green circles, red triangles, and black squares correspond to fires <500 ha, 500–3000 ha, and >3000 ha, respectively. They are plotted in function of soil moisture (SM) and land surface temperature (LST) prior to fire occurrences. Black dashed lines show the median SM and LST during 2013 in unburned pixels. The distribution for both variables in not burned areas during 2013 is shown as the percentage of pixels and days presenting each pair of SM-LST values. White represents <0.1‰.

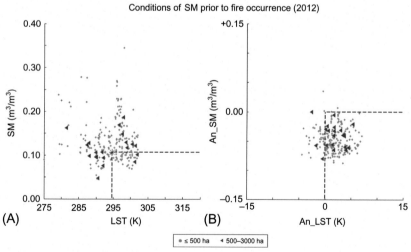

FIG. 3   Prefire conditions on Feb. and Mar. 2012. (A) Absolute values of soil moisture (SM) and land surface temperature (LST), (B) Anomalies for both variables. Green circles and red triangles represent fires <500 ha and 500–3000 ha, respectively. No fires >3000 ha occurred in this period. Blue dashed lines show the median SM and LST during Feb. and Mar. 2012 in unburned pixels.

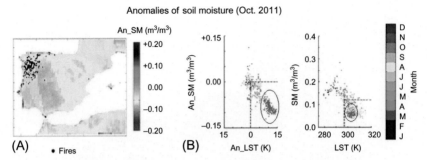

FIG. 4 (A) Soil moisture (SM) anomalies in the Iberian Peninsula during Oct. 2011 showed dry conditions. Fires are plotted in blue. (B) SM and land surface temperature (LST) anomalies (left) and SM and LST absolute values (right) during 2011, plotted by months. Those for Oct. are circled, showing remarkable drought conditions.

to SM absolute values for fire risk assessment, especially when exceptional burning episodes occur during heat waves in out-of-summer periods.

Fig. 4 shows anomalies of SM and LST for Oct. 2011 fires, when soils were very dry and temperatures rose over typical values for the season.

Chaparro et al. (2015) concluded that there was a relation between burned area and SM-LST conditions and that the studied variables could be useful to improve fire risk information services (see Section 5). The authors also reported that the largest fires were concentrated in a few days, mainly during summer 2012, probably due to extreme drought conditions. This is in coherence with results reported in Portugal during early Aug. 2003 (Trigo et al., 2006).

5 FIRE RISK ASSESSMENT IN THE IBERIAN PENINSULA USING SMOS DATA

The fire risk assessment is of great importance in the Iberian Peninsula, a region where fires are frequent (mainly in summer) and occasionally big (especially in the Mediterranean region), endangering environment and, in several occasions, also human lives and goods. The use of remotely sensed data in fire prevention services is interesting as satellite data provides full coverage of the region and provide frequent and accurate information. In the case of SMOS, it allows monitoring soil moisture, a variable of great importance due to its relation with dead fuels water content.

5.1 New Fire Risk Maps Over the Iberian Peninsula Based on SMOS Data

The study presented by Chaparro et al. (2015) led to the configuration of fire risk thresholds associated with fire propagation. This was feasible as drier and hotter soils were generally found prior to the occurrence of big fires, while

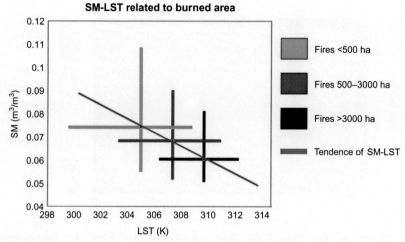

FIG. 5 Green (<500 ha), red (500–3000 ha), and black (>3000 ha) bars represent prefire conditions of soil moisture (SM) and land surface temperature (LST). Cross centers represent median SM-LST conditions. Edges of bars represent first and third quartiles for each variable within each area category.

smaller fires tended to be associated to wetter and colder conditions. The objective of this analysis was to obtain fire risk maps in near real-time to be released daily in the early morning by BEC, so they could be used by Diputació de Barcelona's (DiBa, the provincial government of Barcelona) fire prevention services. Consequently, the study was repeated for SM and LST at 6 am (instead of noon LST) and same conclusions were obtained.

Fig. 5 shows fires occurred in the Iberian Peninsula during 2010–14 grouped by burned area and SM-LST conditions. A random selection of 70% of these fires was used to generate the fire risk maps, while the remaining 30% was allocated for validation.

A novel drought-fire risk index based only on remotely sensed soil moisture and ancillary temperature information was developed using these results. SM, LST, and anomalies (An_SM and An_LST), all of them at 1 km resolution, are combined in a two-step process to obtain the fire risk maps:

1. The third quartile of SM and the first quartile for LST in each category shown in Fig. 5 are established as the minimum thresholds for a pixel to be classified in one of the fire risk categories: risk of ignition (associated to fires <500 ha), risk of a big fire (>500 ha), and risk of a very big fire (>3000 ha).

2. An_SM and An_LST are used as complementary variables. Negative anomalies of SM increase risk category even when cold or wet conditions are found (e.g., an area with $SM > 0.11$ m^3/m^3 and $LST < 300$ K, which is beyond the limits for ignition risk, is anyway included as a risky area if $An\_SM < 0$). Otherwise, $LST\_An > 0$ contribute in determining the high risk level (danger of fire >3000 ha).

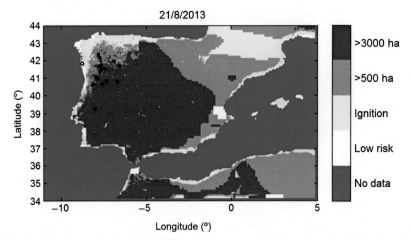

FIG. 6 Fire risk map for Aug. 21, 2013. Four levels are presented (from low risk to high risk, indicating the possibility of fires >3000 ha). Filled circles represent correct predictions and empty circles represent incorrect predictions.

The final risk maps contain four risk levels: one for "low risk" and the other three for each of the categories proposed (Fig. 6 illustrates an example). The product was validated using the remaining, randomly selected, 30% of fires. Results showed that 87% of small fires were predicted, and that the largest fires were detected in almost all the cases (only 9% of fires >500 ha were not predicted, and all fires >3000 ha were expected). Fig. 7 presents the results of the validation analysis, separated per burned area category.

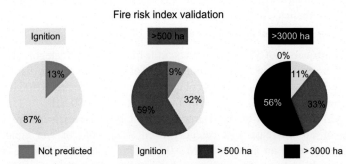

FIG. 7 Three validations are presented, one for each ground truth burned area. Fractions in each pie chart indicate the percentage of fires predicted for each category.

5.2 Fire Risk Maps Availability and Operational Applications

The fire risk maps presented in this section were operationally produced by the BEC in early morning near real-time since Jul. 2015 (available at http://cp34-bec.cmima.csic.es/NRT after registration). Maps are produced at 1 km spatial resolution and cover the Iberian Peninsula, Southern France, and Northern Africa (45°N–34°N, −11°W–5°E).

BEC soil moisture information is routinely used by the Diputació de Barcelona (DiBa, the provincial government of Barcelona) for fire risk assessment and prevention during the forest fire season (Jun.–Sep.). Since Jun. 2012, SMOS-derived fine-scale soil moisture information has been included in DiBa's fire risk prevention service daily bulletin with the purpose of detecting areas with extreme dryness posing a risk of forest fire (Fig. 8).

With the development of the drought-fire risk maps, these are also included in the forest rangers information bulletin from Jul. 2015 (Fig. 9). This implies an important advantage in planning and organization of fire prevention services, as the distribution of vehicles and staff over the territory is in part based on the information provided by the BEC. To continue with this operational application and to foster the use of remotely sensed data in forest fire risk prevention services, BEC has been chosen as a SMAP early adopter.

6 CONCLUSIONS

Climate change is predicted to increase the occurrence and extension of forest fires due to rising global temperatures and an increase of the duration and intensity of droughts. This could reinforce global warming through gases released from combustion and decrease the capability of forests as carbon sinks. Increasing forest fires could also go beyond human extinction and management capabilities, so improvement of prediction tools is essential. In that sense, new soil moisture satellite information can be crucial.

Due to the importance of soil and fuel moisture content in fires, this variable is indirectly included in most fire risk indices. However, soil moisture information is approximated by precipitation data from meteorological stations, which are scarce in some regions. In this regard, SMOS and SMAP global soil moisture maps offer a promising new path for the improvement of main fire risk indices.

The direct use of remotely sensed soil moisture information in fire risk prevention services is a novel and promising field of research. Dry soil conditions has been demonstrated to favor fire propagation, and some studies in the Iberian Peninsula have provided evidence that dry soils and high temperatures increase fire occurrence and extension probabilities. A relationship has been found between remotely sensed SM, LST, and burned area and has set the basis to obtain the first drought-fire risk maps in near real-time. These fire risk maps have

FIG. 8 Soil moisture map from Barcelona Expert Center (BEC) L4 "all-weather" product is provided to forest rangers in the Province of Barcelona through the fire risk prevention service daily bulletin. *(Source: Barcelona Expert Centre.)*

been in BEC operations since Jun. 2015, and are being used by the fire risk assessment team in the provincial government of Barcelona.

Further studies are needed to evaluate the improvement of fire risk indices such as FWI using satellite soil moisture data, and also to take into account lagged effects of soil moisture in different situations such as heat waves or changes in the water content in permafrost soils.

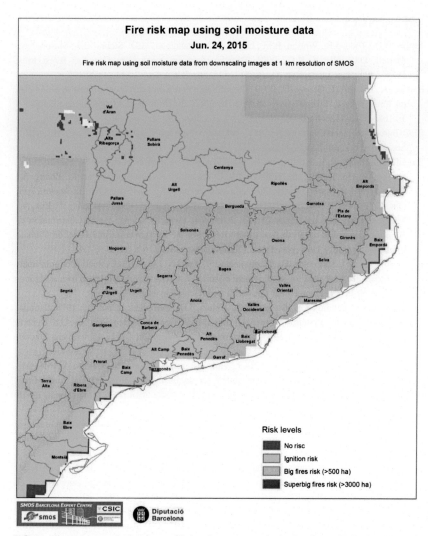

FIG. 9 Fire risk maps product is provided to forest rangers in the Province of Barcelona through the fire risk prevention service daily bulletin. *(Source: SMOS Barcelona Expert Centre.)*

ACKNOWLEDGMENTS

This study was funded by the Spanish government through the project PROMISES: Productos y servicios innovadors con sensores de microondas, SMOS y Sentinels para tierra (ESP2015-67549-C3-1-R), the pre-doctoral grant *Ayudas para contratos predoctorales para la Formación de Doctores*, with reference BES-2013-066240, and the European Regional Development Fund (ERDF).

REFERENCES

Amiro, B.D., Todd, J.B., Wotton, B.M., Logan, K.A., Flannigan, M.D., Stocks, B.J., Mason, J.A., Martell, D.L., Hirsch, K.G., 2001. Direct carbon emissions from Canadian forest fires, 1959–1999. Can. J. For. Res. 31 (3), 512–525. http://dx.doi.org/10.1139/x00-197.

Bartsch, A., Balzter, H., George, C., 2009. The influence of regional surface soil moisture anomalies on forest fires in Siberia observed from satellites. Environ. Res. Lett. 4. http://dx.doi.org/10.1088/1748-9326/4/4/045021.

BEC (Barcelona Expert Centre), 2015. BEC data distribution and visualization services. http://cp34-bec.cmima.csic.es/ (accessed 16.07.15).

Bindoff, N.L., Stott, P.A., Achuta Rao, K.M., Allen, M.R., Gillett, N., Gutzler, D., Hansingo, K., Hegerl, G., Hu, Y., Jain, S., Mokhov, I.I., Overland, J., Perlwitz, J., Sebbar, R., Zhang, X., 2013. 2013: detection and attribution of climate change: from global to regional. In: Stocker, T.F., Qin, D., Plattner, G.-K., Tignor, M., Allen, S.K., Boschung, J., Nauels, A., Xia, Y., Bex, V., Midgley, P.M. (Eds.), Climate Change 2013: The Physical Science Basis. Contribution of Working Group I to the Fifth Assessment Report of the Intergovernmental Panel on Climate Change. Cambridge University Press, Cambridge/New York, NY.

Bourgeau-Chávez, L.L., Leblon, B., Charbonneau, F., Buckley, J.R., 2013. Evaluation of polarimetric Radarsat-2 SAR data for development of soil moisture retrieval algorithms over a chronosequence of black spruce boreal forests. Remote Sens. Environ. 132, 71–85.

Bradshaw, L.S., Deeming, J.E., Burgan, R.E., Cohen, J.D., Compilers, 1984. The 1978 National Fire-Danger Rating System: Technical Documentation. Gen. Tech. Rep. INT-169, U.S. Department of Agriculture, Forest Service, Intermountain Forest and Range Experiment Station, Ogden, UT. 44 pp.

Chaparro, D., Vayreda, J., Martínez-Vilalta, J., Vall-llossera, M., Banqué, M., Camps, A., Piles, M., 2014. SMOS and climate data applicability for analyzing forest decline and forest fires. In: Geoscience and Remote Sensing Symposium (IGARSS), 2014 IEEE International, 13–18 July 2014, pp. 1069–1072.

Chaparro, D., Vall-llossera, M., Piles, M., Camps, A., Rüdiger, C., 2015. Low soil moisture and high temperatures as indicators for forest fire occurrence and extent across the Iberian Peninsula. In: Geoscience and Remote Sensing Symposium (IGARSS), 2015 IEEE International.

Chuvieco, E., Cocero, D., Riaño, D., Martín, P., Martínez-Vega, J., de la Riva, J., Pérez, F., 2004. Combining NDVI and surface temperature for the estimation of live fuel moisture content in forest fire danger rating. Remote Sens. Environ. 92, 322–331.

Dale, V.H., Joyce, L.A., McNulty, S., Neilson, R.P., Ayres, M.P., Flannigan, M.D., Hanson, P.J., Irland, L.C., Lugo, A.E., Peterson, C.J., Simberloff, D., Swanson, F.J., Stocks, B.J., Wotton, B.L., 2001. Climate change and forest disturbances. Bioscience 51 (9), 723–734.

Deeming, J.E., Burgan, R.E., Cohen, J.D., 1977. The National Fire-Danger Rating System—1978. General technical report INT-39, USDA Forest Service, Rocky Mountain Forest and Range Experiment Station, Ogden, UT.

Entekhabi, D., Njoku, E.G., O'Neill, P.E., Kellogg, K.H., Crow, W.T., Edelstein, W.N., Entin, J.K., Goodman, S.D., Jackson, T.J., Johnson, J., Kimball, J., Piepmeier, J.R., Koster, R.D., Martin, N., McDonald, K.C., Moghaddam, M., Moran, S., Reichle, R., Shi, J.-C., Spencer, M.W., Thurman, S.W., Leung Tsang, Van Zyl, J., 2010. The Soil Moisture Active Passive (SMAP) mission. Proc. IEEE 98, 704–716.

ESA, 2015. Climate change initiative. http://www.esa-cci.org/ (accessed 24.07.15).

European Centre for Medium-Range Weather Forecasts (ECMWF), 2015. ERA-Interim project. http://www.ecmwf.int/en/research/climate-reanalysis/era-interim (accessed 12.07.15).

European Comission, 2010. Forest Fires in Europe 2009, EUR 24502 EN, Office for Official Publications of the European Communities, Luxembourg, p. 81.

European Commission, 2015. Fire history, European Forest Fires Information System (EFFIS) applications. http://forest.jrc.ec.europa.eu/effis/applications/fire-history (accessed 16.07.15).

European Environment Agency (EEA), 2006. Corine Land Cover Map.

FAO, 2006. Fire Management-Global Assessment 2006. A Thematic Study Prepared in the Framework of the Global Forest Resources Assessment 2005. Food and Agriculture Organization, Rome.

Fischer, E.M., Seneviratne, S.I., Vidale, P.L., Lüthi, D., Schär, C., 2007. Soil moisture-atmosphere interactions during the 2003 European summer heat wave. J. Climate 20 (20), 5081–5099. http://dx.doi.org/10.1175/JCLI4288.1.

Flannigan, M., Stocks, B., Turestsky, M., Wotton, M., 2009. Impacts of climate change on fire activity and fire mangement in the circumboreal forest. Global Change Biol. 15, 549–560.

Forkel, M., Thonicke, K., Beer, C., Cramer, W., Bartalev, S., Schmullus, C., 2012. Extreme fire events are related to previous-year surface moisture conditions in permafrost-underlain larch forests of Siberia. Environ. Res. Lett. 7, 9. http://dx.doi.org/10.1088/1748-9326/7/4/044021.

García, M., Chuvieco, E., Nieto, H., Aguado, I., 2008. Combining AVHRR and meteorological data for estimating live fuel moisture content. Remote Sens. Environ. 112, 3618–3627.

García-Herrera, R., Díaz, J., Trigo, R.M., Luterbacher, J., Fischer, E.M., 2010. A review of the European summer heat wave of 2003. Crit. Rev. Environ. Sci. Technol. 40 (4), 267–306. http://dx.doi.org/10.1080/10643380802238137.

Gillett, N.P., Weaver, A.J., 2004. Detecting the effect of climate change on Canadian forest fires. Geophys. Res. Lett. 31.

IPCC, 2000. Summary for policymakers: emissions scenarios. A special report of working group III of the intergovernmental panel on climate change. ISBN: 92-9169-113-5. Group 20.

Ireland, G., Petropoulos, G.P., 2015. Exploring the relationships between post-fire vegetation regeneration dynamics, topography and burn severity: a case study from the Montane Cordillera Ecozones of Western Canada. Appl. Geogr. 56, 232–248. http://dx.doi.org/10.1016/j.apgeog.2014.11.016.

Jung, M., Reichstein, M., Ciais, P., Seneviratne, S.I., Sheffield, J., Goulden, M.L., Bonan, G., Cescatti, A., Chen, J., de Jeu, R., Johannes Dolman, A., Eugster, W., Gerten, D., Gianelle, D., Gobron, N., Heinke, J., Kimball, J., Law, B.E., Montagnani, L., Mu, Q., Mueller, B., Oleson, K., Papale, D., Richardson, A.D., Roupsard, O., Running, S., Tomelleri, E., Viovy, N., Weber, U., Williams, C., Wood, E., Zaehle, S., Zhang, K., 2010. Recent decline in the global land evapotranspiration trend due to limited moisture supply. Nature 467 (7318), 951–954. http://dx.doi.org/10.1038/nature09396.

Karamesouti, M., Petropoulos, G.P., Papanikolaou, I.D., Kairis, O., Kosmas, K., 2016. An evaluation of the PESERA and RUSLE in predicting erosion rates at a Mediterranean site before and after a wildfire: comparison & implications. Geoderma.

Kerr, Y.H., Waldteufel, P., Wigneron, J.P., Delwart, S., Cabot, F., Boutin, J., Escorihuela, M.-J., Font, J., Reul, N., Gruhier, C., Juglea, S.E., Drinkwater, M.R., Hahne, A., Martin-Neira, M., Mecklenburg, S., 2010. The SMOS mission: new tool for monitoring key elements of the global water cycle. Proc. IEEE 98, 666–687.

Kerr, Y.H., Waldteufel, P., Richaume, P., Wigneron, J.P., Ferrazzoli, P., Mahmoodi, A., Al Bitar, A., Cabot, F., Gruhier, C., Juglea, S.E., Leroux, D., Mialon, A., Delwart, S., 2012. The SMOS soil moisture retrieval algorithm. IEEE Trans. Geosci. Rem. Sens. 50 (5), 1384–1403. http://dx.doi. org/10.1109/TGRS.2012.2184548.

Liu, Y.Y., Dorigo, W.A., Parinussa, R.M., de Jeu, R.A.M., Wagner, W., McCabe, M.F., Evans, J.P., van Dijk, A.I.J.M., 2012. Trend-preserving blending of passive and active microwave soil moisture retrievals. Remote Sens. Environ. 123, 280–297.

Martínez-Fernández, J., González-Zamora, A., Sánchez, N., Gamuzzio, A., 2015. A soil water based index as a suitable agricultural drought indicator. J. Hydrol. 522, 265–273.

McArthur, A.G., 1967. Fire Behavior in Eucalypt Forests. Leaflet 107, Commonwealth of Australia Forestry and Timber Bureau, Canberra, ACT.

McKee, T.B., Doesken, N.J., Kleist, J., 1993. The relationship of drought frequency and duration to time scales. In: Eighth Conference on Applied Climatology.

Merlin, O., Chehbouni, A., Kerr, Y., Njoku, E., Entekhabi, D., 2005. A combined modeling and multispectral/multiresolution remote sensing approach for disaggregation of surface soil moisture: application to SMOS configuration. IEEE Trans. Geosci. Remote Sens. 43 (9), 2036–2050.

Merlin, O., Walker, J., Chehbouni, A., Kerr, Y., 2008. Towards deterministic downscaling of SMOS soil moisture using MODIS derived soil evaporative efficiency. Remote Sens. Environ. 112 (10), 3935–3946.

Moriondo, M., Good, P., Durao, R., Bindi, M., Giannakopoulos, C., Corte-Real, J., 2006. Potential impact of climate change on fire risk in the Mediterranean area. Clim. Res. 31, 85–95. http://dx.doi.org/10.3354/cr031085.

Oppenheimer, M., Campos, M., Warren, W., Birkmann, J., Luber, G., O'Neill, B., Takahashi, K., 2013. Emergent risks and key vulnerabilities. In: Stocker, T.F., Qin, D., Plattner, G.-K., Tignor, M., Allen, S.K., Boschung, J., Nauels, A., Xia, Y., Bex, V., Midgley, P.M. (Eds.), Climate Change 2014: Impacts, Adaptation, and Vulnerability. Working Group II Contribution to the IPCC 5th Assessment Report. Cambridge University Press, Cambridge/New York, NY.

Padilla, M., Vega-García, C., 2011. On the comparative importance of fire danger rating indices and their integration with spatial and temporal variables for predicting daily human-caused fire occurrences in Spain. Int. J. Wildland Fire 20, 46–58.

Petropoulos, G.P., Ireland, G., Srivastava, P.K., 2015. Evaluation of the soil moisture operational estimates from SMOS in Europe: results over diverse ecosystems. IEEE Sensor. J. http://dx.doi.org/10.1109/JSEN.2015.2427657.

Piles, M., Camps, A., Vall-llossera, M., Corbella, I., Panciera, R., Rüdiger, C., Kerr, Y.H., Walker, J., 2011a. Downscaling SMOS-derived soil moisture using MODIS visible/infrared data. IEEE Trans. Geosci. Remote Sens. 49, 3156–3166.

Piles, M., Camps, A., Vall-llossera, M., Marín, A., Martínez, J., 2011b. SMOS derived soil moisture at 1 km spatial resolution and first results of its application in identifying fire outbreaks. In: Oral Contribution to the 1st SMOS Science Conference, Arles, France, 27–29 September 2011. http://earth.eo.esa.int/workshops/smos_science_workshop/SESSION_6_OTHER_APPLICATIONS/M.Piles_SMOS_Fire%20Outbreaks_.pdf.

Piles, M., Petropoulos, G.P., Ireland, G., Sanchez, N., in press. A novel method to retrieve soil moisture at high spatio-temporal resolution based on the synergy of SMOS and MSG SEVIRI observations. Rem. Sens. Environ.

Piles, M., Vall-llossera, M., Camps, A., Sánchez, N., Martínez-Fernández, J., Martínez, J., Gonzalez-Gambau, V., Riera, R., 2013. On the synergy of SMOS and Terra/Aqua MODIS: high resolution soil moisture maps in near real time. In: Geoscience and Remote Sensing Symposium (IGARSS), Proceedings of the 2013 IEEE International, pp. 3423–3426.

Piles, M., Sánchez, N., Vall-llossera, M., Camps, A., Martínez-Fernández, J., Martínez, J., González-Gambau, V., 2014. A downscaling approach for SMOS land observations: evaluation of high resolution soil moisture maps over the Iberian Peninsula. IEEE J. Sel. Top. Appl. Earth Observ. Remote Sens. 7 (9), 3845–3857. http://dx.doi.org/10.1109/JSTARS.2014.2325398.

Portal, G., Vall-llossera, M., Chaparro, D., Piles, M., Camps, A., Pou, X., Sánchez, N., Martínez-Fernández, J., 2015. SMOS and ECV long term soil moisture comparison: study over the Iberian Peninsula. In: Poster Presented at 2nd SMOS Science Conference, Madrid, Spain, 25–29 May 2015.

Randerson, J.T., Liu, H., Flanner, M.G., Chambers, S.D., Jin, Y., Hess, P.G., Pfister, G., Mack, M.C., Treseder, K.K., Welp, L.R., Chapin, F.S., Harden, J.W., Goulden, M.L., Lyons, E., Neff, J.C., Schuur, E.A.G., Zender, C.S., 2006. The impact of boreal forest fire on climate warming. Science (New York, NY) 314 (5802), 1130–1132. http://dx.doi.org/10.1126/science.1132075.

Ross, M.A., Ponce-Campos, G.E., Barnes, M.L., Hottenstein, J.D., Moran, S., 2014. Response of grassland ecosystems to prolonged soil moisture deficit. Geophysical research abstracts, 16. EGU general assembly 2014.

Sánchez, N., Martínez-Fernández, J., Calera, A., Torres, E.A., Pérez-Gutiérrez, C., 2010. Combining remote sensing and in situ soil moisture data for the application and validation of a distributed water balance model (HIDROMORE). Agric. Water Manag. 98, 69–78.

Sánchez, N., Martínez-Fernández, J., Scaini, A., Pérez-Gutiérrez, C., 2012. Validation of the SMOS L2 soil moisture data in the REMEDHUS network (Spain). IEEE Trans. Geosci. Rem. Sens. 50, 1602–1611.

San-Miguel-Ayanz, J., Moreno, J.M., Camia, A., 2013. Analysis of large fires in European Mediterranean landscapes: lessons learned and perspectives. For. Ecol. Manag. 294, 11–22. http://dx.doi.org/10.1016/j.foreco.2012.10.050.

Scaini, A., Sánchez, N., Vicente-Serrano, S.M., Martínez-Fernández, J., 2014. SMOS-derived soil moisture anomalies and drought indices: a comparative analysis using in situ measurements. Hydrol. Process. 29 (3), 373–383. http://dx.doi.org/10.1002/hyp.10150.

Settele, J., Scholes, R., Betts, R., Bunn, S., Leadley, P., Nepstad, D., Overpeck, J.T., Taboada, M.A., 2014. Terrestrial and inland water systems. In: Field, C.B., Barros, V.R., Dokken, D.J., Mach, K.J., Mastrandrea, M.D., Bilir, T.E., Chatterjee, M., Ebi, K.L., Estrada, Y.O., Genova, R.C., Girma, B., Kissel, E.S., Levy, A.N., MacCracken, S., Mastrandrea, P.R., White, L.L. (Eds.), Climate Change 2014: Impacts, Adaptation, and Vulnerability. Part A: Global and Sectoral Aspects. Contribution of Working Group II to the Fifth Assessment Report of the Intergovernmental Panel on Climate Change. Cambridge University Press, Cambridge/New York, NY.

Shvetsov, E., 2013. Fire danger estimation in Siberia using SMOS data. Geophysical research abstracts, 15. EGU general assembly 2013.

Stoyanova, J.S., Georgiev, C.G., Neytchev, P.N., 2014. Drought extremes on the land surface: operational assessment for agricultural fields over Southeastern Europe. In: The Climate Symposium 2014, Darmstadt, Germany, 13–17 October 2014.

Trigo, R.M., Pereira, J.M.C., Pereira, M.G., Mota, B., Calado, T.J., Dacamara, C.C., Santo, F.E., 2006. Atmospheric conditions associated with the exceptional fire season of 2003 in Portugal. Int. J. Climatol. 26, 1741–1757.

Van Wagner, C.E., 1974. Structure of the Canadian Forest Fire Weather Index. Publication no. 1333, Department of the Environment, Canadian Forestry Service, Ottawa.

Van Wagner, C.E., 1987. Development and structure of the Canadian Forest Fire Weather Index System. Forest technical report 35, Government of Canada, Canadian Forestry Service, Ottawa.

Verdú, F., Salas, J., Vega-García, C., 2012. A multivariate analysis of biophysical factors and forest fires in Spain, 1991–2005. Int. J. Wildland Fire 21, 498–509.

Vicente-Serrano, S.M., Beguería, S., López-Moreno, J.I., 2010. A multiscalar drought index sensitive to global warming: the Standardized Precipitation Evapotranspiration Index. J. Climate 23, 1696–1718.

Wagner, W., Dorigo, W., Dejeu, R., Fernandez, D., Benveniste, J., Haas, E., Ertl, M., 2012. Fusion of active and passive microwave observations to create an Essential Climate Variable data record on soil moisture. In: ISPRS Annals of the Photogrammetry, Remote Sensing and Spatial Information Sciences, vols. I–7. XXII ISPRS Congress, 25 August–01 September 2012, Melbourne, Australia.

Chapter 14

Integrative Use of Near-Surface Satellite Soil Moisture and Precipitation for Estimation of Improved Irrigation Scheduling Parameters

M. Gupta*, P.K. Srivastava*,† and T. Islam‡,§

*NASA Goddard Space Flight Center, Greenbelt, MD, United States, †Banaras Hindu University, Banaras, India, ‡NASA Jet Propulsion Laboratory, Pasadena, CA, United States, §California Institute of Technology, Pasadena, CA, United States

1 INTRODUCTION

Irrigation water management represents the use of a suitable quantity of water at the proper time and it is usually pursued by combining measurements of soil moisture with an optimized irrigation plan (Loew et al., 2006), resulting in maximum crop yields and minimum leaching of water and nutrients to the groundwater. The accurate soil moisture estimation by continuous monitoring of soil moisture for irrigation scheduling will help in improving the management of water resources and reduce pressure on a society facing water scarcity. The irrigation plan is relatively easier to design if water is available; however, having detailed information on the spatial variability of soil moisture is still challenging. The in situ measurement using probes are available but are very costly, and the other limitation, like spatial variability, does not allow a thorough characterization of the whole region. The absence of calibration and validation soil moisture data sets pose a serious challenge for hydrometeorological studies in ungauged basins (Srivastava et al., 2013a), which strongly limit our understanding of ongoing changes and utilization of the resources in water limiting regions (Petropoulos et al., in press). The regions facing the issues of water scarcity need to have improved irrigation scheduling; that is, the efficiency of water use needs to be increased. The water use efficiency can be enhanced by

Satellite Soil Moisture Retrieval. http://dx.doi.org/10.1016/B978-0-12-803388-3.00014-0

271

improving the timing and amounts of water application over the growing season (Dabach et al., 2013; Garg and Hassan, 2007). This requires ensuring optimal water content and maintaining it throughout the growth period through the decision of allocating irrigation water at each time step (Garg and Dadhich, 2014; Garg and Gupta, 2015). Soil water estimation helps in assessing the amount of soil water available for use by the crop. However, due to the spatial heterogeneity and sampling errors, the observational data if available, alone may not be sufficient to provide accurate water budget information. Also, measurement of change in soil storage is difficult as in situ methods are locally restrictive and limited and often nonexistent (Rodell and Famiglietti, 1999). Secondly, inaccurate estimation of rainfall, which is one of the important components in soil water balance, can affect the quantification of soil moisture (Wagner et al., 2003; Gupta, et al., 2014a). Conventionally, rain gauges and gravimetric-based soil moisture measurements have been used for decades to estimate precipitation and soil moisture over a region (Zamora et al., 2011). However, maintenance and establishment of good gauge density network or weather radars in the remote areas has always been a challenge. This problem becomes more complicated in many parts of the world, where the gauges are sparsely distributed, and over the remote/hilly areas, being almost nonexistent and the active radars measurements are rarely available, which complicate the problems (Gupta et al., 2014a). In this context, microwave remote sensing (RS) data can be useful and has demonstrated the possibilities of measuring the soil moisture quantitatively (Srivastava et al., 2013b). Thus, near-surface soil moisture at a regional scale can be observed using RS systems. However, this data still cannot provide information at greater depths below the soil surface (Jackson et al., 1995). As a result, it has not been possible to provide soil moisture information in the root zone, knowledge of which is needed for agricultural and water management applications. Agrohydrological and hydrological models, on the other hand, can simulate soil moisture in the root zone, including spatiotemporal variations (Gupta et al., 2012; Gupta et al., 2014b). Thus, RS data sets along with numerical models have gained considerable interest in many applications including environment, meteorological sciences (Michaelides et al., 2009), and hydrology. Hence, the latest RS products integrated with numerical models provide opportunities for optimum predictive modeling at local/regional or global scales (Gupta and Srivastava, 2010). Numerical models, which have been increasingly utilized to predict unsaturated flow between the upper soil surface and the subsurface environment, are based on the numerical solutions of the Richards equation for variable water movement, involving highly nonlinear processes. Effective management and monitoring of environmental resources requires integrating hydroecologic parameters into biophysical models. However, although models performances have continuously been improved over the years, regional applications for the management of agricultural water are still limited because of the shortage of key input modeling parameters over large areas like the soil-moisture release curve, unsaturated hydraulic conductivity, and soil hydraulic properties, affecting the accuracy of the model in simulating soil water content.

Soils hold different amounts of water depending on their texture and structure. The upper limit of water holding capacity is often called field capacity while the lower limit is called the permanent wilting point. The water available capacity refers to the difference between the field capacity and the wilting point of the soil and represents the soil moisture content at 0.33 bar and 15 bar, respectively. Thus, soil water content at field capacity and wilting point water content is critical information for irrigation scheduling. The two parameters represent part of the soil water retention curve and can be estimated through pressure plate experiments in the laboratory. These curves provide the estimation of soil hydraulic parameters (SHPs), which are the important inputs in a numerical model and affect the model's accuracy in simulating soil water content. The SHPs include van Genuchten parameters (θ_s, θ_r, α, and n). However, laboratory methods being time-consuming and expensive, have resulted in the emergence of indirect methods, pedotransfer functions (PTFs) that take into account soil textural fractions, bulk density, and soil organic carbon content. Empirical relationships between SHPs and routinely measured soil properties, such as the soil textural distribution, organic carbon content, and bulk density, have been established. A number of PTFs are available in the literature (Minasny and McBratney, 2002; Rawls and Brakensiek, 1989; Schaap et al., 2001; Scheinost et al., 1997; Vereecken et al., 1989; Wosten et al., 1999) that relate the descriptive equations (van Genuchten, 1980) with easily measured soil properties. Most efficient and sustainable irrigation scheduling tools obtain field capacity and wilting point water content data from soil texture-based database/PTFs (Jabro et al., 2009). Few studies have suggested that these critical values for determining irrigation timing and amount should be determined on-site as soil texture-based capacity values are not always reliable for irrigation scheduling at field scale. For example, according to Soet and Stricker (2003), the PTF-generated soil hydraulic properties may introduce considerable errors in modeling as no statistical agreement was found between measured and PTF results. Li et al. (2007) confirmed regional limitation of existing PTFs in predicting SHPs, while comparing simulated water contents using directly measured and PTFs-derived soil hydraulic properties and deduced overprediction of water content when simulated using PTFs.

Scientific advances in satellite soil moisture RS such as advanced microwave scanning radiometer for earth observing (AMSR-E), soil moisture and ocean salinity, and the recently launched soil moisture active and passive mission, could solve the previously mentioned problems as recent methods employ inverse modeling and data assimilation with usage of near-surface soil moisture and surface temperature obtained from the satellite data to estimate the SHPs (Entekhabi et al., 1994; Ines and Mohanty, 2008). Inverse modeling, regression techniques, data assimilation methods by utilizing soil-vegetation-atmosphere transfer models, genetic algorithm-based optimization, and uncertainty quantification using ensemble-based approaches under synthetic and field conditions have been developed and adapted over the years. Soil moisture mapping using passive-based microwave RS techniques has proven to be one of the most effective ways of acquiring reliable global soil moisture information on a routine basis. The current study involves making advances toward improving the SHPs for improving soil moisture simulation, especially focusing on the potential integration of microwave near-surface soil moisture (AMSR-E) time series with the physically based model (HYDRUS-1D) and precipitation missions (tropical rainfall measuring mission, TRMM) for effective irrigation scheduling. HYDRUS-1D has been used in the present study as it produces very promising results for soil water content simulations using rain gauge-based measurements as well as TRMM-based rainfall measurements (Gupta et al., 2012).

This work addresses the improvement of field capacity and wilting point by developing a technique for estimating SHPs through the utilization of satellite-based near-surface soil moisture. The method involves the usage of Monte Carlo analysis to utilize the historical RS soil moisture data available from the AMSR-E with the hydrological model. The main hypothesis used in this study is that near-surface RS soil moisture data contain useful information that can describe the effective hydrologic conditions of the basin such that when appropriately inverted would provide a set of SHPs representative of that basin. Thus, a method is followed in this study where hydraulic parameters are derived directly from information on the soil moisture state at the AMSR-E footprint scale and the available water capacity is derived for the root zone by coupling of AMSR-E soil moisture with the physically based hydrological model. Once the SHPs are estimated, the integration of precipitation mission like TRMM/global precipitation measurement can be utilized to estimate the irrigation amount, which could solve the problem for the regions devoid of gauges and radar networks (Hou et al., 2008). The study would help to know how well the derived soil moisture products perform when compared to ground-based measurements and whether it is possible to use the derived data for ungauged catchments and also pave the way for improving the efficiency of irrigation scheduling in the agricultural areas.

2 MATERIAL AND METHODS

2.1 Study Area

The Little Washita River watershed covers parts of Caddo, Comanche, and Grady counties in southwestern Oklahoma, USA, with an area of 600 km^2. The soil texture mainly ranges from fine sand to silt loam with a mean annual precipitation of 760 mm. Hydrological and meteorological measurements of the watershed have been conducted by Grazing Lands Research Laboratory (USDA ARS, http://ars.mesonet.org/), providing a data source to study soil and water conservation and basin hydrology. Also, currently, the ARS monitors the environmental conditions of the Little Washita watershed with a 20-station network called the ARS Micronet. The locations of the stations are shown in Fig. 1. The land use/land cover is primarily composed of range, pasture, forest, cropland, oil waste land, quarries, urban/highways, and water.

2.2 Ground Validation Data

The data sets from the 20 stations were considered for the year 2007, where Jan. 2007 through Aug. 2007 data were used for calibration, and Sep. 2007 through Dec. 2007 data were used for validation purposes. The available data set consists of hourly volumetric soil moisture at 5, 25, and 45 cm soil depths, rainfall data and other meteorological data including air temperature, relative humidity, and solar radiation. Other ancillary data were also obtained such as soil information (soil water retention data, bulk density, saturated hydraulic conductivity) from State Soil Geographic database (Schwarz and Alexander, 1995).

2.3 Satellite Data

A series of quasi-global, near-real time, TRMM-based precipitation estimates has been obtained for the 3-year period from http://mirador.gsfc.nasa.gov. The Level-3 data product TRMM-3B42, a 3-h 0.25 degree product based on multi-satellite precipitation analysis with grid over the latitude band 50 degree N-S (Huffman et al., 2007), is utilized in this study. The moderate resolution imaging spectroradiometer (MODIS) was also chosen for this study, which is available from the NASA ECHO portal (http://reverb.echo.nasa.gov/reverb/) and has

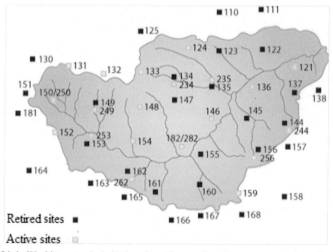

FIG. 1 Little Washita watershed site location. (Source: *http://ars.mesonet.org/sites/*.)

high temporal resolution data of leaf area index (LAI). The LAI from MODIS 8-day composite (MOD15A2) has spatial resolution of 1 km. For testing and evaluation, LAI data for 2007 was processed. The extracted LAI profiles were further processed (smoothed, then averaged) to arrive at an aggregate LAI time series for the study region. Remotely sensed data have a great potential for providing estimates of soil moisture, and AMSR-E data have been used in this study. The AMSR-E is a passive microwave instrument onboard NASA's Earth Observing System Aqua satellite, which measures brightness temperatures at frequencies of 6.9, 10.7, 19, 37, and 89 GHz, and many products are derived from these observations. In this project, we use the Level-3 land surface product available through National Snow and Ice Data Center. The daily soil moisture estimates used here are derived from AMSR-E descending passes. In this study images acquired for 2007 have been used to test the proposed approach. The root mean square error (RMSE) between the AMSR-E soil moisture estimates and in situ data at a depth of 5 cm was found to be 0.04.

2.4 Numerical Modeling

The transport of soil water content was numerically simulated by solving the Richards equation using HYDRUS-1D (Simunek et al., 2005), which solves the following one-dimensional Richards equation for the nonsteady unsaturated water flow:

$$\frac{\partial \theta}{\partial t} = \frac{\partial}{\partial x}\left[K(h)\left(\frac{\partial h}{\partial x}\right)\right] - S \tag{1}$$

where h is the water pressure head (cm), θ is the volumetric water content ($cm^3\, cm^{-3}$), t is time (days), x is the spatial coordinate (cm) (positive upward), K is the unsaturated hydraulic conductivity ($cm\, day^{-1}$), and S is the sink term ($cm^3\, cm^{-3}\, day^{-1}$), representing the root water extraction rate for which the model, Feddes et al. (1978), is used, which defines S as

$$S(h, z) = \alpha_1(h)S_{max}(h, z) \tag{2}$$

where S_{max} ($cm^3\, cm^{-3}\, s^{-1}$) is the maximum possible root water extraction rate when soil water is not limiting, z is the soil depth (cm), and α_1 is a dimensionless water stress reduction factor as a function of pressure head h (cm). α_1 can be expressed as

$$\alpha_1(h) = \begin{cases} 0 & h \geq h_4 \text{ or } h \leq h_1 \\ \dfrac{h - h_1}{h_2 - h_1} & h_1 \leq h \leq h_2 \\ 1 & h_2 \leq h \leq h_3 \\ \dfrac{h_w - h}{h_w - h_3} & h_3 \leq h \leq h_w \end{cases} \tag{3}$$

where h_1 is the pressure head below which roots start to extract water from the soil (cm), h_2 and h_3 are pressure heads between which optimal water uptake exist (cm), and h_w is the permanent wilting point pressure head, below which root water uptake ceases (cm). The values of the parameters are taken from a database provided by Wesseling et al. (1991). Evaporation is calculated based on the differences between precipitation, storage changes, and percolation. The water balance formula is as follows:

$$\frac{dS}{dt} = P - \left(E_p + T_p\right) - L \tag{4}$$

where P is precipitation, L is percolation, dS/dt is the changes of water storage in the soil, E_p is the potential soil evaporation rate, and T_p is the potential transpiration rate. The details are available in the manual (Simunek et al., 2005). Potential evapotranspiration (ET_p) can be estimated indirectly from meteorological data as shown for the study period in Fig. 2.

The selection of a particular method for the determination of ET_p depends upon the type of meteorological data available for the given region and the accuracy desired in the computation of water needs. Several researchers have studied the reliability of the Penman-Monteith method for estimating ET_p (Allen et al., 1998). Thus, ET_p was estimated from meteorological data using the Penman-Monteith method (Allen et al., 1998), which is used to obtain both T_p and potential soil evaporation rate (E_p) according to Eqs. (5) and (6), respectively.

$$E_p = \exp\left(-\alpha_2 LAI\right) ET_0 \tag{5}$$

$$T_p = [1 - \exp\left(-\alpha_2 LAI\right)] ET_0 \tag{6}$$

where LAI is the leaf area index and α is an extinction coefficient of radiation (Feddes et al., 1978). As discussed in Section 2.3, LAI data was obtained from MODIS data.

2.4.1 Initial and Boundary Conditions

The initial condition was taken to be the initial soil moisture in the study domain at time $t = 0$,

$$\theta(z, t) = \theta_i(z, 0), \quad \text{for} - L \leq z \leq 0 \tag{7}$$

where θ_i is the initial moisture content ($cm^3 \, cm^{-3}$) and L is the depth of soil profile (cm). The upper boundary represents the soil-air interface, where water flux depends on the precipitation, evaporation, and the moisture conditions in the soil (Simunek et al., 2005). Thus, system-dependent atmospheric boundary condition was implemented as follows:

$$K(h)\left(\frac{\partial h}{\partial z} + 1\right) \leq E \quad \text{at} \quad z = 0, \text{for} \, t > 0 \tag{8}$$

$$h_A \leq h \leq h_s, \quad \text{at} \quad z = 0, \text{for} \, t > 0 \tag{9}$$

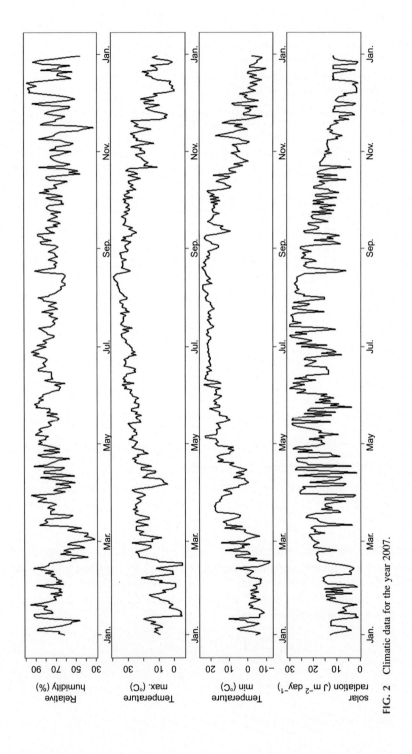

FIG. 2 Climatic data for the year 2007.

where E is the maximum potential rate of infiltration or evaporation under the current atmospheric conditions (cm day^{-1}), and h_A and h_S are, respectively, minimum and maximum pressure head at the soil surface allowed under the prevailing soil conditions (cm). Free drainage was set as the bottom boundary condition for water flow.

$$\frac{\partial h}{\partial z} = 0, \quad z = -L \tag{10}$$

Finite element is used for the spatial variation and finite difference is used for the temporal variation in solving the Richards equation. Weak formulation using the Galerkin method is applied to derive the finite element equations and Crank-Nicholson implicit finite difference scheme is utilized for the temporal approximation.

2.4.2 Soil Parameters

The van Genuchten model (1980), along with K_s, was used to determine the unsaturated hydraulic conductivity (K). The nonlinear unsaturated flow Eq. (1) is solved using an explicit expression between the dependent variable h and the nonlinear terms K and θ. The following closed form θ-h relationship is used,

$$\theta = \begin{cases} \theta_s & h \geq 0 \\ \theta_r + \dfrac{(\theta_s - \theta_r)}{[1 + (\alpha h)^n]^m} & h < 0 \end{cases}, \quad m = 1 - \frac{1}{n} \tag{11}$$

where θ_s is saturated soil water content (cm^3 cm^{-3}), θ_r is residual soil water content (cm^3 cm^{-3}), and θ is soil water content (cm^3 cm^{-3}) at soil matrix potential (h). n and α are curve-fitting SHPs. van Genuchten (1980) used the statistical pore size distribution model (Mualem, 1976) along with Eq. (14) and determined K to yield the K-h relationship,

$$K = \begin{cases} K_s & h \geq 0 \\ K_s \dfrac{\left\{1 - (\alpha h)^{n-1}[1 + (\alpha h)^n]^{-m}\right\}^2}{\{1 + (\alpha h)^n\}^{ml}} & h < 0 \end{cases} \tag{12}$$

where l is the tortuosity parameter. As l is a nonsensitive parameter, it was assumed to be equal to 0.5 and has been similarly incorporated by others (Ines and Droogers, 2002; Pollacco et al., 2013). The SHPs in Eq. (13) were obtained using the Monte Carlo approach. In this study, the method involves extracting target (SHPs) characteristics from remotely sensed (e.g., AMSR-E soil moisture) data. In practice, the probability distributions can be considered to represent either the imprecise knowledge regarding the true value of the parameter, the natural variability of the parameter, or a combination of both. In the procedure, the inference about the set of SHPs is obtained after integrating all possible combinations of the SHPs in the full joint probability posterior distribution. In this study the integration is performed on the set of parameters using a Monte Carlo-based numerical method. The initial ranges of SHPs are taken in correspondence with the values available from the literature. Then, on the basis of the Monte Carlo simulation, the ranges are narrowed in the region where simulation shows a good match between predicted and near-surface soil moisture from AMSR-E. Thus, the model forecasting will be combined with the observational information to optimize the model state and the SHPs simultaneously. The optimization process can be divided into two steps during one time interval: the state variable is first optimized through the state filter. The optimal parameter values are then transferred for retrieving soil moisture. However, soil moisture from sensors such as AMSR-E can only be retrieved for the top couple of centimeters of soil at most. So, for the present study the soil layers have been considered homogeneous. The objective with multidata analysis was to minimize the overall absolute difference between the observed RS near-surface soil moisture and the simulated near-surface soil moisture across time, for all modeling conditions and for all data sources is the total number of data sources/replicates. Here the likelihood function is the time series of AMSR-E-derived soil moisture data at a particular grid point. The 10,000 simulations are performed for the root zone depth to sample the parameters space with the help of LHS. The acceptable or behavioral parameters comprise around 1% of the total number of simulations.

2.5 Evaluation Criteria

To evaluate the performance of the system in helping improve simulation accuracy and whether the system can be used to obtain soil moisture profiles at poorly gauged catchments along with the available water capacity, the RMSE and mean bias error (MBE) are used to measure the performance of the simulations (Petropoulos et al., 2015a, Petropoulos et al., 2015b). It can be calculated as follows.

RMSE was calculated for a comparison between estimated and measured soil water content values. RMSE is defined as

$$\text{RMSE} = \sqrt{\text{MSE}} \tag{13}$$

$$\text{MSE} = \frac{1}{\text{nmsd}} \sum_{i=1}^{\text{nmsd}} (M_i - P_i)^2 \tag{14}$$

where MSE is the mean square error, $\bar{M}$ is the measured mean value, M_i is the measured soil water content, P_i is simulated soil water content corresponding to the measured value, and nmsd is the number of measured data points. Smaller values of RMSE indicate that simulated values match the measured values.

3 RESULTS AND DISCUSSION

In this study, the SHPs were obtained from the PTFs and were compared with the effective estimates obtained from optimization using the AMSR-E soil moisture product. The SHPs were determined for four models of Rosetta PTF (Schaap et al., 2001), given in Table 1, and optimized SHPs methods, which were used as the input parameters in HYDRUS-1D model to assess their suitability by comparing the simulated and measured soil water content values to validate the robustness of the parameters obtained by optimization. The obtained parameters were then also used to calculate the field capacity and wilting point for the root soil zone, which was validated with the available in situ data.

3.1 Comparison of TRMM Rainfall With Ground-Based Rainfall Measurements

The daily TRMM rainfall and ground-based rainfall is shown in Fig. 3A with a scatter plot between ground-based rainfall and TRMM rainfall in Fig. 3B. Fig. 3A shows the trends to be similar for the data extracted from the TRMM

TABLE 1 PTF for Determining Soil Hydraulic Parameters

PTF	Input Requirement
Rosetta (i)	Sand, silt, and clay content
Rosetta (ii)	Rosetta (i) input and bulk density
Rosetta (iii)	Rosetta (ii) input and soil water content at field capacity
Rosetta (iv)	Rosetta (iii) input and soil water content at permanent wilting point

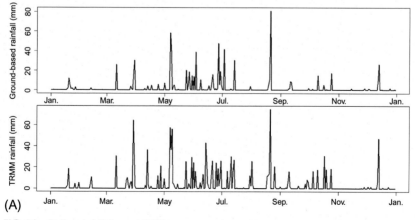

(A)

FIG. 3A   Daily rainfall for year 2007.

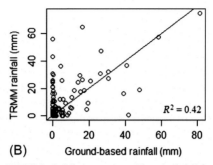

(B)

FIG. 3B   Scatter plot of TRMM rainfall versus ground-based rainfall for year 2007.

to the station-based rainfall data with Nash-Sutcliffe efficiency (NSE) (Nash and Sutcliffe, 1970) being 0.91. The rain events have been well captured by the TRMM; however, the scatter plot demonstrates that there is slight overestimation with TRMM rainfall, which is consistent with other studies (Gupta et al., 2014a).

3.2   Soil Hydraulic Parameters

The SHPs for the current study were optimized using 6 months of data of AMSR-E soil moisture for the period Jan. 2007 to Jul. 2007. The relationship between soil water content (θ) and soil water matric potential (h) is described through the soil water retention curve, which correspond to a continuous van Genuchten curve generated using the SHPs, obtained through both Rosetta PTF and Monte Carlo optimization given in Table 2 and represented in Fig. 4. The optimized parameters based on AMSR-E soil moisture and SHPs based on soil texture are compared with the available in situ data sets of the region. It can be seen from Fig. 4 that optimized parameters are closer to the in situ values as compared to the parameters based on PTF. The accuracy of SHPs obtained through the two approaches was evaluated using NSE, RMSE,

TABLE 2 Soil Hydraulic Parameters

Method		θ_r	θ_s	α	n
Rosetta-PTF based soil hydraulic parameters	i	0.08	0.45	0.01	1.5
	ii	0.08	0.46	0.01	1.56
	iii	0.06	0.44	0.12	1.41
	iv	0.09	0.45	0.03	1.37
Optimized soil hydraulic parameters		0.08	0.38	0.02	1.65
In situ soil hydraulic parameters		0.09	0.40	0.05	1.50

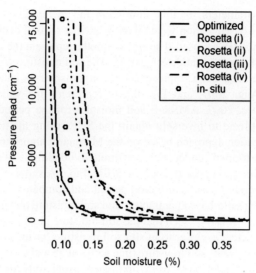

FIG. 4 Soil water retention curves.

TABLE 3 Performance Evaluation for Soil Hydraulic Parameters

	RMSE (Root Mean Square Error)	NSE (Nash-Sutcliffe Efficiency)	MBE (Mean Bias Error)
Rosetta (i)	0.08	−0.001	0.45
Rosetta (ii)	0.09	0.06	0.39
Rosetta (iii)	0.02	0.80	−0.14
Rosetta (iv)	0.06	0.58	0.36
Optimized parameters	0.02	0.90	−0.05

and MBE with the available data set. The optimized parameters show an improvement as compared with texture-based parameters (Table 3). Thus, the optimized hydraulic parameters were utilized in HYDRUS-1D to obtain the surface and root zone soil moisture.

3.3 Soil Moisture Calibration and Validation

The basin average of observed soil moisture was compared with the simulated soil water content from HYDRUS-1D utilizing the optimized parameters and texture-based parameters. In this simulation the meteorological parameters from the ground-based station were taken, while the rainfall data set is utilized when optimized SHPs are used. The first 6 months represent the calibration and the next 4 months are the validation period. The soil moisture was compared at 5, 25, and 45 cm. Fig. 5A–C represents the soil moisture for calibration period Jan. 2007 to Aug. 2007, and Fig. 6A–C represents the validation period, Sep. 2007 to Dec. 2007. AMSR-E soil moisture for the period Jan. 2007 to Aug. 2007 is utilized to inversely obtain the SHPs using the nonlinear least-squares optimization approach based on the Marquardt algorithm. The optimization was performed under the constrained conditions with physically meaningful lower and upper bounds on SHPs. Fig. 5A shows the surface soil moisture time series for the calibration period and compares the soil moisture obtained using Rosetta-based parameters and soil moisture obtained using optimized parameters. It can be seen that the soil moisture obtained from using the optimized parameters is closer to in situ soil moisture for the basin. Similarly, Fig. 5B and C represents soil moisture time series at 25 and 45 cm depth of soil. Even though the SHPs have been optimized using only near-surface soil moisture and the soil depth has been assumed homogeneous for the current study, it can be seen the soil moisture simulated is closer to the in situ soil moisture at both the depths (i.e., 25 and 45 cm) as compared to soil moisture obtained using Rosetta-based SHPs. Although, the bias is high for the root zone, the trend is captured. Moreover, as compared to the pedotransfer SHPs, the RMSE and

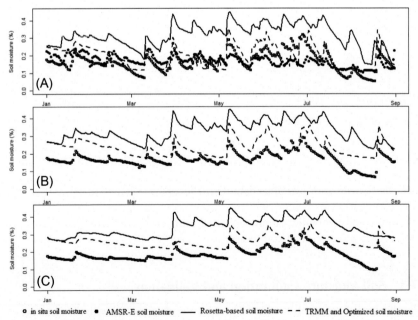

FIG. 5   Soil moisture for calibration period at depths (A) 0–5 cm, (B) 25 cm, and (C) 45 cm.

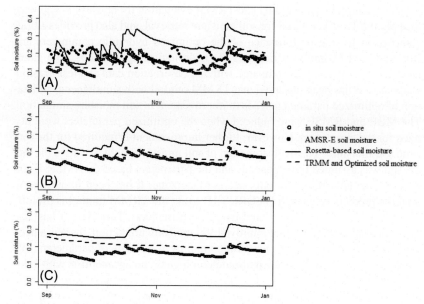

FIG. 6   Soil moisture for validation period at depths (A) 0–5 cm, (B) 25 cm, and (C) 45 cm.

TABLE 4 Performance Evaluation of Soil Moisture

		Calibration Period (cm)			Validation Period (cm)		
	Soil Depth	*0–5*	*25*	*45*	*0–5*	*25*	*45*
Optimized SHPs	RMSE	0.05	0.07	0.07	0.03	0.04	0.05
	MBE	0.25	0.44	0.41	0.03	0.23	0.31
Rosetta SHPs	RMSE	0.15	0.16	0.15	0.11	0.11	0.11
	MBE	1.03	1.01	0.82	0.77	0.71	0.69

MBE have been greatly reduced for both surface layer and the root zone as seen in Table 4. The same trend is also visible in the validation period Sep. 2007 to Dec. 2007. The optimized SHPs perform better as compared with the Rosetta-based SHPs. The RMSE and MBE values are given in Table 4.

4   CONCLUSIONS

A numerical model such as the HYDRUS-1D is able to predict the soil moisture data in a space and time suitable for hydrometeorological modeling. This study provides comprehensive model and TRMM evaluation for new users who wish to apply the TRMM rainfall for soil moisture retrieval, and also provides a brief comparison for the performance of the optimized hydraulic parameters and Rosetta-based hydraulic parameters for hydrological applications. The simulated results obtained are compared with the observation data at a test Washita watershed. It has been observed that TRMM rainfall with simultaneous utilization of optimized parameters leads to improvement in soil moisture simulation. The MBE and RMSE were reduced when the optimized parameters were utilized for the simulation. However, further improvement is required for the root zone soil moisture. The overestimation of soil moisture could be due to the fact the soil was assumed homogeneous and the parameters based on AMSR-E soil moisture actually represent only near-surface zone. It has been found that in total the prediction through HYDRUS 1D using the TRMM rainfall is promising in the absence of any ground-based measurements. However, it has also been concluded that for more improved soil moisture prediction, there is a need of some more efficient algorithms for an improved estimate of parameters. We believe this study will provide hydrometeorologists and agronomists with valuable information on soil moisture prediction. Future research will focus on assimilating soil moisture and simultaneously deriving parameters for the heterogeneous soil column.

ACKNOWLEDGMENTS

The authors would like to acknowledge the validation data of the ARS Micronet stations in the Little Washita River Watershed, provided by Agricultural Research Service, Grazinglands Research Laboratory, El Reno, OK.

REFERENCES

Allen, R.G., Pereira, L.S., Raes, D., Smith, M., 1998. Crop Evapotranspiration—Guidelines for Computing Crop Water Requirements. FAO Irrigation and drainage paper 56, FAO, Rome. 300.

Dabach, S., Lazarovitch, N., Šimůnek, J., Shani, U., 2013. Numerical investigation of irrigation scheduling based on soil water status. Irrig. Sci. 31, 27–36.

Entekhabi, D., Nakamura, H., Njoku, E.G., 1994. Solving the inverse problem for soil moisture and temperature profiles by sequential assimilation of multifrequency remotely sensed observations. IEEE Trans. Geosci. Remote Sens. 32, 438–448.

Feddes, R., Kowalik, P., Zarangy, H., 1978. Simulation of Field Water Use and Crop Yield. Wiley, New York, NY.

Garg, N.K., Dadhich, S.M., 2014. Integrated non-linear model for optimal cropping pattern and irrigation scheduling under deficit irrigation. Agric. Water Manag. 140, 1–13.

Garg, N.K., Gupta, M., 2015. Assessment of improved soil hydraulic parameters for soil water content simulation and irrigation scheduling. Irrig. Sci. 33, 247–264.

Garg, N.K., Hassan, Q., 2007. Alarming scarcity of water in India. Curr. Sci. 93, 932–941.

Gupta, M., Srivastava, P.K., 2010. Integrating GIS and remote sensing for identification of groundwater potential zones in the hilly terrain of Pavagarh, Gujarat, India. Water Int. 35 (2), 233–245.

Gupta, M., Garg, N., Joshi, H., Sharma, M., 2012. Persistence and mobility of 2, 4-D in unsaturated soil zone under winter wheat crop in sub-tropical region of India. Agric. Ecosyst. Environ. 146, 60–72.

Gupta, M., Srivastava, P.K., Islam, T., Ishak, A.M.B., 2014a. Evaluation of TRMM rainfall for soil moisture prediction in a subtropical climate. Environ Earth Sci. 71, 4421–4431.

Gupta, M., Garg, N., Joshi, H., Sharma, M., 2014b. Assessing the impact of irrigation treatments on thiram residual trends: correspondence with numerical modelling and field-scale experiments. Environ. Monit. Assess. 186, 1639–1654.

Hou, A.Y., Skofronick-Jackson, G., Kummerow, C.D., Shepherd, J.M., 2008. Global precipitation measurement. In: Precipitation: Advances in Measurement, Estimation and Prediction. Springer, Berlin Heidelberg, pp. 131–169.

Huffman, G.J., Bolvin, D.T., Nelkin, E.J., Wolff, D.B., Adler, R.F., Gu, G., Hong, Y., Bowman, K.P., Stocker, E.F., 2007. The TRMM multisatellite precipitation analysis (TMPA): Quasi-global, multiyear, combined-sensor precipitation estimates at fine scales. J Hydrometeorol. 8 (1), 38–55.

Ines, A.V., Droogers, P., 2002. Inverse modelling in estimating soil hydraulic functions: a Genetic Algorithm approach. Hydrol. Earth Syst. Sci. Discuss. 6, 49–66.

Ines, A.V., Mohanty, B.P., 2008. Near-surface soil moisture assimilation for quantifying effective soil hydraulic properties under different hydroclimatic conditions. Vadose Zone J. 7, 39–52.

Jabro, J., Evans, R., Kim, Y., Iversen, W., 2009. Estimating in situ soil—water retention and field water capacity in two contrasting soil textures. Irrig. Sci. 27, 223–229.

Jackson, T.J., Le Vine, D.M., Swift, C.T., Schmugge, T.J., Schiebe, F.R., 1995. Large area mapping of soil moisture using the ESTAR passive microwave radiometer in Washita'92. Rem. Sens. Environ. 54 (1), 27–37.

Li, Y., Chen, D., White, R., Zhu, A., Zhang, J., 2007. Estimating soil hydraulic properties of Fengqiu County soils in the North China plain using pedo-transfer functions. Geoderma 138, 261–271.

Loew, A., Ludwig, R., Mauser, W., 2006. Derivation of surface soil moisture from ENVISAT ASAR wide swath and image mode data in agricultural areas. IEEE Trans. Geosci. Remote Sens. 44 (4), 889–899.

Michaelides, S., Levizzani, V., Anagnostou, E., Bauer, P., Kasparis, T., Lane, J., 2009. Precipitation: measurement, remote sensing, climatology and modeling. Atmos. Res. 94 (4), 512–533.

Minasny, B., McBratney, A., 2002. The method for fitting neural network parametric pedotransfer functions. Soil Sci. Soc. Am. J. 66, 352–361.

Mualem, Y., 1976. New model for predicting hydraulic conductivity of unsaturated porous-media. Water Resour. Res. 12, 513–522.

Nash, J., Sutcliffe, J., 1970. River flow forecasting through conceptual models part I—a discussion of principles. J. Hydrol. 10, 282–290.

Petropoulos, G.P., Ireland, G., Barrett, B., in press. Surface soil moisture retrievals from remote sensing: evolution, current status, products & future trends. Phys. Chem. Earth. http://dx.doi.org/10.1016/j.pce.2015.02.009.

Petropoulos, G.P., North, M.R., Ireland, G., Srivastava, P.K., Rendall, D.V., 2015a. Validating a 1D SVAT model in global ecosystems: a tool to study our climate and earth system interactions. Geosci. Model. Dev. http://dx.doi.org/10.5194/gmdd-8-2437-2015.

Petropoulos, G.P., Ireland, G., Srivastava, P.K., 2015b. Evaluation of the soil moisture operational estimates from SMOS in Europe: results over diverse ecosystems. IEEE Sens. J. http://dx.doi.org/10.1109/JSEN.2015.2427657.

Pollacco, J.A., Mohanty, B.P., Efstratiadis, A., 2013. Weighted objective function selector algorithm for parameter estimation of SVAT models with remote sensing data. Water Resour. Res. 49, 6959–6978.

Rawls, W., Brakensiek, D., 1989. Estimation of soil water retention and hydraulic properties. In: Morel-Seytoux, H.J. (Ed.), Unsaturated Flow in Hydrologic Modeling. Volume 275 of the series NATO ASI Series. Springer, Netherlands, pp. 275–300.

Rodell, M., Famiglietti, J., 1999. Detectability of variations in continental water storage from satellite observations of the time dependent gravity field. Water Resour. Res. 35, 2705–2723.

Schaap, M.G., Leij, F.J., van Genuchten, M.T., 2001. ROSETTA: a computer program for estimating soil hydraulic parameters with hierarchical pedotransfer functions. J. Hydrol. 251, 163–176.

Scheinost, A., Sinowski, W., Auerswald, K., 1997. Regionalization of soil water retention curves in a highly variable soilscape, I. Developing a new pedotransfer function. Geoderma 78, 129–143.

Schwarz, G.E., Alexander, R.B., 1995. State Soil Geographic (STATSGO) Data Base for the Conterminous United States. No. 95–449.

Simunek, J., Sejna, M., Van Genuchten, M.T., 2005. The HYDRUS-1D Software Package for Simulating the One-Dimensional Movement of Water, Heat, and Multiple Solutes in Variably-Saturated Media. University of California, Riverside, CA. Research reports:240.

Soet, M., Stricker, J., 2003. Functional behaviour of pedotransfer functions in soil water flow simulation. Hydrol. Process. 17, 1659–1670.

Srivastava, P.K., Han, D., Ramirez, M.R., Islam, T., 2013a. Appraisal of SMOS soil moisture at a catchment scale in a temperate maritime climate. J. Hydrol. 498, 292–304.

Srivastava, P.K., Han, D., Rico-Ramirez, M.A., Islam, T., 2013b. Sensitivity and uncertainty analysis of mesoscale model downscaled hydro-meteorological variables for discharge prediction. Hydrol. Process. 28, 4419–4432. http://dx.doi.org/10.1002/hyp.9946.

Van Genuchten, M.T., 1980. A closed-form equation for predicting the hydraulic conductivity of unsaturated soils. Soil Sci. Soc. Am. J. 44, 892–898.

Vereecken, H., Maes, J., Feyen, J., Darius, P., 1989. Estimating the soil-moisture retention characteristic from texture, bulk-density, and carbon content soil. Science 148, 389–403.

Wagner, W., Scipal, K., Pathe, C., Gerten, D., Lucht, W., Rudolf, B., 2003. Evaluation of the agreement between the first global remotely sensed soil moisture data with model and precipitation data. J. Geophys. Res. 108 (D19), 4611.

Wesseling, J.G., Elbers, J.A., Kabat, P., van den Broek, B.J., 1991. SWATRE: Instructions for Input. Internal Note, Winand Staring Centre, Wageningen.

Wosten, J.H.M., Lilly, A., Nemes, A., Le Bas, C., 1999. Development and use of a database of hydraulic properties of European soils. Geoderma 90, 169–185.

Zamora, R.J., Ralph, F.M., Clark, E., Schneider, T., 2011. The NOAA hydrometeorology testbed soil moisture observing networks: design, instrumentation, and preliminary results. J. Atmos. Ocean. Technol. 28 (9), 1129–1140.

Chapter 15

A Comparative Study on SMOS and NLDAS-2 Soil Moistures Over a Hydrological Basin— With Continental Climate

L. Zhuo*, D. Han*, Q. Dai[†] and T. Islam[‡,§]

**University of Bristol, Bristol, United Kingdom,* [†]*Nanjing Normal University, Nanjing, China,* [‡]*NASA Jet Propulsion Laboratory, Pasadena, CA, United States,* [§]*California Institute of Technology, Pasadena, CA, United States*

1 INTRODUCTION

Soil moisture has been identified as one of the essential climate variables in the Global Climate Observing System (GCOS) (ESA, 2010). Soil is vital to our ecosystems, playing a key role in the carbon cycle, storing and filtering water, and improving resilience to floods and droughts (U.N., 2015). The existence of soil moisture especially at the land surface and atmosphere interface has a strong influence on the water and energy balance (Kerr et al., 2001) because wet soil evaporates more water into the atmosphere. In this context, the important role of soil moisture in various feedback loops of the Earth system cannot be overstated. At a regional scale, accurate soil moisture measurement is important for water resources management and agricultural monitoring. In addition, soil moisture is of great value for weather and climate modeling, and thus significant for preventing extreme events such as floods and droughts, at a global scale (Ottlé and Vidal-Madjar, 1994; Ridler et al., 2014; Srivastava et al., 2013b,c, 2014; Wagner et al., 2007; Wanders et al., 2012, 2014; Zhuo et al., 2015).

A considerable number of quantitative methods have been employed to monitor the spatial and temporal distribution of soil moisture over a wide range of scales (Petropoulos et al., 2015; Vereecken et al., 2014). Traditionally, soil moisture is mainly measured by ground instruments (Petropoulos et al., 2013), such as by using the rather popular time domain reflectometry and the more recent ground penetrating radar system. In-situ measurements have certain advantages, such as the ability to provide soil moisture information at various

Satellite Soil Moisture Retrieval. http://dx.doi.org/10.1016/B978-0-12-803388-3.00015-2

depths with relatively acceptable accuracy, as well as their maturity. However they are expensive, labor extensive, and normally disturb the soil structure (Rahimzadeh-Bajgiran et al., 2013; Zhang et al., 2014). More importantly, in-situ measurements are currently limited to discrete measurements at particular locations and are too sparse to represent the spatial soil moisture distribution (SMD). Therefore, they are not suitable for basin-scale studies (Al-Shrafany et al., 2013; Engman and Gurney, 1991; Srivastava et al., 2013c; Walker et al., 2004; Wang and Qu, 2009).

The advent of satellite remote sensing techniques over the last few decades has produced a brand new field of surface soil moisture observation from space. Nowadays, the advanced Earth observation technology has already shown that soil moisture measurement can be achieved by using almost all regions of the electromagnetic radiation (EMR) spectrum (Petropoulos et al., 2015). Various satellites have been launched to realize this purpose and the most progress has been made by employing the microwave domain of the EMR. This is because microwave owns longer wavelength and thus has better penetration capability through clouds and some vegetation than other bands (e.g., visible and thermal infrared bands) (Petropoulos et al., 2015). Such satellites include the passive scanning multichannel microwave radiometer at 6.6 (C-band) and 10 (X-band) GHz (Reichle et al., 2004); the passive special sensor microwave/imager at 19 (Ku-band) and 37 (Ka-band) GHz (Paloscia et al., 2001); the passive advanced microwave scanning radiometer for Earth observing system from 6.9 to 89.0 GHz (Njoku et al., 2003); the passive WindSat from 6.8 to 37 GHz (Parinussa et al., 2012); the active advanced scattermeter at 5.255 GHz (Brocca et al., 2010); the passive Soil Moisture and Ocean Salinity (SMOS) at 1.4 GHz (Kerr et al., 2001); and the recently launched active soil moisture active and passive at 1.2–1.4 GHz (Entekhabi et al., 2010). Among which SMOS is the first microwave satellite dedicated to monitoring the soil moisture variations over the land (Kerr et al., 2001). In addition, SMOS has a relatively long period of data records since its launch in 2009; hence, it has been selected in this study.

Because all of the aforementioned microwave soil moisture products are obtained at a relatively coarse resolution (around tens of km), evaluating them through comparison with the ground measurement is clearly not appropriate (Srivastava et al., 2014). Although some studies have attempted to validate the satellite soil moisture observations by using the in-situ soil moisture networks (Al Bitar et al., 2012; Jackson et al., 2012; Leroux et al., 2014a), not all basins have these facilities as has been discussed previously. Furthermore, in comparison with the weather and climate modeling fields, there is a lack of research on the satellite soil moisture products in hydrological modeling. Hence, in this study we carry out a comprehensive basin-scale evaluation of the SMOS Level-3 soil moisture products against a widely used hydrological model Xinanjiang (XAJ) as a benchmark, in the central United States. Such an approach provides an opportunity to assess the potential of using the SMOS soil moisture product in hydrological modeling. In land surface models (LSMs), soil moisture is also a key state variable in calculating the evaporative fraction at the surface and the infiltration in the root zone (Leroux et al., 2014a). Therefore the soil moisture information simulated from the four North American Land Data Assimilation System-Phase 2 (NLDAS-2) LSMs, Noah, variable infiltration capacity (VIC), Mosaic, and Sacramento (SAC) are also employed in this study for the comparison. This is to provide a comprehensive assessment of the SMOS soil moisture observations at a basin.

The following section describes the comparison of the SMOS soil moisture observations and the four NLDAS-2 LSMs soil moisture, through the XAJ model derived SMD over the Pontiac basin. The data sets as well as the methodology used in this study are shown in Section 2. The comparison results are analyzed in Section 3. Finally, some concluding points are summarized in Section 4.

2 DATA AND METHODOLOGY

2.1 Study Area and Data Sets

In this study, the mid-sized Pontiac basin (1500 km^2) is chosen. The basin is part of the Vermilion River domain located in mid-Illinois, USA (40.878°N, 88.636°W). The basin is mainly influenced by hot summer continental climate (Peel et al., 2007) and is primarily used for agriculture purposes (Bartholomé and Belward, 2005; Hansen et al., 1998). The majority of the basin area is covered by Mollisols soil type (Webb et al., 2000). The average altitude of the catchment is 188 m mean sea level and the average annual rainfall is 867 mm. The layout of the Pontiac basin is shown in Fig. 1 along with the location of its flow gauge, NLDAS-2 grids, and distribution of the river networks.

The NLDAS-2 (Mitchell et al., 2004) precipitation and potential evapotranspiration at 0.125 degree spatial resolution and daily temporal resolution (converted from hourly resolution) are used to drive the XAJ model. The potential evapotranspiration is derived from the North American Regional Reanalysis (NARR). The precipitation data are derived from the temporal disaggregation of the gauged daily precipitation data from the National Centers for Environmental Prediction (NCEP)/Climate Prediction Center (CPC) with an orographic adjustment based on the monthly climatological precipitation of the Parameter-elevation Regressions on Independent Slopes Model (PRISM) (Daly et al., 1994). The four LSMs soil moisture outputs forced by the same NLDAS-2 meteorological forcings are downloaded from the NLDAS-2 website (Cai et al., 2014) in the period from Jan. 2010 to Dec. 2012, and have been converted from hourly to daily intervals. Furthermore, as shown in Fig. 1, there are a total of 20 NLDAS-2 grids that cover the entire basin. Because XAJ is a lumped hydrological model, each of the NLDAS-2 products (i.e., precipitation, potential evapotranspiration, and the soil moistures) has to be converted into one basin-scale data set by using the weighted average method. The U.S. Geological

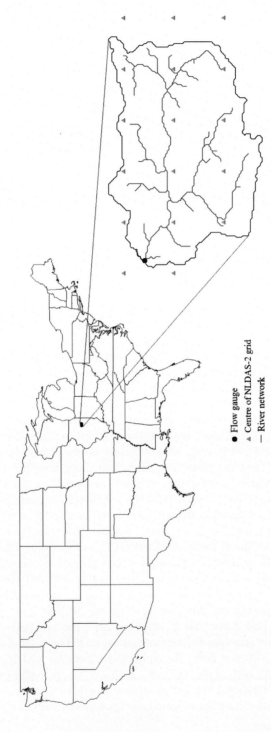

FIG. 1 Location of the Pontiac basin, with the flow gauge and NLDAS-2 grid points over the river network.

Survey (USGS) daily flow data from Jan. 2010 to Apr. 2011 has been used for the calibration of the XAJ model, and the period of May 2011 to Dec. 2011 has been used for validation purposes. The daily air temperature at the nearest meteorological station (i.e., the Pontiac station) is provided by the NOAA National Climatic Data Center (NCDC, 2015). The Level-3 SMOS soil moisture data set evaluated in this study is available for the period from Jan. 2010 to Dec. 2012 and is obtained from the SMOS Barcelona Expert Center (SMOS-BEC).

2.2 SMOS Soil Moisture Products

SMOS is able to provide a global coverage in less than 3 days with a spatial resolution around 35–50 km. It offers global coverage at the equator crossing the times of 6 am (local solar time (LST), ascending) and 6 pm (LST, descending) (Kerr et al., 2012). SMOS acquires brightness temperature at multiple incidence angles from 0 to 55 degrees with full-polarization mode. This feature has been well employed as a key element in the SMOS algorithm in retrieving the soil moisture and the vegetation optical depth (i.e., by minimizing the cost function between the modeled and the observed brightness temperatures) (Kerr et al., 2012).

In this study the Level-3 SMOS soil moisture products generated by the SMOS-BEC (http://cp34-bec.cmima.csic.es) are selected, which provides soil moisture data sets at various temporal resolutions: daily, 3-day, 9-day, monthly, and annual maps. These products are generated in the NetCDF format on two types of grids: first the Icosahedral Snyder Equal Area (ISEA) 4H9 grid projection with aperture 4, resolution 9, with its cell shape in hexagon (Pinori et al., 2008) (the same grid as the Level-2 product), and second the Equal Area Scalable Earth (EASE) grid with a pixel size of $\sim$25 × 25 km. In this study the daily soil moisture product with the EASE grid is used because it is more widely used. The main method implemented to retrieve surface soil moisture is the same as the one employed by the European Space Agency (ESA) operational algorithm for generating the standard Level-2 soil moisture products (Kerr et al., 2012). This study separates the two SMOS passes to see if there are any differences between them and to determine which pass is more suitable for hydrological applications.

2.3 NLDAS-2 LSMs Soil Moisture Simulations

The overall modeling strategy of NLDAS-2 is to generate surface meteorological and hydrological products using the observed gauge precipitation and the bias-corrected reanalysis forcing to drive the four NLDAS-2 LSMs in an offline mode (Xia et al., 2014). More details regarding the setup, parameters, and forcing data can be found in Mitchell et al. (2004), Xia et al. (2012a,b), and Zhuo et al. (2015). The data sets started from Jan. 1, 1979 (Xia et al., 2014).

Four LSMs were selected in NLDAS-2 (Noah, Mosaic, VIC, and SAC). As described in Zhuo et al. (2015), these four models apply different mechanisms in land surface modeling, giving a cross section of different comparison aspects, including small scale versus large scale, coupled versus uncoupled, distributed versus lumped, and others. The Noah LSM is a land surface model of the NCEP operational regional and global weather and climate models (Betts et al., 1997; Chen et al., 1996, 1997; Ek et al., 2003). The Mosaic model is a LSM for the National Aeronautics and Space Administration (NASA) global climate model (Koster and Suarez, 1994, 1996), which has been replaced by the Catchment LSM for the recent upgrade of NASA's GOES-5 system. The VIC model is built as a macroscale semidistributed model, which solves full water and energy balances (Liang et al., 1994). The SAC model is designed as a semidistributed hydrological model (Koren et al., 1999) based on a lumped conceptual hydrological model (Burnash et al., 1973) and has been used widely for small-basin flood forecasting. The first two models emerged within the Soil Vegetation Atmosphere Transfer (SVAT) scheme for coupled atmospheric modeling with a focus on the energy and water flux exchange between the atmosphere and the land, with little calibration; whereas the last two models were originally designed within the hydrological community as uncoupled hydrological models with a focus on flood simulation and considerable calibration (Xia et al., 2014). Through developments, Mosaic, Noah, and VIC have been widely applied as both coupled and uncoupled for all spatial scales. As a result, all three models are considered as both SVATs and semidistributed hydrological models (Mitchell et al., 2004; Zhuo et al., 2015). To compare with the SMOS surface soil moisture observations, only the top layer of the four LSMs' soil moisture is used. Noah, Mosaic, and VIC all have the surface soil layer designed at 0–10 cm, whereas the SAC model is a storage-type hydrological model, so its soil layers are not fixed. A postprocessed soil moisture product with the same soil layer as the Noah model has been developed and can be downloaded from the NOAA/NCEP/EMC website (ftp://nomad6.ncep.noaa.gov/pub/raid2/wd20yx/nldas/Postprocessed_SAC/).

2.4 XAJ Hydrological Model

There are a great variety of hydrological models available globally, ranging from the relatively simple black-box models to the much more complicated physically based, fully distributed models (Zhuo et al., 2014). In this study, a widely used conceptual model XAJ is adopted whose very informative and readable account is given by Zhao (1980). The model has been widely applied to various basins around the world (Chen et al., 2013; Shi et al., 2011; Zhao, 1992; Zhao et al., 1995). The XAJ model is a relatively simple conceptual lumped rainfall-runoff model; its main hypothesis is the runoff generation is on repletion of storage, which means that runoff is not generated until the soil water content of the aeration zone reaches the field capacity. The structure of the XAJ model includes an evapotranspiration unit, a runoff production unit, and a runoff concentration unit. The model includes three soil layers (upper, lower, and deep) which represent the three soil moisture storage components. The runoff component is also known as a water balance module that simulates lumped values of runoff with given rainfall and potential evapotranspiration data sets. The simulated effective rainfall (runoff) is then routed as flow through a routing model to the basin outlet, among which the Muskingum routing method is applied in this study (Zhuo et al., 2015). The model formulations are well suited for automatic parameter estimations. The adopted flowchart of the XAJ model is shown in Fig. 2. The three-layer SMDs are calculated to determine the effect of drying and wetting on the basin soil water storage. In this study only the upper layer's SMD is used for the comparison because it is more scale-matched. The SMD can be calculated using the following equation (Srivastava et al., 2013b):

$$SMD = FC - SMC \tag{1}$$

where FC is the field capacity, which is considered as the upper limit in hydrological modeling for soil moisture and SMC is the soil moisture content.

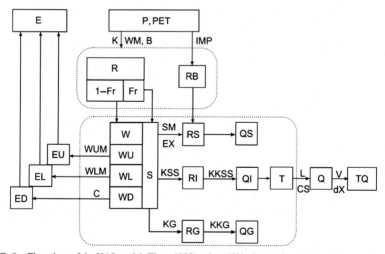

FIG. 2 Flowchart of the XAJ model (Zhao, 1992), where W is the areal mean tension water storage with three components WU, WL, and WD representing upper, lower, and deep soil layers respectively; S is the areal mean free water storage; Fr is the factor of runoff producing area related to W; IMP is the factor of impervious area in a basin; RB is the direct runoff produced from the small portion of impervious area; R is the total runoff generated from the model with surface runoff (RS), interflow (RI), and groundwater runoff (RG) components, respectively. These three runoff components are then transferred into QS, QI, and QG and combined as the total subbasin inflow (T) to the channel network. The flow outputs Q from each subbasin are then routed to the basin outlet to produce the final flow result (TQ).

3 RESULTS AND DISCUSSION

3.1 XAJ Simulations for SMD Estimation

First, the XAJ modeled surface SMD is computed for the Pontiac basin by optimizing the model's 16 parameters. The optimization is based on minimizing the difference between the modeled and the observed flow. For the calibration of the XAJ model, the period of Jan. 2010 to Apr. 2011 is used for calibration and the period of May 2011 to Dec. 2011 is selected for validation. The calibration procedure focuses especially on modeling the actual evapotranspiration and the distribution of the total runoff (e.g., surface runoff, interflow, and groundwater). The optimal values of the model-calibrated parameters are shown in Table 1 (along with their initialization values used in this study). The performance of the model is judged by the Nash-Sutcliffe Efficiency

TABLE 1 The XAJ Model Parameters Used in the Pontiac Basin

Symbol	Model Parameters	Unit	Optimal Value	Range
SM	Areal mean free water capacity of the surface soil layer, which represents the maximum possible deficit of free water storage	mm	31.48	10–50
KG	Outflow coefficients of the free water storage to groundwater relationships	[–]	0.10	0.10–0.70
KSS	Outflow coefficients of the free water storage to interflow relationships	[–]	0.19	0.10–0.70
KKG	Recession constants of the groundwater storage	[–]	0.31	0.01–0.99
KKSS	Recession constants of the lower interflow storage	[–]	0.01	0.01–0.99
CS	Recession constant in the lag and route method for routing through the channel system with each subbasin	[–]	0.26	0.10–0.70
WUM	Averaged soil moisture storage capacity of the upper layer	mm	46.96	30–50
WLM	Averaged soil moisture storage capacity of the lower layer	mm	39.08	20–150

TABLE 1 The XAJ Model Parameters Used in the Pontiac Basin—cont'd

Symbol	Model Parameters	Unit	Optimal Value	Range
WDM	Averaged soil moisture storage capacity of the deep layer	mm	30.11	30–400
IMP	Percentage of impervious and saturated areas in the catchment	%	0.00	0.00–0.10
B	Exponential parameter with a single parabolic curve, which represents the nonuniformity of the spatial distribution of the soil moisture storage capacity over the catchment	[−]	0.70	0.10–0.90
C	Coefficient of the deep layer that depends on the proportion of the basin area covered by vegetation with deep roots	[−]	0.49	0.10–0.70
EX	Exponent of the free water capacity curve influencing the development of the saturated area	[−]	1.93	1.10–2.00
L	Lag in time	[−]	0.00	0.00–6.00
V	Parameter of the Muskingum method	m/s	0.43	0.40–1.20
dX	Parameter of the Muskingum method	[−]	0.18	0.00–0.40

(NSE) (Nash and Sutcliffe, 1970) because it is the most common and important performance measure used in hydrology. The overall performance of the XAJ model shows a NSE value of 0.81 in the calibration and 0.80 in the validation. This result reveals that the calibrated model is hydrologically acceptable. In this study the surface SMD generated from XAJ is employed as a benchmark given the fact that the XAJ model is capable of simulating the hydrological processes in the basin, even though the model is calibrated using the river flow (Lacava et al., 2012; Zhuo et al., 2015). Therefore the XAJ simulated SMD is a hydrologically reliable indicator to surface soil water content, albeit in an inverse relationship. The time series plots of rainfall and flow during the calibration and validation periods are illustrated in Fig. 3. The modeling result indicates that the XAJ model tends to match the measured flow rather well while there is a slight overestimation of low flows during the calibration. On the other hand, during the

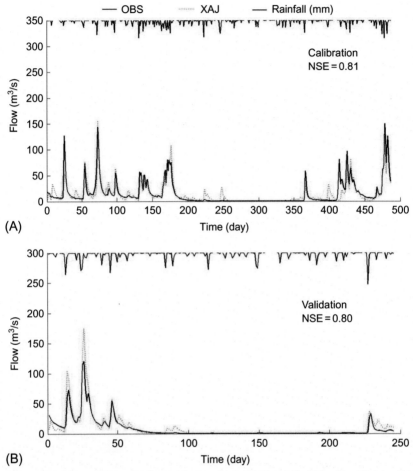

FIG. 3 Time series of daily rainfall and daily flow for the Pontiac basin, during the calibration (A) and validation (B) periods.

validation period there is an overestimation of the overall flow simulation. Presumably, this could be largely because of the methods used for model parameter identification or the deficiency of the model structure itself. For this comparison study, the relevant state variable in the XAJ model is the SMD. The SMD plays an important role in distributing rainfall into different runoff components. If SMD is small, then more surface runoff will be generated after a rainfall event and the remainder to interflow and ground water. It is clear that at some parts of the flow simulation, the model is incapable of calculating the nonlinear behavior of the hydrological processes, and this is particularly evident in the recession curves and low flows. Nevertheless, during most of the monitoring periods, the XAJ model has a relatively good performance and both NSE values are sufficiently high (NSE ≥ 0.8) for an acceptable hydrological model.

3.2   Comparison Soil Moisture Products With the XAJ SMD

The time series of SMOS and NLDAS-2 LSMs soil moisture products, over the period between Jan. 2010 and Dec. 2012 (in the unit of m^3/m^3) are shown in Fig. 4, along with the XAJ SMD simulations (in the unit of m). In addition, complementary information (air temperature, potential evapotranspiration (PET), and precipitation) is added to the time series. It is noted that both the SMOS morning and afternoon soil moisture products are taken into consideration.

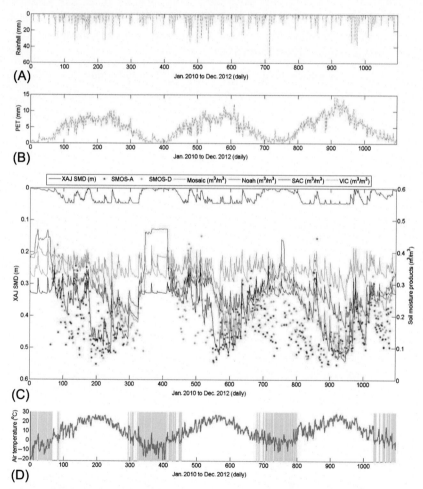

FIG. 4   Time series plots of (A) precipitation, (B) potential evapotranspiration (PET), (C) SMOS (both the ascending and descending orbits) and the NLDAS-2 Noah, Mosaic, VIC, and SAC soil moisture products along with the XAJ simulated SMD, (D) observed air temperature with potential freezing days (≤ 0 °C) shown in the blue shadows. SMOS-A means the SMOS ascending pass, and SMOS-D stands for the SMOS descending pass.

Table 2 and Fig. 5 show the statistical scores of the comparison between the six soil moisture data sets and the XAJ SMD simulations over the Pontiac basin. The Pearson product moment correlation coefficient (r) and the Spearman rank correlation coefficient (r_{sp}) are computed.

The Pontiac basin is mainly covered by cropland subjected to frequent frozen soil events during winter periods. Based on the air temperature (the blue-shaded area in Fig. 4D), the surface soil starts to freeze at the beginning

TABLE 2 Comparison Results Between the SMOS (Both the Ascending and Descending Orbits), NLDAS-2 Mosaic, Noah, SAC, and VIC Soil Moisture Products and the XAJ Simulated SMD

	Unfrozen Period		Frozen Period	
	r	r_{sp}	r	r_{sp}
SMOS-A	−0.56	−0.61	−0.30	−0.41
SMOS-D	−0.68	−0.71	−0.31	−0.39
Mosaic	−0.57	−0.60	−0.59	−0.37
Noah	−0.65	−0.67	−0.58	−0.53
SAC	−0.73	−0.79	−0.72	−0.52
VIC	−0.57	−0.64	−0.09	−0.14

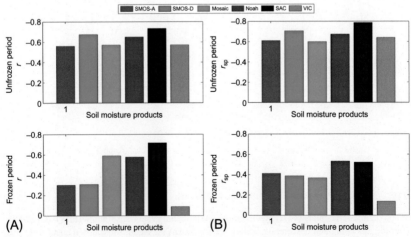

FIG. 5 Statistics scores regarding the comparison between soil moisture from SMOS and NLDAS-2 LSMs and XAJ simulated SMD. (A) r under both unfrozen and frozen periods; (B) r_{sp} for both unfrozen and frozen periods.

of Nov. with the soil moisture value falling suddenly from above 0.2 to below 0.1 m³/m³ and thaw at the end of Mar.. It can be seen that due to frozen soil, SMOS is not able to retrieve valid observations until late Mar.; and the case is even worse for the ascending overpass in 2011. This is clearly an issue in the current microwave soil moisture sensing technology, which has also been reported in other previous studies (Leroux et al., 2014b; Panciera et al., 2009). For this reason, data that are potentially affected by frost/snow cover (air temperature $\leq 0°C$) are separated as the frozen data sets; whereas the rest are indicated as the unfrozen data set (air temperature $> 0°C$) (Wagner et al., 2014).

As shown in Fig. 4C, all soil moisture products (except for VIC during the whole monitoring period and both SMOS orbits during the 2011 winter period) follow the seasonal trend of soil moisture fluctuations and the XAJ SMD very well. So, during the hot summer period (when PET demand is higher than the precipitation replenishment as presented in Fig. 4A and B), soil becomes considerably drier and SMD rises significantly. Whereas when soil gets wetter during rainy seasons, SMD tends to be around zero. Specifically, all soil moisture products are able to capture individual rainfall events rather responsively. In comparison with the other three LSMs, the SAC model gives the largest variations in soil moisture. A similar result was also found in a US basin covered by forest land (Zhuo et al., 2015). On the other hand, VIC presents less seasonal variation when compared with all the other soil moisture products. One possible reason is due to its inappropriate setting of bare soil fraction within a grid cell, as discussed in Xia et al. (2014). As a result, this leads to relatively weaker correlations with the XAJ SMD, as revealed in Table 2. Furthermore, during the 2010 winter (days 300–400), both Noah and SAC models exhibit abrupt soil moisture variations. The reason could be due to the relatively inaccurate forcing data (e.g., rain gauges covered by snow) during the winter time, which have particularly higher impacts on these two models. To further check the linear and nonlinear relationships (Srivastava et al., 2013c), the Spearman rank correlation coefficients (r_{sp}) are also generated, which yield nearly similar trends and values. The similar results reveal that there are no strong nonlinearities between the six soil moisture products and the XAJ modeled SMD. Among the four NLDAS-2 LSMs, the best correlation is achieved by the SAC model at $r = -0.73$ for the unfrozen data set and $r = -0.72$ for the frozen data set. Interestingly both the XAJ model and the SAC model are originated from the hydrological modeling community. They are very effective in modeling water balance-based variables such as soil moisture. Generally speaking, there is no distinct difference in statistical performances between frozen and unfrozen data sets for all the four LSMs (except the VIC model).

The scatter plots between the six soil moisture products and the XAJ SMD are also generated and presented in Fig. 6. As seen from the figure, all soil moisture products (except VIC) range between less than 0.1 m³/m³ to slightly over 0.45 m³/m³. In addition the lower right-hand part of the six scatter plots

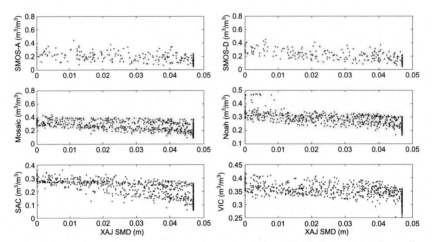

FIG. 6   Scatterplots of the six soil moisture products against the XAJ SMD, during the whole monitoring period of 2010–2012.

represent the wilting points of the six soil moisture data sets and the higher range represents the saturation points of soil. Their values depend on soil porosity (for saturation) and vegetation cover (for the wilting point), as well as the algorithms used in different retrieval models. It is noted that they cannot be either near 0 m^3/m^3 or close to 1 m^3/m^3.

For SMOS, it is expected that soil moisture measurements are more accurate in the hours near dawn when the soil profile has the most time to return to an equilibrium state from the previous day's fluxes (Jackson, 1980). Hence, based upon this hypothesis, it is more likely to be true that the ascending soil moisture would have a better performance than the descending one (Jackson et al., 2012). However, the statistical results reveal that the performance of descending retrievals is better than the ascending's for the whole monitoring period. In addition it is evident from Fig. 4C that at many oscillations, higher soil moisture values are often observed in the descending pass and lower values in the ascending pass. However, this result contradicts the supposition that soil moisture should be at its driest in the late afternoon as SMOS registers the opposite; similar results were discovered in Collow et al. (2012). This phenomenon could be partly explained by the fact that this area is highly affected by radio frequency interference (RFI), which preferentially affects the ascending retrievals because of the SMOS antenna pattern (Collow et al., 2012). The RFI increases the brightness temperature and hence artificially reduces the measured soil moisture (Collow et al., 2012). It is believed that RFI is the main source of errors in the SMOS soil moisture products (Chen et al., 2013). If the RFI emissions are relatively constant, useful corrections could be made in the future (Collow et al., 2012). The SMOS retrievals are then compared with the reference soil moisture data (XAJ SMD). As shown in Fig. 5, the correlations observed by the unfrozen data set are remarkably better than the frozen data set. In addition both SMOS overpasses show good correlations with the XAJ-generated surface SMD under unfrozen condition. Therefore, it is logical to conclude that the SMOS-retrieved soil moisture is only valuable under unfrozen conditions.

4 CONCLUSIONS

Soil moisture is a key variable in many disciplines, especially in the hydrological and meteorology fields. For example, in real-time flood forecasting, the antecedent soil moisture can significantly influence runoff generation by precipitation. Thus, accurate estimation of soil moisture has a huge potential to improve flood forecasting performance. Before any estimated antecedent soil moisture is to be incorporated into a hydrological model, it is important to understand how the soil moisture products can be matched with the corresponding component in a hydrological model (Zhuo et al., 2015). This study investigates the potential of SMOS for capturing the soil moisture variability by a hydrological model (i.e., XAJ SMD), in a medium-sized basin in the central United States. SMOS and NLDAS-2 Noah, Mosaic, VIC, and SAC soil moisture products are compared with the XAJ simulated SMD.

It is found that both SMOS ascending and descending overpasses demonstrate high variability with seasons and follow strong seasonal cycles. They correlate reasonably well with the XAJ SMD for the unfrozen data set. However none of the overpasses show reliable estimates for the frozen data set. This leads to the conclusion that SMOS retrieval is not reliable for hydrological usage when there is frozen soil. Generally speaking, the descending orbit shows a stronger potential for the improved hydrological relevance. This outcome contradicts the previous hypothesis from other studies that the ascending soil moisture would have a better performance than the descending one because at dawn soil is often in near-hydraulic equilibrium. The reason could be explained by the existence of RFI in this region, which preferentially affects the ascending observations.

On the other hand, for LSMs' soil moisture outputs, the SAC model has a significant strong correlation with the XAJ soil moisture information. This is because both XAJ and SAC models are originated from the similar hydrological concept. Although their model structures are relatively simple, they are very effective in modeling water balance-based variables such as soil water storage. Unlike the microwave soil moisture retrieving technique, LSMs' soil moisture simulations show no distinct difference under frozen and unfrozen conditions (except VIC). Another advantage of LSMs soil moisture generation over satellite soil moisture monitoring is that they continue in time. Although LSMs have overall better correlations with the XAJ SMD, they have some limitations. For example, model-based soil moisture tends to suffer from a time drifts problem (e.g., accumulation of errors), while satellite-based soil moisture is based on instantaneous observation of the Earth, so it is not affected by the time drifting issues.

As more and more soil moisture products are becoming available from a wide range of sources, it is important to understand and improve the quality of the current soil moisture products for their relevance in hydrology. For example, it is known that frequent revisit time is important for hydrological applications, especially to obtain adequate measurements of surface wetting and drying between precipitation events (Njoku et al., 2003). Hence it might be valuable to use data fusion technology among different soil moisture products for a wider range of hydrological applications. A useful data source that could be considered is the remote sensing soil moisture products by visible light/infrared bands (Carlson et al., 1994; Piles et al., 2011). Although these techniques are highly influenced by weather conditions (e.g., clouds and rainfall), they can provide soil moisture information at a much finer resolution. Moreover, they monitor land surface temperature information, which has proven to be valuable in downscaling the microwave banded observations by various studies (Kim and Hogue, 2012; Piles et al., 2011; Srivastava et al., 2013a). In addition, data fusion based on multiple active and passive microwave sensors is also a potential solution, such as the ESA CCI project (Wagner et al., 2012).

ACKNOWLEDGMENTS

We acknowledge the U.S. Geological Survey for making available daily streamflow records (http://waterdata.usgs.gov/nwis/rt). The NLDAS-2 data sets used in this article can be obtained through the http://ldas.gsfc.nasa.gov/nldas/NLDAS2-forcing.php website, and the SMOS Level-3 soil moisture is from the SMOS-BEC at http://cp34-bec.cmima.csic.es.

REFERENCES

Al Bitar, A., Leroux, D., Kerr, Y.H., Merlin, O., Richaume, P., Sahoo, A., Wood, E.F., 2012. Evaluation of SMOS soil moisture products over continental US using the SCAN/SNOTEL network. IEEE Trans. Geosci. Remote Sens. 50, 1572–1586.

Al-Shrafany, D., Rico-Ramirez, M.A., Han, D., Bray, M., 2013. Comparative assessment of soil moisture estimation from land surface model and satellite remote sensing based on catchment water balance. Meteorol. Appl. 21, 521–534.

Bartholomé, E., Belward, A., 2005. GLC2000: a new approach to global land cover mapping from Earth observation data. Int. J. Remote Sens. 26, 1959–1977.

Betts, A.K., Chen, F., Mitchell, K.E., Janjic, Z.I., 1997. Assessment of the land surface and boundary layer models in two operational versions of the NCEP Eta Model using FIFE data. Mon. Weather Rev. 125, 2896–2916.

Brocca, L., Melone, F., Moramarco, T., Wagner, W., Hasenauer, S., 2010. ASCAT soil wetness index validation through in situ and modeled soil moisture data in central Italy. Remote Sens. Environ. 114, 2745–2755.

Burnash, R.J.C., Ferral, R.L., McGuire, R.A., McGuire, R.A., U.S.J.F.-S.R.F. Center, 1973. A Generalized Streamflow Simulation System: Conceptual Modeling for Digital Computers. U.S. Department of Commerce, National Weather Service, and State of California, Department of Water Resources, Sacramento, CA.

Cai, X., Yang, Z.L., David, C.H., Niu, G.Y., Rodell, M., 2014. Hydrological evaluation of the Noah-MP land surface model for the Mississippi River Basin. J. Geophys. Res. Atmos. 119, 23–38.

Carlson, T.N., Gillies, R.R., Perry, E.M., 1994. A method to make use of thermal infrared temperature and NDVI measurements to infer surface soil water content and fractional vegetation cover. Remote Sens. Rev. 9, 161–173.

Chen, F., Mitchell, K., Schaake, J., Xue, Y., Pan, H.L., Koren, V., Duan, Q.Y., Ek, M., Betts, A., 1996. Modeling of land surface evaporation by four schemes and comparison with FIFE observations. J. Geophys. Res. Atmos. 101, 7251–7268.

Chen, T.H., Henderson-Sellers, A., Milly, P., Pitman, A., Beljaars, A., Polcher, J., Abramopoulos, F., Boone, A., Chang, S., Chen, F., 1997. Cabauw experimental results from the project for intercomparison of land-surface parameterization schemes. J. Clim. 10, 1194–1215.

Chen, X., Yang, T., Wang, X., Xu, C.-Y., Yu, Z., 2013. Uncertainty intercomparison of different hydrological models in simulating extreme flows. Water Resour. Manag. 27, 1393–1409.

Collow, T.W., Robock, A., Basara, J.B., Illston, B.G., 2012. Evaluation of SMOS retrievals of soil moisture over the central United States with currently available in situ observations. J. Geophys. Res. Atmos. 117.

Daly, C., Neilson, R.P., Phillips, D.L., 1994. A statistical-topographic model for mapping climatological precipitation over mountainous terrain. J. Appl. Meteorol. 33, 140–158.

Ek, M., Mitchell, K., Lin, Y., Rogers, E., Grunmann, P., Koren, V., Gayno, G., Tarpley, J., 2003. Implementation of Noah land surface model advances in the National Centers for Environmental Prediction operational mesoscale Eta model. J. Geophys. Res. Atmos. 108. http://dx.doi.org/10.1029/2002JD003296.

Engman, E.T., Gurney, R.J., 1991. Remote Sensing in Hydrology. Chapman and Hall Ltd., London.

Entekhabi, D., Njoku, E.G., O'Neill, P.E., Kellogg, K.H., Crow, W.T., Edelstein, W.N., Entin, J.K., Goodman, S.D., Jackson, T.J., Johnson, J., 2010. The soil moisture active passive (SMAP) mission. Proc. IEEE 98, 704–716.

ESA, 2010. Soil Moisture Essential Climate Variable. ESA. http://www.esa-soilmoisture-cci.org/ (accessed 09.06.15).

Hansen, M., DeFries, R., Townshend, J.R.G., Sohlberg, R., 1998. UMD global land cover classification. 1 Kilometer. Department of Geography, University of Maryland, College Park, MD, pp. 1981–1994.

Jackson, T.J., 1980. Profile soil moisture from surface measurements. J. Irrig. Drain Eng. ASCE 106, 81–92.

Jackson, T.J., Bindlish, R., Cosh, M.H., Zhao, T., Starks, P.J., Bosch, D.D., Seyfried, M., Moran, M.S., Goodrich, D.C., Kerr, Y.H., 2012. Validation of soil moisture and ocean salinity (SMOS) soil moisture over watershed networks in the US. IEEE Trans. Geosci. Remote Sens. 50, 1530–1543.

Kerr, Y.H., Waldteufel, P., Wigneron, J.-P., Martinuzzi, J., Font, J., Berger, M., 2001. Soil moisture retrieval from space: the Soil Moisture and Ocean Salinity (SMOS) mission. IEEE Trans. Geosci. Remote Sens. 39, 1729–1735.

Kerr, Y.H., Waldteufel, P., Richaume, P., Wigneron, J.-P., Ferrazzoli, P., Mahmoodi, A., Al Bitar, A., Cabot, F., Gruhier, C., Juglea, S.E., 2012. The SMOS soil moisture retrieval algorithm. IEEE Trans. Geosci. Remote Sens. 50, 1384–1403.

Kim, J., Hogue, T.S., 2012. Improving spatial soil moisture representation through integration of AMSR-E and MODIS products. IEEE Trans. Geosci. Remote Sens. 50, 446–460.

Koren, V., Schaake, J., Mitchell, K., Duan, Q.Y., Chen, F., Baker, J., 1999. A parameterization of snowpack and frozen ground intended for NCEP weather and climate models. J. Geophys. Res. Atmos. 104, 19569–19585.

Koster, R.D., Suarez, M.J., 1994. The components of a 'SVAT' scheme and their effects on a GCM's hydrological cycle. Adv. Water Resour. 17, 61–78.

Koster, R., Suarez, M., 1996. Energy and water balance calculations in the Mosaic LSM. NASA Tech. Memo., 104606, 59.

Lacava, T., Matgen, P., Brocca, L., Bittelli, M., Pergola, N., Moramarco, T., Tramutoli, V., 2012. A first assessment of the SMOS soil moisture product with in situ and modeled data in Italy and Luxembourg. IEEE Trans. Geosci. Remote Sens. 50, 1612–1622.

Leroux, D.J., Kerr, Y.H., Al Bitar, A., Bindlish, R., Jackson, T.J., Berthelot, B., Portet, G., 2014a. Comparison between SMOS, VUA, ASCAT, and ECMWF soil moisture products over four watersheds in US. IEEE Trans. Geosci. Remote Sens. 52, 1562–1571.

Leroux, D.J., Kerr, Y.H., Wood, E.F., Sahoo, A.K., Bindlish, R., Jackson, T.J., 2014b. An approach to constructing a homogeneous time series of soil moisture using SMOS. IEEE Trans. Geosci. Remote Sens. 52, 393–405.

Liang, X., Lettenmaier, D.P., Wood, E.F., Burges, S.J., 1994. A simple hydrologically based model of land surface water and energy fluxes for general circulation models. J. Geophys. Res. Atmos. 99, 14415–14428.

Mitchell, K.E., Lohmann, D., Houser, P.R., Wood, E.F., Schaake, J.C., Robock, A., Cosgrove, B.A., Sheffield, J., Duan, Q., Luo, L., 2004. The multi-institution North American Land Data Assimilation System (NLDAS): utilizing multiple GCIP products and partners in a continental distributed hydrological modeling system. J. Geophys. Res. Atmos. 109. http://dx.doi.org/10.1029/2003JD003823.

Nash, J., Sutcliffe, J., 1970. River flow forecasting through conceptual models part I—a discussion of principles. J. Hydrol. 10, 282–290.

NCDC, 2015. NOAA National Climatic Data Center. http://www.ncdc.noaa.gov/ (accessed 09.06.15).

Njoku, E.G., Jackson, T.J., Lakshmi, V., Chan, T.K., Nghiem, S.V., 2003. Soil moisture retrieval from AMSR-E. IEEE Trans. Geosci. Remote Sens. 41, 215–229.

Ottlé, C., Vidal-Madjar, D., 1994. Assimilation of soil moisture inferred from infrared remote sensing in a hydrological model over the HAPEX-MOBILHY region. J. Hydrol. 158, 241–264.

Paloscia, S., Macelloni, G., Santi, E., Koike, T., 2001. A multifrequency algorithm for the retrieval of soil moisture on a large scale using microwave data from SMMR and SSM/I satellites. IEEE Trans. Geosci. Remote Sens. 39, 1655–1661.

Panciera, R., Walker, J.P., Kalma, J.D., Kim, E.J., Saleh, K., Wigneron, J.-P., 2009. Evaluation of the SMOS L-MEB passive microwave soil moisture retrieval algorithm. Remote Sens. Environ. 113, 435–444.

Parinussa, R.M., Holmes, T.R., de Jeu, R.A., 2012. Soil moisture retrievals from the WindSat spaceborne polarimetric microwave radiometer. IEEE Trans. Geosci. Remote Sens. 50, 2683–2694.

Peel, M.C., Finlayson, B.L., McMahon, T.A., 2007. Updated world map of the Köppen-Geiger climate classification. Hydrol. Earth Syst. Sci. Discuss. 4, 439–473.

Petropoulos, G.P., Griffiths, H., Dorigo, W., Xaver, A., Gruber, A., 2013. Surface soil moisture estimation: significance, controls, and conventional measurement techniques. Remote Sensing of Energy Fluxes and Soil Moisture Content. CRC Press, Boca Raton, FL, p. 29.

Petropoulos, G.P., Ireland, G., Barrett, B., 2015. Surface soil moisture retrievals from remote sensing: current status, products & future trends. Phys. Chem. Earth, Parts A/B/C 83–84, 36–56.

Piles, M., Camps, A., Vall-Llossera, M., Corbella, I., Panciera, R., Rudiger, C., Kerr, Y.H., Walker, J., 2011. Downscaling SMOS-derived soil moisture using MODIS visible/infrared data. IEEE Trans. Geosci. Remote Sens. 49, 3156–3166.

Pinori, S., Crapolicchio, R., Mecklenburg, S., 2008. Preparing the ESA-SMOS (soil moisture and ocean salinity) mission-overview of the user data products and data distribution strategy. In: Microwave Radiometry and Remote Sensing of the Environment, 2008. MICRORAD 2008. IEEE, Firenze, pp. 1–4.

Rahimzadeh-Bajgiran, P., Berg, A.A., Champagne, C., Omasa, K., 2013. Estimation of soil moisture using optical/thermal infrared remote sensing in the Canadian Prairies. ISPRS J. Photogramm. Remote Sens. 83, 94–103.

Reichle, R.H., Koster, R.D., Dong, J., Berg, A.A., 2004. Global soil moisture from satellite observations, land surface models, and ground data: implications for data assimilation. J. Hydrometeorol. 5, 430–442.

Ridler, M.E., Madsen, H., Stisen, S., Bircher, S., Fensholt, R., 2014. Assimilation of SMOS-derived soil moisture in a fully integrated hydrological and soil-vegetation-atmosphere transfer model in Western Denmark. Water Resour. Res. 50, 8962–8981.

Shi, P., Chen, C., Srinivasan, R., Zhang, X., Cai, T., Fang, X., Qu, S., Chen, X., Li, Q., 2011. Evaluating the SWAT model for hydrological modeling in the Xixian watershed and a comparison with the XAJ model. Water Resour. Manag. 25, 2595–2612.

Srivastava, P.K., Han, D., Ramirez, M.R., Islam, T., 2013a. Machine learning techniques for downscaling SMOS satellite soil moisture using MODIS land surface temperature for hydrological application. Water Resour. Manag. 27, 3127–3144.

Srivastava, P.K., Han, D., Rico-Ramirez, M.A., Al-Shrafany, D., Islam, T., 2013b. Data fusion techniques for improving soil moisture deficit using SMOS satellite and WRF-NOAH land surface model. Water Resour. Manag. 27, 5069–5087.

Srivastava, P.K., Han, D., Rico Ramirez, M.A., Islam, T., 2013c. Appraisal of SMOS soil moisture at a catchment scale in a temperate maritime climate. J. Hydrol. 498, 292–304.

Srivastava, P.K., Han, D., Ramirez, M.A., O'Neil, P., Islam, T., Gupta, M., 2014. Assessment of SMOS soil moisture retrieval parameters using tau-omega algorithms for soil moisture deficit estimation. J. Hydrol. 519, 574–587.

U.N., 2015. Spotlighting humanity's 'silent ally', UN launches 2015 International Year of Soils. UN News Centre. http://www.un.org/apps/news/story.asp?NewsID=49520#.VWsf8zHF9ag (accessed 09.06.15).

Vereecken, H., Huisman, J., Pachepsky, Y., Montzka, C., Van Der Kruk, J., Bogena, H., Weihermüller, L., Herbst, M., Martinez, G., Vanderborght, J., 2014. On the spatio-temporal dynamics of soil moisture at the field scale. J. Hydrol. 516, 76–96.

Wagner, W., Naeimi, V., Scipal, K., de Jeu, R., Martínez-Fernández, J., 2007. Soil moisture from operational meteorological satellites. Hydrogeol. J. 15, 121–131.

Wagner, W., Dorigo, W., de Jeu, R., Fernandez, D., Benveniste, J., Haas, E., Ertl, M., 2012. Fusion of active and passive microwave observations to create an essential climate variable data record on soil moisture. In: XXII ISPRS Congress, Melbourne, Australiapp. 315–321.

Wagner, W., Brocca, L., Naeimi, V., Reichle, R., Draper, C., de Jeu, R., Ryu, D., Su, C.-H., Western, A., Calvet, J.-C., 2014. Clarifications on the "Comparison between SMOS, VUA, ASCAT, and ECMWF soil moisture products over four watersheds in US" IEEE Trans. Geosci. Remote Sens. 52, 1901–1906.

Walker, J.P., Willgoose, G.R., Kalma, J.D., 2004. In situ measurement of soil moisture: a comparison of techniques. J. Hydrol. 293, 85–99.

Wanders, N., Karssenberg, D., Bierkens, M., Parinussa, R., de Jeu, R., van Dam, J., de Jong, S., 2012. Observation uncertainty of satellite soil moisture products determined with physically-based modeling. Remote Sens. Environ. 127, 341–356.

Wanders, N., Karssenberg, D., Roo, A.d., de Jong, S., Bierkens, M., 2014. The suitability of remotely sensed soil moisture for improving operational flood forecasting. Hydrol. Earth Syst. Sci. 18, 2343–2357.

Wang, L., Qu, J.J., 2009. Satellite remote sensing applications for surface soil moisture monitoring: a review. Front. Earth Sci. China 3, 237–247.

Webb, R.W., Rosenzweig, C.E., Levine, E.R., 2000. Global Soil Texture and Derived Water-Holding Capacities (Webb et al.). Data set Oak Ridge National Laboratory Distributed Active Archive Center, Oak Ridge, TN. http://www. daac. ornl. gov.

Xia, Y., Mitchell, K., Ek, M., Cosgrove, B., Sheffield, J., Luo, L., Alonge, C., Wei, H., Meng, J., Livneh, B., 2012a. Continental-scale water and energy flux analysis and validation for North American Land Data Assimilation System project phase 2 (NLDAS-2): 2. Validation of model-simulated streamflow. J. Geophys. Res. Atmos. 117. http://dx.doi.org/10.1029/2011JD016051.

Xia, Y., Mitchell, K., Ek, M., Sheffield, J., Cosgrove, B., Wood, E., Luo, L., Alonge, C., Wei, H., Meng, J., 2012b. Continental-scale water and energy flux analysis and validation for the North American Land Data Assimilation System project phase 2 (NLDAS-2): 1. Intercomparison and application of model products. J. Geophys. Res. Atmos. 117. http://dx.doi.org/10.1029/2011JD016048.

Xia, Y., Sheffield, J., Ek, M.B., Dong, J., Chaney, N., Wei, H., Meng, J., Wood, E.F., 2014. Evaluation of multi-model simulated soil moisture in NLDAS-2. J. Hydrol. 512, 107–125.

Zhang, J., Zhou, Z., Yao, F., Yang, L., Hao, C., 2014. Validating the modified perpendicular drought index in the north china region using in situ soil moisture measurement. Geosci. Remote Sens. 12, 542–546.

Zhao, R.-J., 1980. The Xinanjiang model. In: Hydrological Forecasting Proceedings Oxford Symposium, IASH 129, pp. 351–356.

Zhao, R.-J., 1992. The Xinanjiang model applied in China. J. Hydrol. 135, 371–381.

Zhao, R.-J., Liu, X.R., 1995. The Xinanjiang model. In: Singh, V. (Ed.), Computer Models of Watershed Hydrology. Water Resources Publications, Colorado, pp. 215–232.

Zhuo, L., Dai, Q., Han, D., 2014. Meta-analysis of flow modeling performances—to build a matching system between catchment complexity and model types. Hydrol. Process. 29, 2463–2477.

Zhuo, L., Han, D., Dai, Q., Islam, T., Srivastava, P.K., 2015. Appraisal of NLDAS-2 multi-model simulated soil moistures for hydrological modelling. Water Resour. Manage., 1–15. http://dx.doi.org/10.1007/s11269-015-1011-1.

Chapter 16

Continental Scale Monitoring of Subdaily and Daily Evapotranspiration Enhanced by the Assimilation of Surface Soil Moisture Derived from Thermal Infrared Geostationary Data

N. Ghilain

Royal Meteorological Institute of Belgium, Brussels, Belgium

1 INTRODUCTION

Monitoring evapotranspiration (ET) over large areas is required in some environmental applications that need regularly updated information about water loss from the soil. Such information, if delivered daily, could benefit different sectors relying on warning systems of drought or flood occurrence (Sepulcre-Canto et al., 2014; Petropoulos et al., in press). But not only that, as some energy, transport (Wade et al., 2006; Dierickx, 2014), and recreational infrastructures (Throssell et al., 2009) also rely on this information to take appropriate measures. Daily ET estimations over continents can already be obtained by the time integration over the day of short-term forecasts by numerical weather prediction (NWP) models. However, the limited timely access to observation data for the surface characterization and the coarse resolution of the final products constrains their use in the aforementioned applications.

A monitoring system based on more Earth observation satellite data seems a valuable option. In the EUMETSAT (European Organization for the exploitation of Meteorological Satellites), Satellite Application Facility (LSA-SAF) (Trigo et al, 2011), an algorithm delivering operationally half-hourly and daily ET maps in near-real time using a combination of numerical weather forecasts and surface variables (mainly surface radiation) derived from the SEVIRI

Satellite Soil Moisture Retrieval. http://dx.doi.org/10.1016/B978-0-12-803388-3.00016-4

sensors onboard the European Meteosat Second Generation (MSG) geostationary satellites (Ghilain et al., 2011).

Its capabilities have been demonstrated in several studies (Ghilain et al., 2011; Petropoulos et al., 2015; Hu et al., 2015; Jovanovic et al., 2014). However, difficulties of this monitoring system arise in some places from the strong dependence of the ET model parameterization to some forcing variables from the numerical weather forecasts, and especially on the modeled soil water status (Ghilain et al., 2011). Therefore, we have explored ways to further constrain the model by assimilating more available satellite data. We included, among others, a daily variation of the vegetation status, and an estimation of daily averaged soil moisture using the satellite-derived land surface temperature, wherever it was possible to adequately relate both variables in a near-real time operation mode. The special focus on soil moisture is justified by its strong connection with ET in semiarid regions: a strong reduction in soil moisture generally implies a restriction on plant transpiration (Williams and Albertson, 2004). Therefore, soil moisture is found to be one of the major components limiting ET in semiarid regions (Teuling et al., 2009).

In this chapter, we review the proposed strategy to monitor ET over Europe and Africa in the LSA-SAF context (half-hourly/daily at 3–5 km spatial resolution). The impact on surface heat fluxes of assimilating vegetation state and surface soil moisture from Earth observation (EO) thermal infrared (TIR) data is assessed by comparing the new proposed prototype with ground observations of surface heat fluxes, and we discuss possibilities for further research lines.

2 OPERATIONAL PRODUCTION OF EVAPOTRANSPIRATION MAPS USING GEOSTATIONARY SATELLITE OBSERVATIONS IN NEAR-REAL TIME

To face the demand from the different regional, national, and international organizations for natural hazard management, and especially in agriculture and hydrometeorology, the EUMETSAT LSA-SAF has set up a service to deliver products related to continental land surface, exploiting the capabilities of the European meteorological satellites. Evapotranspiration is one of the variables that is so far drawing the most interest among public and private sectors (http://landsaf.meteo.pt/viewDocument.jsp?id=599), as it provides an indicator of early drought warning due to its high sensitivity to soil moisture deficit. In providing a near-real time service with freely available products, LSA-SAF has therefore launched an almost unique service (Romaguera et al., 2010, 2012, 2014a,b) (http://aramis.obspm.fr/~jimenez/Docs/WACMOSET/WACMOSET_WP3300_approved.pdf). With the daily ET product, continuous in time, available in near-real time with continental coverage, and a spatial resolution of 3–5 km, LSA-SAF is therefore complementing nicely and partially filling the gap in current scientific services.

The uniqueness of the product finds its roots in the exploitation of continuous satellite data, in comparison to products delivered by integrated forecast systems, like the European Center for Medium-Range Weather Forecasts (ECMWF) Integrated Forecast System (IFS). The products are designed for land surface processes with a spatial sampling that is better than the worldwide forecasts. On the other end, LSA-SAF ET product is complementary to the different ET data sets that have been worked out for climate analyses, usually at a coarser spatial resolution (e.g., Miralles et al., 2010; Balsamo et al., 2015). LSA-SAF ET also responds to several of the requirements from the agriculture sector (http://aramis.obspm.fr/~jimenez/Docs/WACMOSET/WACMOSET_WP1100_approved.pdf).

2.1 Physical Principles

The ET algorithm developed in the framework of the LSA-SAF (Trigo et al., 2011) targets the estimation of the flux of water vapor and heat release from the ground surface into the atmosphere using physical drivers retrieved from MSG satellites. Satellite remote sensing (SRS) stays as the only method capable of providing wide area coverage of environmental variables at economically affordable costs. However, a major difficulty to the use of SRS for monitoring ET at regional and global scale is that the phase change of water molecules produces neither emission nor absorption of an electromagnetic signal. Therefore, the ET process cannot directly be quantitatively observed from satellites: it has to be modeled, taking advantage of information retrieved from the satellite observations about surface variables having a direct influence on ET (Choudhury, 1991). Most of the proposed methods use SRS-derived data combined into models with different degrees of complexity, ranging from empirical direct methods to complex deterministic models based on land surface schemes that compute the different components of the surface energy budget (Courault et al., 2005; Verstraeten et al., 2008). The simplest methods are only applicable locally and should be calibrated again when applied elsewhere. Most methods can only be applied for clear sky conditions, as necessary observations from EO satellites cannot be retrieved in cloudy situations.

The methodological approach selected in the framework of the LSA-SAF intends to be applicable at regional to global scales, to be able to provide a monitoring at short time step, as it allows to follow the diurnal cycle evolution, and to obtain results continuously for all nebulosity conditions. In addition, it has been conceived to ingest an increasing number of remote sensing-derived products as they are released to the community through time. The adopted method can be described as a soil vegetation and atmosphere transfer scheme modified to work in the LSA-SAF operational chain and accept input data from external sources (Ghilain et al., 2011). It is based on the physical model of the Tiled ECMWF surface scheme for exchange processes over land (TESSEL) (Viterbo and Beljaars, 1995; van den Hurk et al., 2000). Main modifications operated to the initial model allow the model to run uncoupled from the atmospheric model and to use data from external sources like SRS-derived data, NWP output and recent information about land-cover characteristics. The adopted methodology combines the forces of remote sensing approaches and NWP models.

In this approach, the area for which ET has to be assessed is divided into independent pixels, in a one-to-one correspondence with the pixels of a satellite image. Each pixel is in turn considered as being a mix of homogeneous *tiles*, each tile representing a particular soil surface: bare soil, C3 crops, grass, deciduous broadleaf forest, and so on. The global pixel value is obtained through the weighted contribution of each tile. Theoretically, ET can be derived in near-real time at the time resolution of MSG satellite images. In practice, the generation of ET will be limited by the availability of input data.

The ET model is conceived to solve an energy balance at the surface for each tile composing each pixel, taking into account the specificity of each tile by a dedicated set of parameters (Ghilain et al., 2011). Once a solution has been found by iteration for each tile, a value of the surface heat fluxes is computed as a weighted average of each tile. Evapotranspiration is directly computed from latent heat flux. At the end of the day, the individual values are summed to compose a daily estimate, without requiring further assumptions.

The needed drivers of the ET model are short and longwave radiation reaching the surface, the near-surface air temperature and humidity, the wind speed at a few meters above the surface, soil moisture and temperature at various depths, the surface albedo, the type of surface and associated characteristics: fraction of vegetation, leaf area index (LAI), roughness lengths, and emissivity.

2.2 Earth Observation Satellite Data

The suite of Meteosat meteorological satellites is interesting for a close monitoring of fast evolving processes in the atmosphere, but also at the surface of the Earth, because of their strategic position in geosynchronous orbit. The SEVIRI instrument carried on board, delivers information in 12 spectral channels in the optical, and near to thermal infrared range, for an observation of the different essential atmospheric and land surface processes (http://www.eumetsat.int). Of interest for land surface, channels operating in the optical, near infrared, and thermal infrared spectral ranges present an excellent opportunity to monitor land surface at every full scanning time (i.e., every 15 min). Meteosat satellites are good candidates for monitoring as they are operational satellites committed in continuous service that has not been interrupted since the early stage by the launch of the Meteosat satellites of the first generation in the 1980s. After the suite of the four second-generation satellites, EUMETSAT is preparing actively the launch of the next series of the third generation, providing with the flexible combined imager instrument, continuity to MVIRI and SEVIRI sensors aboard previous satellite generations, but complemented with new technology allowing new observation capabilities in terms of spatial and temporal sampling.

For ET monitoring, the Meteosat satellites provide possibilities to have access to a continental coverage every 15 min. Within the LSA-SAF, a whole operational production chain has been set up to derive variables of interest for land surface monitoring. Products related to surface energy budget, vegetation, water stress, and fires are being generated mostly from Meteosat data, but also from the polar orbiters MetOp-A and -B (see Fig. 1).

In a first algorithm, delivering ET maps operationally since 2009, only radiation products from LSA-SAF have been exploited: the half-hourly surface short and longwave downwelling radiation components (Geiger et al., 2008a; Trigo et al., 2010) and the daily surface albedo (Geiger et al., 2008b). For the surface state characterization, and vegetation parameterization, the ECO-CLIMAP database (Masson et al., 2003) projected and aggregated to the variable resolution of the Meteosat view is used: it gives the needed information on land cover type, its decomposition into a few plant functional types, a classification useful for SVAT models, a monthly varying vegetation parameterization, including LAI, fraction of vegetation cover, the surface roughness, and the surface albedo in case LSA-SAF product is missing. The database is simple to use

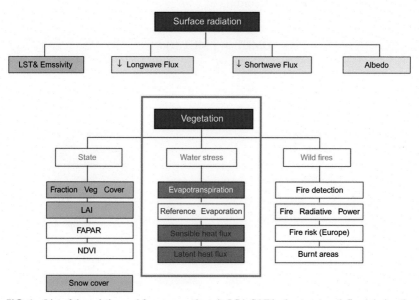

FIG. 1 List of the existing and foreseen products in LSA-SAF in the structure defined during the continuous development and operation phase 2 (2012–2017). Evapotranspiration products are part of a broad class gathering information on vegetation and characterizing water stress. Sensible and latent heat flux are foreseen to be delivered at the same time as evapotranspiration. In the operational evapotranspiration algorithm, three LSA-SAF products are used *(light blue)*. In a later version, LSA-SAF ET v2, the use of four additional products is foreseen *(darker blue)*.

as it refers to a monthly varying look-up table. Only the three dominant plant functional types are retained per observing surface unit for computation efficiency.

The other input information is obtained from ECMWF IFS. Forecasts between 12 and 24 h departing from the initial analysis of wind speed at 10 m, air temperature and humidity at 2 m, soil moisture and temperature at four depths, initially at 0.25 degree spatial resolution and temporal sampling of 3 h, are disaggregated to 1 h and to the spatial resolution of MSG using a bilinear interpolation, and a topographical correction on air temperature (De Bruin et al., 2010).

2.3 Operational Production and Dissemination

The ET algorithm is run by the operational chain of LSA-SAF every 30 min for each of four districts of interest composing the MSG field of view: Europe (Euro), Africa-North (NAfr), Africa-South (SAfr), and South America (SAme). The ET rates are stored in mm h^{-1} in a hierarchically formatted file together with a quality flag, informing the user on the expected accuracy of the estimates and the possible causes of degradation (see Fig. 2).

Daily ET are obtained by temporal integration of instantaneous values generated by the ET algorithm according to

$$\text{DMET} = \int_{h_1}^{h_2} \text{MET}_i(t)dt \tag{1}$$

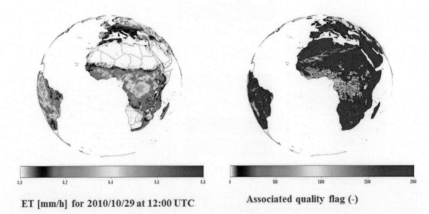

ET [mm/h] for 2010/10/29 at 12:00 UTC   Associated quality flag (-)

FIG. 2 Visualization of an example of half-hourly ET product. The MSG field of view has been recomposed from the four districts of interest. The instantaneous evapotranspiration rate of 12 coordinated universal time (UTC) for Oct. 29, 2010, is accompanied by its quality flag, showing pixels of reduced quality *(in yellow and in green)*.

where MET_i are the instantaneous ET estimates provided by the ET algorithm every 30 min for a given day. The integration limits (h_1, h_2) correspond to the first (theoretically at 00:30 UTC) and last (theoretically at 24:00 UTC) existing slots for a given day. In the best situation, a total of 48 images are generated by day what means that for each MSG pixel, 48 ET values have to be integrated. It happens that some images are missing for a given day. To provide a daily consistent value for each pixel, a linear interpolation between the closest slots is applied according to

$$MET_j = MET_{j-1} + 0.5\left(MET_j + MET_{j-1}\right)N \qquad (2)$$

where MET_{j-1} and MET_j are previous and next existing ET values for the same pixel in considered day and N is the number of time steps between previous and next existing ET values.

The daily product consists of three layers in file: the daily ET per pixel, the percentage of values used for time integration, and the number of missing slots (Arboleda et al., 2011).

Once generated, the products are stored, disseminated via satellite telecommunication to the users via EUMETCAST service, the LSA-SAF website (http://landsaf.meteo.pt), from where it feeds other dissemination tools owned by third-parties for targeted users after some transformations (e.g., Water Observation and Information System, Guzinski et al., 2014).

2.4 Accuracy and Caveats of the Operational Retrieval

The accuracy of the LSA-SAF ET product has been reported in several studies (see Fig. 3). Comparisons have been made with observations from the ground measurements networks (e.g., FLUXNET, RMI-AWS (Gellens-Meulenberghs, 2005)), either on a reprocessed archive (Ghilain et al, 2011) or directly on the disseminated product (Hu et al., 2015; Jovanovic et al., 2014). Also comparisons at continental scale or basin scales with other models have been reported (Hofste, 2014; Ghilain et al., 2011; Helman et al., 2015), and usage of it for other applications (Sepulcre-Canto et al., 2014; Romaguera et al., 2012, 2014a,b; Cammalleri et al., 2014). The studies have reported good accuracies over Europe, with an ability to represent well the seasonal cycle of vegetation, the daily variations, and effects of specific events, like vegetation fires (Hu et al., 2015). In Africa, contrasted accuracies have been reported, but mostly with a better accuracy over temperate regions of Africa (Jovanovic et al., 2014) and low accuracies over semidesert areas (Hofste, 2014; Ghilain et al., 2011; Petropoulos et al., in press). After careful examination, it was concluded that soil moisture input used by the model could be inadequate in those areas (Ghilain et al., 2011). Also, the vegetation representation in the model could play a role in adjusting the transpiration to the actual annual vegetation cycle (Ghilain et al., 2012).

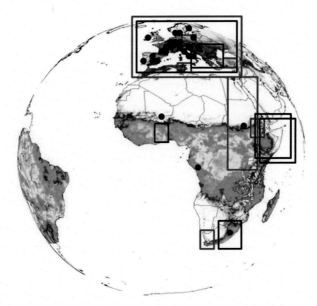

FIG. 3 Areas where validation *(black dots and red frames)* or usage *(black frames)* of LSA-SAF ET products have been reported in the scientific literature are mostly localized in Europe, but also cover parts of the African continent.

Stability over time of the incoming information is a crucial element in setting up and maintaining a proper continuous near-real time operational production. The ET algorithm of LSA-SAF relies on various sources of information with a strong dependency. A change of satellite, for example, replacing the exploitation of MSG-3 by MSG-4, might have an impact on all the primary and secondary satellite products, due to calibration problems, faulty onboard devices, or in the numerous processing software packages. For meteorological fields, ECMWF IFS has been selected for its stability and high performance, providing very high quality meteorological variables at an ever-increased spatial resolution. However, to maintain this, changes of ECMWF forecast system cycles can bring a discontinuity in the supplied fields, triggering large efforts to adapt the processing of application products downstream. Over the years, several updates have been applied to the land surface scheme TESSEL used in operational weather forecasts at ECMWF (see Table 1), a few of them having heavy consequences on the nominal generation of the ET products of the LSA-SAF, and especially related to soil moisture.

Both arguments, degraded quality over semiarid Africa and strong dependence in ECMWF soil moisture, pushed further development in relying less on ECMWF IFS soil moisture and more on remote sensing data, if at all possible. This initiative was eased by the increasing accessibility and maturity of soil moisture products derived from satellites. In addition to soil moisture, the new developments have considered vegetation-related products obtained through remote sensing.

TABLE 1 List and Dates of Implementation of Changes in the IFS at ECMWF, Relevant to Land Surface

Cycle	Date	Changes
32r3	2007-11-06	New soil hydrology scheme
35r2	2009-03-10	Revised snow scheme
35r3	2009-09-08	Routine monitoring of soil moisture observations from Metop-A ASCAT
36r1	2010-01-26	Deterministic forecast and analysis horizontal resolution is increased (16 km)
36r4	2010-11-09	Monthly varying climatology of LAI based on MODIS data
		Change of parameterization of the bare soil evaporation
		Recalibration of vegetation parameters

Source: http//www.ecmwf.int.

3 A NEW PROTOTYPE OF THE EVAPOTRANSPIRATION ALGORITHM CONSTRAINED BY MORE EO SATELLITE DATA

3.1 Surface Soil Moisture Index Derived From Land Surface Temperature

Soil moisture is one of the important environmental factors constraining ET rates in semiarid environments. Finding a source for soil moisture is constrained by the operational timeliness of the monitoring system, but also by the continuity of the generation of such data and products. A potential source of information for soil moisture from EO are products derived from the microwave scanning aboard polar orbiters (Bartalis et al., 2007; Kerr et al., 2010; Petropoulos et al., in press). In the framework of EUMETSAT, an easy and fast transmission of data is possible between EUMETSAT centralized (headquarters) and decentralized (SAF) production facilities. Therefore, the products based on MetOp-ASCAT sounders are on the first line, as they are processed at EUMETSAT headquarters. However, as the surface soil moisture product is provided at a spatial resolution coarser ($\sim$25 km) than the ET product from MSG, an alternative proposed in the early stage of development was to first exploit the TIR signal, and especially the diurnal variation of the land surface temperature, from Meteosat-SEVIRI. It had the advantage of avoiding any delay in the production by relying on LSA-SAF-derived products only, with the desired spatial resolution of a few kilometers.

The exploitation of the land surface temperature for soil moisture monitoring has been studied for decades by scientists showing an increasing interest over the years and yielding promising results mostly at regional scale (Verstraeten et al., 2006; Stisen et al., 2008; Garcia et al., 2014). Within the framework of LSA-SAF production, a stable methodology had to be developed for continental scale application in near-real time. As proposed in earlier studies (Stisen et al., 2008; Garcia et al., 2014), the daily morning heating rate computed from LST at a specific location is used to estimate a daily surface soil moisture estimate using a low-order model.

However, this combination of TIR observations from a geostationary satellite and of a simple low-order model is expected to render meaningful surface soil moisture estimates only in a set of geographical zones depending on the surface characteristics (vegetation and soils) and climates. A simple model has therefore been developed, applied with the land surface temperature product from LSA-SAF (Trigo et al., 2008; Freitas et al., 2010; Ermida et al., 2014) and an extensive validation has been carried out sampling climates of Europe and Africa (Ghilain et al., 2015) revealing the accuracy and limitations in applying this methodology. The results showed acceptable accuracy in soil moisture retrieval using this technique over semiarid and arid areas, including Mediterranean regions and Sahelian regions. Mitigated to insignificant performances were found for temperate climates of Europe and humid climates of Africa. A further analysis showed that the new indicator could be better than, or as accurate as, ECMWF IFS soil moisture short-term forecasts for semiarid regions of Africa and Southern Europe (see examples in Fig. 4) for days with available satellite observations.

In semiarid regions, the TIR signal exploitation brings additional information: wetland extents and irrigation can be easily detected, as shown in Fig. 5, because of their clear contrasts with their neighboring environment.

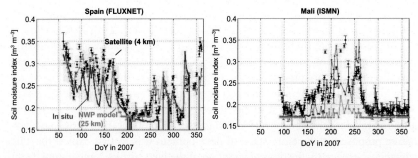

FIG. 4 Examples of validation of the surface soil moisture retrieved from land surface temperature *(black)* with in situ observations *(red)* and comparison with the ECMWF IFS soil moisture input of the current operational ET product *(green)*. The satellite-derived soil moisture relates better than the numerical weather prediction (NWP) soil moisture for the African site.

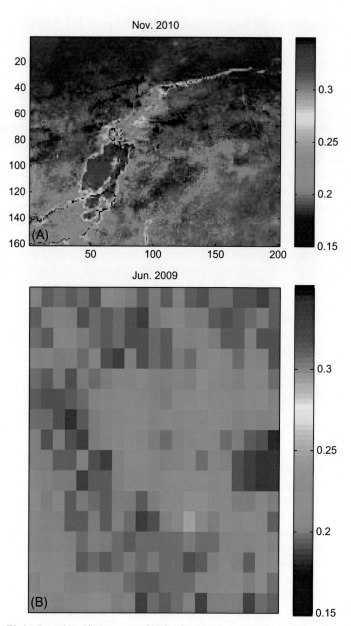

FIG. 5 Wetlands are identified as areas of high soil moisture, as here for the inner Niger delta in Nov. 2010. Irrigation patterns are also recognizable as zones of higher soil moisture even at the coarse resolution of the satellite, as in the area of Barrax-Albacete, Spain, in Jun. 2009. Soil moisture represented is rescaled between 0.171 and 0.35 m^3 m^{-3}.

3.2 Assimilation of Vegetation Products From EO Satellites

LAI is a variable characterizing the vegetation vertical abundance of leaves, quantifying the vertically integrated area of green material per surface unit on the ground. The LAI used in this context is produced daily by LSA-SAF at a spatial resolution of a few kilometers in the SEVIRI projection (Garcia-Haro et al., 2005). Using that product presents several advantages resulting from an up-to-date follow-up of the vegetation state pixelwise: it allows the detection of short-term local fluctuations to interannual variability of vegetation health and productivity. Those are expected to bring a clear improvement to the ET model, as its vegetation information is varies monthly in the operational context, based on past data and does not vary from year to year.

After having implemented the change from the ECOCLIMAP customized database to the LSA-SAF LAI for use in the ET algorithm (Ghilain et al., 2012), we have reprocessed archives for the year 2007 over Europe and Africa. The result of the comparison between the new version and ground observations shows a clear improvement over semiarid areas especially in Europe: correlation scores have been improved, and variability is comparable to the observations (see Fig. 6). However, the scores at Sahelian sites have been found to be still very low, showing that the model is quite more sensitive to soil water availability than vegetation input and that the current soil moisture information was not adequate for those regions.

3.3 Prototype Usage of Soil Moisture Products

Land surface temperature is not a universal indicator of soil moisture. Past studies (Prigent et al., 2005) discovered that the amplitude of the land surface temperature diurnal cycle was related to soil moisture in regions controlled by thermal inertia, not in transpiration-driven regions (Teuling et al., 2009). The validation study presented in Ghilain et al. (2015) exhibits the same conclusions: results from the comparison show an overall good performance over semiarid regions and degradation toward more wet and vegetated areas. As an operational application should not be of degraded quality by a change in the model or in its settings, a comparison of the performances of the two sources was undergone over Africa and Europe to select the best source of soil moisture data in function of climate zones. One of the indicators used in this comparison with observational ground-truthing data sets was the correlation coefficient, which is one of the measures of soil moisture products quality requested by the EUMETSAT users (http://hsaf.meteoam.it/documents/docs/20130100/SAF_HSAF_CDOP2_PRD_1_0.pdf). By grouping the results of the statistical comparison for the different sites within a climate zone, it was possible to state the actual performance of the two sources, and to define where the soil moisture derived from Meteosat satellite data could better characterize soil moisture than

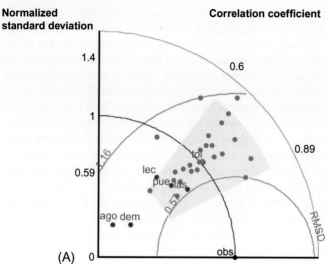

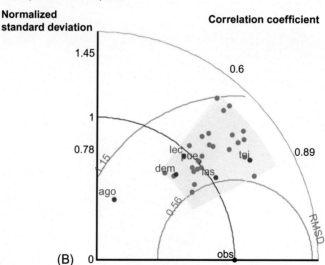

FIG. 6 Statistical scores of the comparison of LSA-SAF ET in its nominal configuration (MET 4.0.3), or with the use of LSA-SAF vegetation products (MET 4.1.0) with observations from the FLUXNET sites over Europe and Africa. Each point (*green* in unstressed and *red* (with label) in stress soil moisture conditions) in the Taylor diagram (Taylor, 2001) represents the comparison of 1 year time series. Sites in Africa are Agoufou (ago) and Demokeya (dem).

ECMWF IFS forecasts for days with not completely overcast conditions (see Fig. 7). The interpretation of the diurnal variation of LST in terms of soil moisture is further hampered by the specific viewing geometry of the SEVIRI sensor and its relation to terrain orography. Areas with complex terrain are assigned to ECMWF IFS soil moisture, as more research would be needed to disentangle the signal from the soil moisture from its perturbation due to shadows and facing south observation. In the Mediterranean region, scores of the ECMWF IFS are slightly higher than the thermal signal from the satellite. But, as irrigation is a common agricultural practice in that zone, and ECMWF IFS does not perform up to now to a spatial resolution where irrigation may have an impact on weather, we select the thermal signal for soil moisture primary source in the regions where irrigation is significant (at least 10% based on the cadastral map of agricultural infrastructure (Siebert et al., 2010, 2013) and the available data from the Food and Agriculture Organization).

Finally, for regions where the primary source is the satellite-derived soil moisture, unavoidable gaps due to persistent cloudy conditions or to missing satellite data are filled by ECMWF IFS soil moisture, locally adapted to match the temporal dynamics imposed by the soil moisture derived from the satellite (Liu et al., 2011) (see Fig. 8).

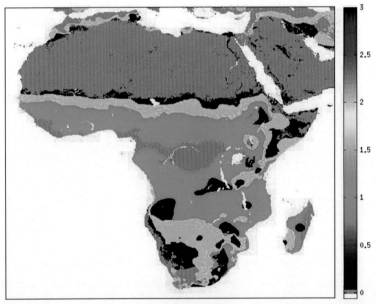

FIG. 7 Map of choice of input data for soil moisture based on the clustering by climate of comparison of both sources to in situ observations: EO derived is best based on correlation *(brown areas)*, ECMWF IFS is best *(light blue)*, equal/slightly better ECMWF IFS (within 3% difference in correlation) *(orange)*, and unknown accuracy *(grey)*. The best source possible is used by the LSA-SAF ET new prototype.

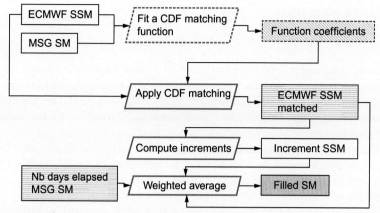

FIG. 8 Flowchart of the process of filling gaps due to persistent cloud coverage in time series of soil moisture derived from LST. ECMWF IFS soil moisture is matched to past series and coefficients are applied to transform it when needed.

Once a surface soil moisture map has been obtained, the root-zone soil moisture is computed for zones with prevalence of the satellite input using an exponential filter (Albergel et al., 2008), differentiated for high and low vegetation types. For zones where ECMWF IFS is selected as the primary source for soil moisture, nothing is altered in the algorithm exposed in the first section.

3.4   A Revised Scheme for Evapotranspiration Modeling

The physical equations and parameterizations behind the algorithm have not been substantially revised. New processes were added to the model: interception of rain and dew by the vegetation canopy and its evaporation, snow sublimation in the case of permanent or temporary snow, and evaporation over small water bodies. Revision of the soil moisture limitation over bare soils and parameters related to vegetation resistance to transpiration was undergone. The representation of the vegetation development in the model has been revised, especially for crops and forests.

The formulation canopy resistance, r_c, of the vegetation to transpiration is based on van den Hurk et al. (2000), while the parameterization for crops and nonperennial grasslands has been slightly changed, such as to introduce more sensitivity of the model for low green biomass conditions.

For crops and grass in semiarid ecosystems:

$$r_c = \left(\frac{r_{s,min}}{0.25.(\exp(LAI) - 0.8)} + 50 \right) f_1(S_{\downarrow}) f_2(\overline{\theta}) f_3(D_a) \tag{3}$$

For the other cases:

$$r_c = \frac{r_{s,min}}{LAI} f_1(S_\downarrow) f_2(\overline{\theta}) f_3(D_a) \tag{4}$$

where $S_\downarrow$ is the surface shortwave radiation, $\overline{\theta}$ is the average unfrozen soil water content, D_a is the atmospheric water pressure deficit, LAI is the leaf area index, $r_{s,min}$ is the minimum stomatal resistance, and $f_1(S_\downarrow), f_2(\overline{\theta}), f_3(D_a)$ are functions parameterizing the effect of environmental factors on the plant transpiration rate.

The minimum stomatal resistance, $r_{s,min}$, scaled by LAI or the effective LAI (crops and nonperennial grasslands) fixes the maximum rate of ET observed for each vegetation type in the absence of any constrains. Values of $r_{s,min}$ were recalibrated.

A separate formulation is used for bare soil. A minimum stomatal resistance is associated with bare soil to represent the minimum soil resistance and the only stress assumed for soil evaporation is due to soil water deficit, via a modified Jarvis function $f_{2,BS}$, which allows a broader range of soil moisture sensitivity. f_{liq} is the fraction of the soil water content that is in liquid phase, θ_{pwp} is the soil water content at wilting point, chosen to be the minimum soil water content possible, and θ_{fc} is the soil water content at field capacity, value above which ET is not restrained anymore by soil water content. Both θ_{pwp} and θ_{fc} act as values to rescale the soil water content of the surface soil layer (0–7 cm) θ_1. In the model, they are set to values prescribed in the TESSEL model for average soils, 0.171 and 0.323 $m^3\ m^{-3}$, respectively.

$$r_c = r_{s,min} \cdot f_2(f_{liq} \cdot \theta_1) \tag{5}$$

$$f_{2,BS}(\overline{\theta})^{-1} = 1 + \exp\left(-50 \cdot \frac{\theta - \theta_{pwp}}{1000 \cdot (\theta_{fc} - \theta_{pwp}) + 1}\right) \tag{6}$$

4 GAIN AND LOSS OF ACCURACY IN CHOOSING A NEW SOURCE FOR SOIL MOISTURE INPUT: VALIDATION OF THE PROTOTYPE BY COMPARISON WITH IN SITU OBSERVATIONS

The revised algorithm was tested over FLUXNET stations in Europe and Africa with soil moisture received from ECMWF IFS and, in addition for Mediterranean and African regions, also with the new soil moisture input derived from land surface temperature retrievals. The result of both implementations was compared to ground measurements. For each site, and each year, statistical measures of agreement have been computed: bias, root mean square difference and correlation index. The difference of each score between both configurations is represented in Fig. 9, highlighting the gain (in green) and loss (in red) of

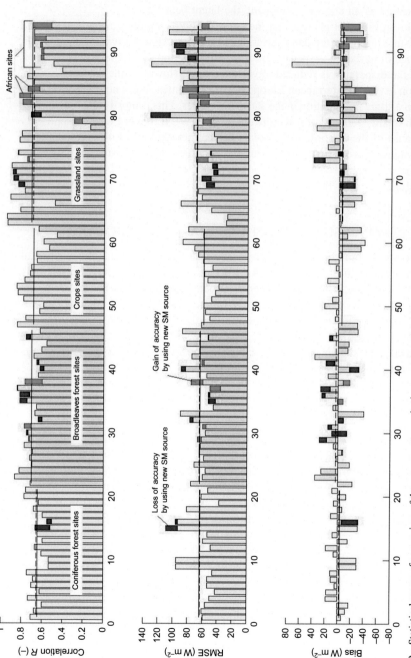

FIG. 9 Statistical scores of comparison of the new evapotranspiration prototype algorithm using and not using EO soil moisture with yearly ground observation data sets of latent heat flux. Gain and loss of accuracy in using EO soil moisture are represented in *green* and *red*, respectively.

accuracy in using satellite-derived soil moisture to derive half-hourly ET, here presented as an energy component, the latent heat flux.

From Fig. 9, it is clear that the assimilation of remote sensing derived soil moisture improves the daily variations of ET over African savannahs with dominance of herbaceous species: the correlation is improved, the bias is reduced, and the dispersion is reduced most of the time. However, the result of assimilation in the Mediterranean European grassland sites shows no improvement or even a slight degradation in the statistical scores. This is mostly attributed to the high-quality forecast of soil moisture by ECMWF IFS over Europe, supposedly because of the assimilation of near-surface meteorological observations from the dense European network. For the Mediterranean forests, no systematic effect can be detected, with some sites presenting gain and loss of accuracy alternatively over the years.

5   COMPARISON WITH THE OPERATIONAL PRODUCT

A comparison with the operational product has been initiated at the location of measurement sites (see Fig. 10). As expected, the ET patterns have been largely improved in the Sahel region, where the operational algorithm has difficulties in accounting for the large seasonality of water fluxes. The combination of assimilating vegetation and soil moisture products from EO seem to work out the actual daily variations, with some exceptions. In Spain, however, the comparison does not show a large improvement, as the operational version gives results already close to the in situ observations.

6   CONCLUSIONS

The LSA-SAF delivers near-real time products derived from Meteosat satellites, as a decentralized production center of EUMETSAT. It is targeting the near-real time provision of land surface variables, such as radiation components at the surface, but also the surface characterization (vegetation monitoring, snow, and fire detection), and finally the heat and water fluxes from the land surface. The operational algorithm for deriving half-hourly ET is based on a model driven by three radiation components produced by LSA-SAF and by short-term forecasts from the ECMWF. When assessing its accuracy, both the development team and the users have reported its accuracy and limitations. One of the limitations in accuracy has been found to be pronounced in semiarid regions, and especially in Africa. After careful examination, the ET algorithm was found to exploit an inappropriate soil moisture source, which is derived from the ECMWF forecasts, caused either by the forecast quality of the soil moisture in those regions or by the regular updates of the forecast system that can cause degradation in downstream applications if they are not well studied beforehand.

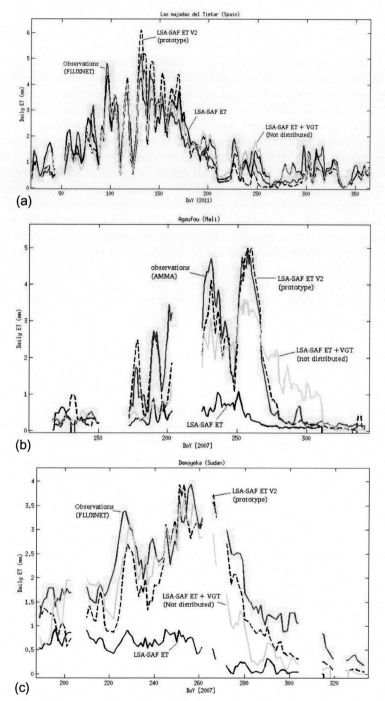

FIG. 10 Comparison of daily evapotranspiration rates at three sites of the operational algorithm (LSA-SAF ET), a modified version using EO-derived vegetation parameters from LSA-SAF (LSA-SAF ET+VGT) and the new proposed algorithm LSA-SAF ET V2. While it is clear that the assimilation of soil moisture derived from land surface temperature and vegetation dynamics does not specifically improve the evapotranspiration rates for the Dehesa site in Spain, the quantification is much more accurate for the tested African semiarid savannahs [Agoufou-Mali (Kergoat et al., 2011; Timouk et al., 2009); Demokeya—Sudan (Merbold et al., 2009; Sjöström et al., 2009)].

To reduce the dependence in ECMWF forecasted soil moisture and improve the quality of the soil moisture input to the ET model, a new strategy has been developed to assimilate the land surface temperature derived from MSG/SEVIRI as a proxy for soil moisture. Jointly, daily vegetation parameters have been assimilated in the model. Assimilation of the soil moisture derived from land surface temperature is only considered in parts of African and Mediterranean regions, where soil water stress is frequent. The evaluation of those changes shows that ET monitoring will definitely gain in accuracy at least in African grassland savannahs if using the land surface temperatures from the satellite, while results do not show a significant improvement, or even sometimes a degradation, in using it in the Mediterranean regions, except for the practical consequences of relying less on the ECMWF system changes. However, the soil moisture from remote sensing allows the detection of irrigation areas and wetlands that cannot be resolved by ECMWF forecasts, and will impact ET monitoring. The new implemented strategy is leading to a new version of ET products in LSA-SAF, which will cope with deficiencies reported in the semiarid regions.

7 FUTURE LINE OF OPERATIONAL RESEARCH

In this study, TIR signal from Meteosat, specifically the land surface temperature product from LSA-SAF, has been exploited to constrain the soil moisture availability for ET processes. From an operation point of view, research should now focus on localizing where the microwave soil moisture products, which are becoming more and more mature, could still further improve the soil moisture constraint, also in cloudy sky conditions, and how it could be achieved in near-real time. One privileged line would be the exploitation of the EUMETSAT H-SAF soil moisture products, from microwave observations delivered by the ASCAT sensors aboard MetOp-A/B polar satellites (Bartalis et al., 2007; Wagner et al., 2007). In addition, the use of the soil moisture delivered by the SM-DAS-2 analysis (Albergel et al., 2012), assimilating jointly synoptic meteorological observations and soil moisture derived from microwave sensors aboard space technology, could be an interesting alternative to the ECMWF IFS, if delivered on time, as it integrates the latest information on the land surface readily accessible using an efficient integrated infrastructure and communication lines in Europe and globally. On the other hand, efforts should carry on to improve the estimation of soil moisture using TIR signal from satellites and on the possibilities of combining the different sources of information (e.g., Zhao and Li, 2013), as precursor studies have undergone for climate studies (Liu et al., 2011).

ACKNOWLEDGMENTS

I would like to thank my collaborators Alirio Arboleda and Françoise Meulenberghs for the fruitful discussions, the partners of the LSA-SAF consortium for providing high-quality data sets, and the flux data providers from FLUXNET and CarboAfrica who shared their data for the validation. The presented work has been done in the context of EUMETSAT LSA-SAF project CDOP-2 and is funded by EUMETSAT and by ESA through the PRODEX program from the Belgian Science Policy.

REFERENCES

Albergel, C., Rüdiger, C., Pellarin, T., Calvet, J.-C., Fritz, N., Froissard, F., Suquia, D., Petitpa, A., Piguet, B., Martin, E., 2008. From near surface to root-zone soil moisture using an exponential filter: an assessment of the method based on in situ observations and model simulations. Hydrol. Earth Syst. Sci. 12, 1323–1337.

Albergel, C., de Rosnay, P., Gruhier, C., Muñoz-Sabater, J., Hasenauer, S., Isaksen, L., Kerr, Y., Wagner, W., 2012. Evaluation of remotely sensed and modelled soil moisture products using global ground-based in situ observations. Remote Sens. Environ. 118, 215–226. http://dx.doi.org/10.1016/j.rse.2011.11.017.

Arboleda, A., Ghilain, N., Gellens-Meulenberghs, F., 2011. EUMETSAT's LSA-SAF evapotranspiration: comparisons of operational product to observations and models at hydrological basins scale. In: Proceedings of the EUMETSAT Conference 2011, Oslo, Norway.

Balsamo, G., Albergel, C., Beljaars, A., Boussetta, S., Brun, E., Cloke, H., Dee, D., Dutra, E., Muñoz-Sabater, J., Pappenberger, F., de Rosnay, P., Stockdale, T., Vitart, F., 2015. ERA-Interim/Land: a global land surface reanalysis data set. Hydrol. Earth Syst. Sci. 19, 389–407. http://dx.doi.org/10.5194/hess-19-389-2015.

Bartalis, Z., Wagner, W., Naeimi, V., Hasenauer, S., Scipal, K., Bonekamp, H., Figa, J., Anderson, C., 2007. Initial soil moisture retrieval from the METOP-A Advanced Scatterometer. Geophys. Res. Lett. 34 (L20). http://dx.doi.org/10.1029/2007GL031088.

Cammalleri, C., Sepulcre-Cantó, G., Vogt, J., 2014. Combining land surface models and remote sensing data to estimate evapotranspiration for drought monitoring in Europe. In: SPIE Remote Sensing, p. 923907.

Choudhury, B.J., 1991. Multispectral satellite data in the context of land surface heat balance. Rev. Geophys. 29, 217–236.

Courault, D., Seguin, B., Olioso, A., 2005. Review on estimation of evapotranspiration from remote sensing data: from empirical to numerical modelling approaches. Irrig. Drain. Syst. 19, 223–249.

de Bruin, H.A.R., Trigo, I.F., Jitan, M.A., Temesgen Enku, N., van der Tol, C., Gieske, A.S.M., 2010. Reference crop evapotranspiration derived from geo-stationary satellite imagery: a case study for the Fogera flood plain, NW-Ethiopia and the Jordan Valley, Jordan. Hydrol. Earth Syst. Sci. 14, 2219–2228. http://dx.doi.org/10.5194/hess-14-2219-2010.

Dierickx, P., 2014. Low flow management in Wallonia, presentation to NHV Symposium. In: International Symposium On Droughts and Low Flows, Maastricht, The Netherlands. (http://bigfiles.nhv.nu/Maastricht2014/2014.10.24_Dierickx.pdf, last accessed 16.11.15).

Ermida, S.L., Trigo, I.F., DaCamara, C.C., Gottsche, F.M., Olesen, F.S., Hulley, G., 2014. Validation of remotely sensed surface temperature over an oak woodland landscape—the problem of viewing and illumination geometries. Remote Sens. Environ. 148, 16–27.

Freitas, S.C., Trigo, I.F., Bioucas-Dias, J.M., Gottsche, F., 2010. Quantifying the uncertainty of land surface temperature retrievals from SEVIRI/Meteosat. IEEE Trans. Geosci. Remote Sens. 48, 523–534.

Garcia, M., Sandholt, I., Ceccato, P., Ridler, M., Mougin, E., Kergoat, L., Timouk, F., Fensholt, R., Domingo, F., 2014. Actual evapotranspiration in drylands derived from in-situ and satellite data: assessing biophysical constraints. Remote Sens. Environ. 131, 103–118.

Garcia-Haro, F.J., Camacho-de Coca, F., Melia, J., Martinez, B., 2005. Operational derivation of vegetation products in the framework of the LSA SAF project. In: Proceedings of the EUMET-SAT Meteorological Satellite Conference. (p. 116).

Geiger, B., Meurey, C., Lajas, D., Franchistéguy, L., Carrer, D., Roujean, J.-L., 2008a. Near real time provision of downwelling shortwave radiation estimates derived from satellite observations. Meteorol. Appl. 15, 411–420.

Geiger, B., Carrer, D., Franchistéguy, L., Roujean, J.-L., Meurey, C., 2008b. Land surface Albedo derived on a daily basis from Meteosat Second Generation observations. IEEE Trans. Geosci. Remote Sens. 46, 3841–3856.

Gellens-Meulenberghs, F., 2005. Sensitivity tests of an energy balance model to choice of stability functions and measurement accuracy. Bound.-Lay. Meteorol. 115 (3), 453–471.

Ghilain, N., Arboleda, A., Gellens-Meulenberghs, F., 2011. Evapotranspiration modelling at large scale using near-real time MSG SEVIRI derived data. Hydrol. Earth Syst. Sci. 15, 771–786. http://dx.doi.org/10.5194/hess-15-771-2011.

Ghilain, N., Arboleda, A., Sepulcre-Canto, G., Batelaan, O., Ardö, J., Gellens-Meulenberghs, F., 2012. Improving evapotranspiration in a land surface model using biophysical variables derived from MSG/SEVIRI satellite. Hydrol. Earth Syst. Sci. 16, 2567–2583.

Ghilain, N., Arboleda, A., Trigo, I.F., Batelaan, O., Gellens-Meulenberghs, F., 2015. Tracking down soil moisture from MSG SEVIRI: fullfilling the promises of geostationary satellites. Remote Sens. Environ. (in review).

Guzinski, R., Kass, S., Huber, S., Bauer-Gottwein, P., Jensen, I.H., Naeimi, V., Tottrup, C., 2014. Enabling the use of earth observation data for integrated water resource management in Africa with the water observation and information system. Remote Sens. 6 (8), 7819–7839.

Helman, D., Lensky, I.M., Givati, A., 2015. Annual evapotranspiration retrieved solely from satellites' vegetation indices for the Eastern Mediterranean. Atmos. Chem. Phys. Discuss. 15, 15397–15429. http://dx.doi.org/10.5194/acpd-15-15397-2015.

Hofste, W.R. 2014. Comparative analysis among near-operational evapotranspiration products for the Nile basin based on Earth observations, first steps towards an ensemble product. Master thesis, Delft University of Technology.

Hu, G., Jia, L., Menenti, M., 2015. Comparison of MOD16 and LSA-SAF MSG evapotranspiration products over Europe for 2011. Remote Sens. Environ. 156, 510–526.

Jovanovic, N., Garcia, C.L., Bugan, R.D., Teich, I., Rodriguez, C.M.G., 2014. Validation of remotely-sensed evapotranspiration and NDWI using ground measurements at Riverlands, South Africa. Water SA 40 (2), 211–220.

Kergoat, L., Grippa, M., Baille, A., Lacaze, R., Mougin, E., Ottlé, C., Pellarin, T., Polcher, J., de Rosnay, P., Roujean, J.-L., Sandholt, I., Taylor, C.M., Zin, I., Zribi, M., 2011. Remote sensing of the land surface during the African Monsoon Multidisciplinary Analysis (AMMA), Atmospheric Science Letters, Special issue: African Monsoon Multidisciplinary Analysis (AMMA): an integrated project for understanding of the West African climate system and its human dimension. Atmos. Sci. Lett. 12, 129–134. http://dx.doi.org/10.1002/asl.325.

Kerr, Y., Waldteufel, P., Wigneron, J.-P., Cabot, F., Boutin, J., Escorihuela, M.-J., Font, J., Reul, N., Gruhier, C., Juglea, S., Delwart, S., Drinkwater, M., Hahne, A., Martin-Neira, M., Mecklenburg, S., 2010. The SMOS mission: new tool for monitoring key elements of the global water cycle. Proc. IEEE 98 (5), 666–687. http://dx.doi.org/10.1109/JPROC.2010.2043032.

Liu, Y.Y., Parinussa, R.M., Dorigo, W.A., De Jeu, R.A.M., Wagner, W., van Dijk, A.I.J.M., McCabe, M.F., Evans, J.P., 2011. Developing an improved soil moisture dataset by blending passive and active microwave satellite-based retrievals. Hydrol. Earth Syst. Sci. 15, 425–436. http://dx.doi.org/10.5194/hess-15-425-2011.

Masson, V., Champeaux, J.L., Chauvin, F., Meriguet, Ch., Lacaze, R.A., 2003. Global database of land surface parameters at 1-km resolution in meteorological and climate models. J. Clim. 16, 1261–1282.

Merbold, L., Ardö, J., Arneth, A., Scholes, R.J., Nouvellon, Y., de Grandcourt, A., Archibald, S., Bonnefond, J.M., Boulain, N., Brueggemann, N., Bruemmer, C., Cappelaere, B., Ceschia, E., El-Khidir, H.A.M., El-Tahir, B.A., Falk, U., Lloyd, J., Kergoat, L., Le Dantec, V., Mougin, E., Muchinda, M., Mukelabai, M.M., Ramier, D., Roupsard, O., Timouk, F., Veenendaal, E.M., Kutsch, W.L., 2009. Precipitation as driver of carbon fluxes in 11 African ecosystems. Biogeosciences 6, 1027–1041. http://dx.doi.org/10.5194/bg-6-1027-2009.

Miralles, D.G., Gash, J.H., Holmes, T.R.H., De Jeu, R.A.M., Dolman, A.J., 2010. Global canopy interception from satellite observations. J. Geophys. Res. D: Atmos. 115, D16122. http://dx.doi.org/10.1029/2009JD013530.

Petropoulos, G., Ireland, G., Cass, A., Srivastava, P. K. 2015. Performance assessment of the SEVIRI evapotranspiration operational product: results over diverse Mediterranean ecosystems. IEEE Sens., http://dx.doi.org/10.1109/jsen.2015.7082390031.

Petropoulos, G.P., Ireland, G., Barrett, B., in press. Surface soil moisture retrievals from remote sensing: evolution, current status, products & future trends. Phys. Chem. Earth. http://dx.doi.org/10.1016/j.pce.2015.02.009.

Petropoulos, G.P., Ireland, G., Lamine, S., Ghilain, N., Anagnostopoulos, V., North, M.R., Srivastava, P.K., Georgopoulou, H., in press. Evapotranspiration estimates from SEVIRI to support sustainable water management. Int. J. Appl. Earth Obs. Geoinf.

Prigent, C., Aires, F., Rossow, W.B., Robock, A., 2005. Sensitivity of satellite microwave and infrared observations to soil moisture at a global scale: relationship of satellite observations to in situ soil moisture measurements. J. Geophys. Res. 110 (D11). http://dx.doi.org/10.1029/2004JD005094.

Romaguera, M., Hoekstra, A.Y., Su, Z., Krol, M.S., Salama, M.S., 2010. Potential of using remote sensing techniques for global assessment of water footprint of crops. Remote Sens. 2 (4), 1177–1196.

Romaguera, M., Krol, M.S., Salama, M., Hoekstra, A.Y., Su, Z., 2012. Determining irrigated areas and quantifying blue water use in Europe using remote sensing Meteosat Second Generation (MSG) products and Global Land Data Assimilation System (GLDAS) data. Photogramm. Eng. Remote Sens. 78 (8), 861–873.

Romaguera, M., Krol, M.S., Salama, M.S., Su, Z., Hoekstra, A.Y., 2014a. Application of a remote sensing method for estimating monthly blue water evapotranspiration in irrigated agriculture. Remote Sens. 6 (10), 10033–10050.

Romaguera, M., Salama, M.S., Krol, M.S., Hoekstra, A.Y., Su, Z., 2014b. Towards the improvement of blue water evapotranspiration estimates by combining remote sensing and model simulation. Remote Sens. 6 (8), 7026–7049.

Sepulcre-Canto, G., Vogt, J., Arboleda, A., Antofie, T., 2014. Assessment of the EUMETSAT LSA-SAF evapotranspiration product for drought monitoring in Europe. Int. J. Appl. Earth Obs. Geoinf. 30, 190–202.

Siebert, S., Burke, J., Faurès, J.-M., Frenken, K., Hoogeveen, J., Döll, P., Portmann, F.T., 2010. Groundwater use for irrigation—a global inventory. Hydrol. Earth Syst. Sci. 14, 1863–1880.

Siebert, S., Henrich, V., Frenken, K., Burke, J. 2013. Update of the global map of irrigation areas to version 5. Project report, 178 p.

Sjöström, M., Ardö, J., Eklundh, L., El-Tahir, B.A., El-Khidir, H.A.M., Hellström, M., Pilesjö, P., Seaquist, J., 2009. Evaluation of satellite based indices for gross primary production estimates in a sparse savanna in the Sudan. Biogeosciences 6, 129–138. http://dx.doi.org/10.5194/bg-6-129-2009.

Stisen, S., Sandholt, I., Noergaard, A., Fensholt, R., Jensen, K.H., 2008. Combining the triangle method with thermal inertia to estimate regional evapotranspiration—applied to MSG-SEVIRI data in the Senegal River basin. Remote Sens. Environ. 110, 1242–1255. http://dx.doi.org/10.1016/j.rse.2007.08.013.

Taylor, K.E., 2001. Summarizing multiple aspects of model performance in a single diagram. J. Geophys. Res. 106 (D7). http://dx.doi.org/10.1029/2000JD900719.

Teuling, A.J., Hirschi, M., Ohmura, A., Wild, M., Reichstein, M., Ciais, P., Buchmann, N., Ammann, C., Montagnani, L., Richardson, A.D., Wohlfahrt, G., Seneviratne, S.I., 2009. A regional perspective on trends in continental evaporation. Geophys. Res. Lett. 36, L02404. http://dx.doi.org/10.1029/2008GL036584.

Throssell, C.S., Lyman, G.T., Johnson, M.E., Stacey, G.A., 2009. Golf course environmental profile measures water use, source, cost, quality, and management and conservation strategies. Appl. Turf. Sci. http://dx.doi.org/10.1094/ATS-2009-0129-01-RS.

Timouk, F., Kergoat, L., Mougin, E., Lloyd, C.R., Ceschia, E., Cohard, J.-M., De Rosnay, P., Hiernaux, P., Demarez, V., Taylor, C.M., 2009. Response of surface energy balance to water regime and vegetation development in a Sahelian landscape. J. Hydrol. 375, 178–189.

Trigo, I.F., Monteiro, I.T., Olesen, F., Kabsch, E., 2008. An assessment of remotely sensed land surface temperature. J. Geophys. Res. D17. http://dx.doi.org/10.1029/2008JD010035.

Trigo, I.F., Barroso, C., Viterbo, P., Freitas, S.C., Monteiro, I.T., 2010. Estimation of downward long-wave radiation at the surface combining remotely sensed data and NWP data. J. Geophys. Res. 115, D24118. http://dx.doi.org/10.1029/2010JD013888.

Trigo, I.F., DaCamara, C.C., Viterbo, P., Roujean, J.-L., Olesen, F., Barroso, C., de Coca, F.C., Carrer, D., Freitas, S.C., Garcia-Haro, J., Geiger, B., Gellens-Meulenberghs, F., Ghilain, N., Melia, J., Pessanha, L., Siljamo, N., Arboleda, A., 2011. The satellite application facility on land surface analysis. Int. J. Remote Sens. 32, 2725–2744.

van den Hurk, B.J.J.M., Viterbo, P., Beljaars, A.C.M., Betts, A.K. 2000. Offline validation of the ERA40 surface scheme. ECMWF Technical Memorandum, 295, 41pp.

Verstraeten, W.W., Veroustraete, F., van der Sande, C.J., Grootaers, I., Feyen, J., 2006. Soil moisture retrieval using thermal inertia, determined with visible and thermal spaceborne data, validated for European forests. Remote Sens. Environ. 101, 299–314.

Verstraeten, W.W., Veroustraete, F., Feyen, J., 2008. Assessment of evapotranspiration and soil moisture content across different scales of observation. Sensors 8, 70–117.

Viterbo, P., Beljaars, A.C.M., 1995. An improved surface parameterization scheme in the ECMWF model and its validation. J. Clim. 8, 2716–2748.

Wade, S.D., Jones, P.D., Osborn, T. 2006. The impacts of climate change on severe drought, implications for decision making. Environmental Agency Science Report SC040068/SR3, 94pp.

Wagner, W., Naeimi, V., Scipal, K., de Jeu, R., Martinez-Fernandez, J., 2007. Soil moisture from operational meteorological satellites. Hydrogeol. J. 15, 121–131.

Williams, C.A., Albertson, J.D., 2004. Soil moisture controls on canopy-scale water and carbon fluxes in an African savanna. Water Resour. Res. 40, W09302. http://dx.doi.org/10.1029/2004WR003208.

Zhao, W., Li, A., 2013. A downscaling method for improving the spatial resolution of AMSR-E derived soil moisture product based on MSG SEVIRI data. Remote Sens. 5, 6790–6811.

Chapter 17

Soil Moisture Deficit Estimation Through SMOS Soil Moisture and MODIS Land Surface Temperature

P.K. Srivastava*,¶, T. Islam†,‡, S.K. Singh§, M. Gupta¶, G.P. Petropoulos@, D.K. Gupta**, W.Z. Wan Jaafar†† and R. Prasad**

*Banaras Hindu University, Banaras, India, †NASA Jet Propulsion Laboratory, Pasadena, CA, United States, ‡California Institute of Technology, Pasadena, CA, United States, §University of Allahabad, Allahabad, India, ¶NASA Goddard Space Flight Center, Greenbelt, MD, United States, @Department of Geography & Earth Sciences, Ceredigion, Wales, United Kingdom, **Indian Institute of Technology (BHU), Varanasi, India, ††University of Malaya, Kuala Lumpur, Malaysia*

1 INTRODUCTION

Soil moisture deficit (SMD) is an important variable in the exchange of water and energy that occurs at the interface of land surface/atmosphere. In real conditions, the soil has less water than at field capacity because of physiological processes such as crop suction and meteorological losses due to soil surface evaporation, which is described here by a parameter called SMD (Srivastava et al., 2014). Every soil has critical SMD level; if the soil is maintained at a dry level beyond this critical value, there will be a resultant decrease in yield and quantity (Martin et al., 1992). Therefore, quantification and prediction of SMD in agriculture are most important for better food security. Furthermore, SMD monitoring will act as an alternative method for irrigation. This represents the proper use of water at the required time to avoid any agricultural losses (Srivastava et al., 2013a) and could help in preventing natural disasters; for example, flood and drought (Srivastava, 2013).

Soil moisture works as a land surface parameter that has an important control over several hydrological and atmospheric processes. The exchange of latent heat and sensible heat is controlled due to the changes of evapotranspiration (ET) processes and latent heat between land surface and atmosphere interface. Soil moisture is an essential parameter at different spatial scales for many applications; for example, flood forecasting (Aubert et al., 2003),

Satellite Soil Moisture Retrieval. http://dx.doi.org/10.1016/B978-0-12-803388-3.00017-6

meteorology (Mahfouf, 1991; Srivastava et al., 2013a), agriculture (Gupta et al., 2014), and global and regional climate models (Liang et al., 1994; Petropoulos et al., 2015b). The significant development of remote sensing has given considerable scope for further studies to determine the soil moisture and land surface temperature (LST) over earth (Piles et al., 2016; Srivastava et al., 2013b; Wan et al., 2004). Previous scientific studies in this field have revealed that LST has a strong relationship with SMD, which could be measured by moderate resolution imaging spectroradiometer (MODIS). In Nov. 2009, the soil moisture and ocean salinity (SMOS) satellite was launched to provide near-surface soil moisture with targeted accuracy of 4% (Kerr et al., 2012) and has also been observed to have a significant relationship with SMD.

Most of the hydrological models require the integration of data with the help of conceptual models or statistical tools such as regression and curve fitting (Srivastava et al., 2015). Few scientific attempts have been made in this field in recent years to develop techniques based on sophisticated algorithms for various hydrological applications (Islam et al., 2014a). Adaptive neuro fuzzy inference systems (ANFIS) are one of the most powerful mathematical tools and have been successfully applied in hydrology for tackling problems such as flow forecasting, evaporation estimation, forecasting of precipitation, and so on (Islam et al., 2013, 2014a,b; Terzi et al., 2006).

So far, limited studies have been performed for the estimation of SMD using SMOS soil moisture information and/or MODIS LST. Therefore, this study can be considered as a first-time comprehensive evaluation of LST from MODIS and SMOS soil moisture for hydrological SMD prediction using the sophisticated ANFIS technique.

In purview of the above, the foremost objective of this chapter is to focus on the potential application of ANFIS for evaluating its capabilities toward SMD estimation using the LST from MODIS and the SMOS soil moisture. The benchmark SMD estimated from the Probability Distributed Model (PDM) over the Brue catchment, southwest of England, United Kingdom, is applied for all the validation. The analysis of the satellite products (viz. SMOS soil moisture and MODIS LST) towards SMD prediction is a crucial step for successful hydrological modeling, agriculture and water resource management, and can provide important assistance in policy and decision making.

2 MATERIALS AND METHODOLOGY

2.1 Study Area and Data Sets

The Brue catchment ($135.5 \, \text{km}^2$) is chosen as the study area, located in the southwest of England, 51.11°N and 2.47°W (see Fig. 1). It is a good experimental site for rainfall-runoff modeling, satellite and mesoscale model-based studies because of low vegetation cover, maintained meteorological and flow station, moderate topography, and availability of nearly all required data sets

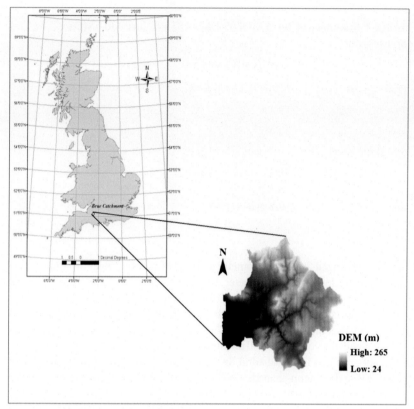

FIG. 1 Geographical location of the study area.

generally used for hydrological applications. All the ground observed data for this study are obtained from the Natural Environment Research Council (NERC) and the British Atmospheric Data Centre (BADC). The most of the area is comprised of clayey type (49%) soil followed by coarse loam 29% and silt 21%. This catchment is predominantly rural catchment with pastureland (94.34%), patches of forest (3.48%) in the higher eastern part, and urban areas (2.18%). The data sets used in this study are provided in Table 1.

2.2 SMOS Satellite Soil Moisture

The SMOS mission is a joint venture of the European Space Agency (ESA), the National Centre for Space Studies, and the Industrial Technological Development Centre with the main goal to precisely estimate soil moisture. The onboard microwave imaging radiometer with aperture synthesis (MIRAS) instrument in the SMOS satellite acquires data of emitted microwave radiation at the frequency of 1.4 GHz (L-band) (Petropoulos et al., 2014; Petropoulos et al., 2015a).

TABLE 1 Summary of Terra MODIS Daily LST Product and SMOS Soil Moisture

	MODIS LST Level 3	SMOS Level 2
Product name	MOD11C1 Daily CMG	SMOS.MIRAS.MIR_SMUDP2
Type of product	Global	Swath-based
Stated accuracy	1.0 K	0.04 m^3 m^{-3}
Spatial resolution	~ 5.6 km	~ 40 km
Algorithm principle	Physics-based day and night algorithm	Iterative Bayesian and physically based models
References	Wan and Li (1997)	Kerr et al., 2012

It is a dual polarized 2-D interferometer and is the first ever, polar orbiting, spaceborne, 2-D interferometric radiometer designed to provide global information on surface soil moisture with an accuracy of 4% (Kerr et al., 2001). In this study the Level 2 products (soil moisture swath-based maps) are employed for preparation of time series over the Brue catchment. The spatial resolution of the SMOS products is ~40 km with the volumetric soil moisture retrieval unit in m^3 m^{-3}, defined on the ISEA 4H9 grid; that is, icosahedral Snyder equal area projection with aperture 4, resolution 9, and hexagon-shaped cells (Pinori et al., 2008). The instrument provides records of brightness temperatures over incidence angles from 0 degree to 55 degrees across a 900 km swath (Pinori et al., 2008). Each point (or node) of this grid is known as a Discrete Global Grid (DGG) that has fixed coordinates and is assigned with an identificator: the "DGG Id." For the comparison between the catchment and SMOS soil moisture (SMOS SM), the SMOS pixel with its centroid over the catchment is extracted and considered for the subsequent analysis. The Beam 4.9 package with SMOS 2.1.3 plugin is used for the extraction.

2.3 Land Surface Temperature From MODIS Satellite

The MODIS Terra satellite LST obtained through MODIS Land Processes Distributed Active Archive Center (LP DAAC) (https://lpdaac.usgs.gov/) is used in this study. The MODIS LST products are archived in Hierarchical Data Format-Earth Observing System (HDF-EOS) format files. The LST product files contain global attributes (metadata) and scientific data sets arrays with local attributes (Wan, 2006). It has a viewing swath width of 2330 km and measuring signals in 36 spectral bands between 0.405 and 14.385 μm. Due to its daily temporal resolution and free near-real time data availability, it has been selected among other operational optical/infrared satellites. The MODIS Level 3 (~5.6 km) LST data are gridded uniformly across the globe and used in this study for LST estimation. The Brue boundary is used for subsetting the MODIS LST global data sets using the environment for visualizing images ITT version 4.8. A total of five pixels are extracted from the MODIS global image covering the whole Brue catchment. To use the MODIS LST it is averaged and a time series is created which is then used as an input in ANFIS for the SMD estimation. As both SMOS and MODIS come under low Earth orbiting satellites, they have local equatorial crossing times of approximately 6:00 am/6:00 pm in the case of SMOS, and 10:30 pm/10:30 am for Terra, in ascending/descending nodes. In this study, SMOS descending passes are combined with MODIS Terra ascending passes with the assumption that the soil moisture pattern is spatially persistent for a few hours before and after the SMOS overpass (Sánchez-Ruiz et al., 2014).

2.4 Fuzzy Logic System

Zadeh (1975) implemented fuzzy set theory to produce domain-specific outputs, when the variables under question represent nonlinear relationships and enhanced complex interactions. Fuzzy sets can trace and quantify uncertainties within the data efficiently overriding intrinsic subjectivity within the data domain by means of fuzzification, where crisp variables are altered into fuzzy or continuous inputs. The ANFIS is a Sugeno fuzzy model with a systematic approach for generating fuzzy rules from a given input-output data set (Sugeno and Kang, 1988; Takagi and Sugeno, 1985); thus, facilitating learning and adaption of a complex dynamical system by integrating the principles of neural networks and fuzzy logic. The ANFIS is generally composed of five layers, namely, a fuzzification layer, a rule layer, a normalization layer, a defuzzification layer, and an output layer. The first layer contains an adaptive node whose outputs are of the fuzzy membership grade of the inputs, and can be estimated by the equations:

$$O_i^1 = \mu_{A_i}(x) \tag{1}$$

$$O_i^1 = \mu_{B_{i-2}}(y) \tag{2}$$

where μ is the weight obtained through a fuzzy membership function. For μ_{Ai} and μ_{Bi} a Gaussian membership function is assumed:

$$\mu_{A_i}(x) = \exp\left[-\left(\frac{x - b_i}{a_i}\right)^2\right] \tag{3}$$

where a_i and b_i are the premise parameters of the membership functions (MFs). The second layer contains a fixed node calculated using a simple multiplier:

$$O_i^2 = w_i = \mu_{A_i}(x)\mu_{B_i}(y) \tag{4}$$

The third layer operated through a fixed node and indicates the normalized firing strength of each rule from the previous layer:

$$O_i^3 = \overline{w_i} = \frac{w_i}{w_1 + w_2} \tag{5}$$

The fourth layer contains an adaptive node following the contribution of the ith rule through the product of first-order Sugeno model and the normalized firing strength:

$$O_i^4 = \overline{w_i}f_i = \overline{w_i}(p_i x + q_i y + r_t) \tag{6}$$

where p_i, q_i, and r_i are the consequent parameters.

The last layer is represented by only one node and indicates the overall outputs by performing the summation of all incoming signals:

$$O_i^5 = \sum \overline{w_i}f_i = \frac{\sum_i w_i f_i}{\sum_i w_i} \tag{7}$$

The extension of the mountain clustering method (Chiu, 1994) (i.e., subtractive clustering method) is employed for training with the advantage that it can reduce the computational complexity of the model. An unsupervised method is used for measuring the potential of data points in the featured space with the assumption that each data point is a potential cluster. The first cluster center with the highest potential (based on the density of surrounding points), is selected while the rest of the data points are destroyed. The unsupervised method determines the number of fuzzy rules and MFs using the ANFIS model. The numbers of domain-specific variables are proportional to each and every MF, predefined by individual input/output. Fuzzy sets recognize 1 and 0 as defined MF, where 0 is false and 1 is true. Whereas applying the criterion of ANFIS for SMD modeling, the crisp inputs include SMOS and MODIS LST, while the crisp output is the predicted SMD.

2.5 Probability Distributed Model

The PDM is a lumped rainfall-runoff model capable of representing a variety of catchment-scale hydrological behaviors and has been widely applied throughout the globe for runoff prediction (Bell and Moore, 1998). PDM requires only rainfall and reference ET for discharge prediction, and the river flow is the model output. PDM is applied over this catchment by using the flow, rainfall, and ET provided by BADC, NERC, and Environment Agency to estimate the benchmark SMD. In this study, data sets were divided as (1) two-year data sets that were used for the calibration period (Feb. 1, 2009, to Jan. 31, 2011) and (2) one-year data sets used for validation (Feb. 1, 2011, to Jan. 31, 2012). The SMD obtained during the validation is considered for all the performance comparisons. The overall performance of PDM indicated a Nash-Sutcliffe Efficiency (NSE) value of 0.84 and 0.81 during the calibration and validation process, respectively. In this study, the PDM is used for SMD estimation through its moisture deficit routine (Moore, 2007):

$$\frac{E_i'}{E_i} = 1 - \left\{ \frac{(S_{max} - S(t))}{S_{max}} \right\}^{b_e} \tag{8}$$

where $\frac{E_i'}{E_i}$ is the ratio of actual ET to potential ET; $(S_{max} - S(t))$ is SMD; b_e is an exponent in the actual evaporation function; S_{max} is the total available storage; and $S(t)$ is storage at a particular time t. Detailed information on PDM calibration, validation, sensitivity, and uncertainty analysis over Brue catchment is reported in Srivastava et al. (2013b).

2.6 Statistical Performances

In the present study the MODIS LST and SMOS-derived SMD are compared against the PDM SMD. Two correlation analyses (parametric and nonparametric) are employed to study the existing relationship between the SMOS soil moisture, MODIS LST, and SMD data sets. These tests are useful to understand the linear and nonlinear behavior of the data sets. The first method is the Pearson correlation coefficient (r), which is a type of parametric test, used to measure the strength of linear relationship in normally distributed variables. The second one is the Spearman correlation coefficient (rs), which is a nonparametric test usually used for nonnormal data sets.

The performances between two patterns are quantified in terms of NSE (Nash and Sutcliffe, 1970), percentage of bias (%Bias), and root mean square error (RMSE). NSE is based on the sum of the absolute squared differences between the simulated and observed values, normalized by the variance of the observed values during the investigation period. The %Bias can be calculated using the simulated and observed data sets. Positive values of %Bias indicate an overestimation, whereas negative values indicate an underestimation by the model. The RMSE is employed to calculate the sample standard deviation of the differences between simulated and observed data sets.

3 RESULTS AND DISCUSSION

3.1 MODIS LST and SMOS Soil Moisture Variation With SMD

The data sets used in this study are comprised of PDM SMD, LST estimated from MODIS satellite, and SMOS soil moisture. As mentioned, both the parametric and nonparametric test are taken into account for quantifying the degree of relationship (i.e., r and the rs, respectively) represented through Fig. 2. The analysis suggests that a strong linearity existed between the MODIS LST and SMD, indicated by similar r (0.84) and rs (0.84) correlation, while the SMOS

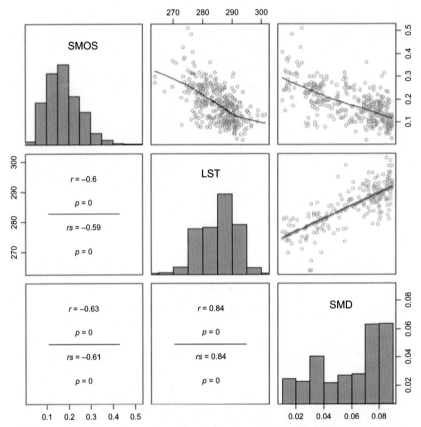

FIG. 2 Correlation matrix plots between SMOS, MODIS LST, and PDM SMD (where r and rs are Pearson and Spearman correlation coefficients, respectively).

soil moisture indicated a moderate correlation with SMD with nearly equal r (-0.63) and rs (-0.61) coefficients. Correlation analysis showed that MODIS LST has a strong relation with the SMD; that means if LST is higher there will be more loss of soil moisture that leads to a higher SMD in the soil. On the other hand, SMOS soil moisture indicated a negative correlation with the SMD, which is expected as high soil moisture will lead to less development of SMD. Similar theory is also valid for MODIS LST and SMOS soil moisture, as high LST will lead to more evaporative losses and, hence, less soil moisture, which is the reason a negative correlation is observed between the two variables.

To understand the temporal relationship between MODIS LST, SMOS soil moisture, and SMD, the time series depicting the variations for the complete monitoring period is plotted and indicated through Fig. 3. From the pattern of SMOS and SMD, we decipher that if the SMOS and SMD have negative

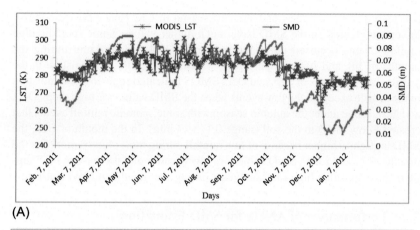

(A)

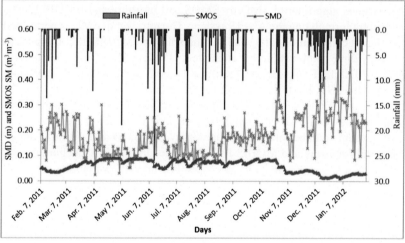

(B)

FIG. 3   Time series depicting the relationship of PDM SMD and rainfall with (A) MODIS LST and (B) SMOS soil moisture.

correlation it means that in the month of Feb. the higher SMOS values linked with the low SMD values. The time series exhibited a high spatial variability during the complete monitoring period. At the start of Feb., temperature is low, therefore high soil moisture values are recorded with correspondingly low values of LST and SMD, which indicate that the evaporative losses during these months are lower than other months. From Mar. onwards up to Jun. low values of SMOS and higher values of SMD are recorded. From Mar. to May, temperature gradually increases except there are few incidences in the month of May, when the peak rainfall is matching with the valley of SMD due to some rainfall events. Higher temperature starts from Jun. to Aug., these months are considered as the warmest and sunniest season in UK. Hence the evaporative losses are greater, so the SMD have much higher values. Some sporadic high rainfall events occurred up to the maximum 20.4 and 15.8 mm during the months of Jul. and Aug., respectively, which causes high soil moisture in the profile. In the months of Sep. and Oct., SMOS indicated some low soil moisture in the profile except at the few events when the SMD values are high. Sep., Oct., and Nov. come under the autumn season with some sporadic rainfall events that created a low SMD in the soil (range 0.1–31.3 mm). In the month of Oct., the SMD is slightly higher because of low rainfall during most of the period. For the months of Nov. to Jan., SMD values are low because of a decrease in LST supplemented by some high rainfall events during the months of Nov. and Dec.

3.2 Performance of ANFIS for SMD Estimation

The retrieval algorithm of SMD by SMOS soil moisture and MODIS LST is realized by formulating the ANFIS system. The retrieval of SMD is performed by different combinations of SMOS soil moisture and MODIS LST as inputs. In this study three combinations are attempted: (1) using SMOS soil moisture, (2) using MODIS LST, and (3) SMD retrieval by both inputs SMOS soil moisture and MODIS LST. First, the subtractive clustering algorithm applied with different combinations of SMOS soil moisture and MODIS LST. The subtractive clustering method divides the input data sets into a number of clusters, which gives the same number of rules for the retrieval of SMD at different levels of SMOS soil moisture and MODIS LST. More accurately, it is categorizing the entire data sets into different clusters depending upon the variability of SMOS soil moisture and MODIS LST. The number of data clusters generated using subtractive clustering depends upon the Euclidean distance between cluster center and data points, which are also known as *radii* of the cluster. The small value of *radii* accounts for a greater number of clusters while the higher value of *radii* accounts for less number of clusters. For a satisfactory performance, ANFIS required optimum cluster radii. The cluster radii provide the range of influence of a cluster when the data space is considered as a unit hypercube. The optimization of cluster radii in the data set is an important task to generate the optimum result from the ANFIS. Herein, the optimum value of radii is generated by running the ANFIS for the different radii values from 0 to 1 with steps of 0.05. At each step of radii, the performance is monitored to obtain the best results. The analysis indicates that the *radii* value of 0.5 is found suitable for all the combinations.

The Gaussian membership function is used to represent the inputs (see Fig. 4). The fuzzy rules are generating using 235 training data sets. In total, three fuzzy models are generated using different combinations of input data sets to check the capability of ANFIS for SMD prediction. Here, two clusters are generated for each combination of input-output data sets with cluster *radii* of 0.5. The first combination of input (SMOS soil moisture)-output (SMD) has cluster

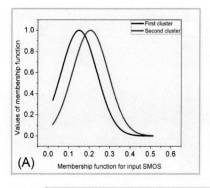

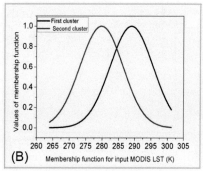

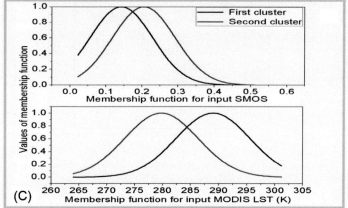

FIG. 4 The relationship between values of membership function and different combination of inputs (A) The values of membership function with input SMOS soil moisture, (B) The values of membership function with input MODIS LST, and (C) The values of membership function with both inputs (SMOS soil moisture and MODIS LST).

centers at 0.15, 0.21, while for the output (SMD) they are obtained as 0.08, 0.04 with sigma values 0.09 and 0.01, respectively. The second combination of input (MODIS LST) and output (SMD) has cluster centers at 288.98, 279.66, whereas the cluster centers for output (SMD) are obtained as 0.08, 0.03 with sigma values 6.58 and 0.01, respectively. The third combination of inputs (SMOS soil moisture and MODIS LST)-output (SMD) has cluster centers for the first input (SMOS soil moisture) at 0.14, 0.21 and for second input (MODIS LST) found at 288.51, 280.21 with the sigma values 0.87 and 6.58, respectively, and the cluster centers for the output (SMD) are 0.08, 0.04 with the sigma value of 0.01. The sigma values here specify the range of influence of a cluster center in each of the data dimensions.

The goodness of fit of SMD predicted from ANFIS using different input parameters—SMOS and MODIS LST—are indicated in Fig. 5. As compared

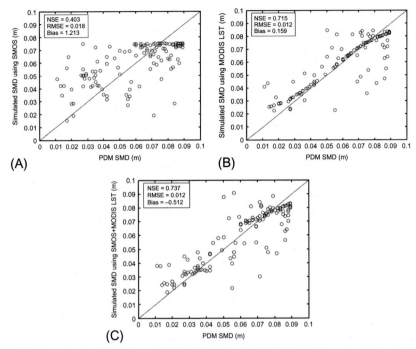

FIG. 5 Scatterplots depicting the ANFIS performance for SMD using different approaches: (A) Using SMOS; (B) Using MODIS LST; and (C) Using both MODIS LST and SMOS.

to SMOS, the MODIS indicate improved performance statistics between the simulated and observed SMD. The validation of MODIS LST-based SMD reported a much lower RMSE than SMOS, which indicates that the performance of MODIS for SMD is better than the SMOS soil moisture. To estimate the bias in the simulated data sets to understand the under/overestimation, % Bias is taken into account. The analysis indicates that the SMOS is showing a very high %Bias of 1.213% as compared to the MODIS, which has the value of 0.159%. In terms of RMSE, MODIS (RMSE = 0.012) also is giving a much lower value than SMOS (RMSE = 0.018). NSE also indicated a better performance by MODIS in comparison to SMOS with much higher value depicted by MODIS (NSE = 0.745) than SMOS (NSE = 0.403).

To evaluate the combined performance of SMOS and MODIS, an ANFIS system is designed with both the inputs. The combined performance indicates comparable values with MODIS only data sets indicating that MODIS alone is sufficient to predict SMD with higher performances. It is also worth mentioning that SMOS and MODIS have different scanning strategy resulting in different resolutions. Therefore, during the development of the ANFIS algorithm, the SMOS and the MODIS data sets are collocated to a single footprint domain. However, there is always some error in footprint matching, and some lower performance of SMOS could be attributed to difference in spatial resolution; for example, SMOS are coarse resolution data sets with spatial resolution of ∼40 km whereas MODIS data sets used in this study are high-resolution products extracted accordingly to the catchment used in this study. The spatial mismatch of SMOS and catchment used in this study could be one reason that even after the use of a high-performance technique like ANFIS, the performance of SMOS is a bit lower than the MODIS data sets.

4 CONCLUSIONS

SMOS is an advanced and specific mission to measure the soil moisture parameter over the globe. This study, to our knowledge for the first time, provides a very comprehensive evaluation of SMD estimation using the data sets from SMOS soil moisture and MODIS LST. High-performance tools like ANFIS could be employed for SMD using the data sets from SMOS and MODIS, and can update hydrologists with valuable information on applicability of SMOS soil moisture and LST for SMD estimations. Findings suggest that the ANFIS is suitable for SMD estimation using both the MODIS LST and SMOS soil moisture satellite data sets. The ANFIS model provides better estimates of SMD with MODIS LST data sets in comparison to SMOS. The performance of SMOS is lower, but it is useful when there is high atmospheric noise such as clouds, aerosols, and so on, as MODIS is generally not available in those conditions because of infrared sensors that have only partial penetration power through atmospheric obstacles. Opposite to this, low-frequency microwave sensors like as L-band SMOS have the advantage that they have high penetration through the atmosphere and provide soil moisture data in both daytime and nighttime conditions (Entekhabi et al., 2010). Furthermore, exploration of this potentially valuable data source by the scientific community is encouraged so that useful understanding and knowledge could be accumulated in the literature for different geographical areas. Future research will focus on the application of the abovementioned scheme for drought forecasting integrated with uncertainty analysis.

ACKNOWLEDGMENTS

The authors would like to thank the Commonwealth Scholarship Commission, British Council, United Kingdom, and the Ministry of Human Resource Development, Government of India, for providing the necessary support and funding for this research. The authors are thankful to the European Space Agency for providing the SMOS data. The first author would also like to thank Yann Kerr for providing the training over SMOS at CESBIO, Toulouse, France. The authors would like to acknowledge the British Atmospheric Data Centre, United Kingdom, for providing the ground observation data sets. The authors also acknowledge the Advanced Computing Research Centre at the University of Bristol for providing the access to its supercomputer facility (The Blue Crystal).

REFERENCES

Aubert, D., Loumagne, C., Oudin, L., 2003. Sequential assimilation of soil moisture and streamflow data in a conceptual rainfall-runoff model. J. Hydrol. 280, 145–161.

Bell, V., Moore, R., 1998. A grid-based distributed flood forecasting model for use with weather radar data: Part 2. Case studies. Hydrol. Earth Syst. Sci. 2, 283–298.

Chiu, S.L., 1994. Fuzzy model identification based on cluster estimation. J. Intell. Fuzzy Syst. 2, 267–278.

Entekhabi, D., et al., 2010. The soil moisture active passive (SMAP) mission. Proc. IEEE 98, 704–716.

Gupta, M., Srivastava, P.K., Islam, T., Ishak, A.M.B., 2014. Evaluation of TRMM rainfall for soil moisture prediction in a subtropical climate. Environ. Earth Sci. 71 (10), 4421–4431. http://dx.doi.org/10.1007/s12665-013-2837-6.

Islam, T., Rico-Ramirez, M.A., Han, D., Bray, M., Srivastava, P.K., 2013. Fuzzy logic based melting layer recognition from 3 GHz dual polarization radar: appraisal with NWP model and radio sounding observations. Theor. Appl. Climatol. 112, 317–338.

Islam, T., Srivastava, P.K., Gupta, M., Zhu, X., Mukherjee, S., 2014a. Computational Intelligence Techniques in Earth and Environmental Sciences. Springer-Verlag, Netherlands, ISBN:978-94-017-8642-3.

Islam, T., Srivastava, P.K., Rico-Ramirez, M.A., Dai, Q., Han, D., Gupta, M., 2014b. An exploratory investigation of an adaptive neuro fuzzy inference system (ANFIS) for estimating hydrometeors from TRMM/TMI in synergy with TRMM/PR. Atmos. Res. 145, 57–68.

Kerr, Y.H., Waldteufel, P., Wigneron, J.P., Martinuzzi, J., Font, J., Berger, M., 2001. Soil moisture retrieval from space: the soil moisture and ocean salinity (SMOS) mission. IEEE Trans. Geosci. Remote Sens. 39, 1729–1735.

Kerr, Y.H., et al., 2012. The SMOS soil moisture retrieval algorithm. IEEE Trans. Geosci. Remote Sens. 50, 1384–1403.

Liang, X., Lettenmaier, D.P., Wood, E.F., Burges, S.J., 1994. A simple hydrologically based model of land surface water and energy fluxes for general circulation models. J. Geophys. Res.-Atmos. 99, 14415–14428.

Mahfouf, J.-F., 1991. Analysis of soil moisture from near-surface parameters: a feasibility study. J. Appl. Meteorol. 30, 1534–1547.

Martin, R., Jamieson, P., Wilson, D., Francis, G., 1992. Effects of soil moisture deficits on yield and quality of 'Russet Burbank' potatoes. N. Z. J. Crop. Hortic. Sci. 20, 1–9.

Moore, R., 2007. The PDM rainfall-runoff model. Hydrol. Earth Syst. Sci. 11, 483–499.

Nash, J.E., Sutcliffe, J., 1970. River flow forecasting through conceptual models part I—A discussion of principles. J. Hydrol. 10, 282–290.

Petropoulos, G.P., Ireland, G., Srivastava, P.K., Ioannou-Katidis, P., 2014. An appraisal of soil moisture operational estimates accuracy from SMOS MIRAS using validated in-situ observations acquired at a Mediterranean environment. Int. J. Rem. Sens. 35 (13), 5239–5250. http://dx.doi.org/10.1080/2150704X.2014.933277.

Petropoulos, G.P., Ireland, G., Srivastava, P.K., 2015a. Evaluation of the soil moisture operational estimates from SMOS in Europe: results over diverse ecosystems. IEEE Sens. J. http://dx.doi.org/10.1109/JSEN.2015.2427657.

Petropoulos, G.P., Ireland, G., Barrett, B., 2015b. Surface soil moisture retrievals from remote sensing: Evolution, Current status, Products & Future trends. Physics and Chemistry of the Earth. http://dx.doi.org/10.1016/j.pce.2015.02.009.

Piles, M., Petropoulos, G.P., Ireland, G., Sanchez, N., 2016. A novel method to retrieve soil moisture at high spatio-temporal resolution based on the synergy of SMOS and MSG SEVIRI observations. Rem. Sens. Environ. [in press].

Pinori, S., Crapolicchio, R., Mecklenburg, S., 2008. Preparing the ESA-SMOS (soil moisture and ocean salinity) mission-overview of the user data products and data distribution strategy. In: Microwave Radiometry and Remote Sensing of the Environment, 2008. MICRORAD 2008. IEEE, pp. 1–4.

Sánchez-Ruiz, S., Piles, M., Sánchez, N., Martínez-Fernández, J., Vall-llossera, M., Camps, A., 2014. Combining SMOS with visible and near/shortwave/thermal infrared satellite data for high resolution soil moisture estimates. J. Hydrol. 516, 273–283.

Srivastava, P.K., 2013. Soil moisture estimation from SMOS satellite and mesoscale model for hydrological applications. PhD Thesis, University of Bristol, UK.

Srivastava, P.K., Han, D., Ramirez, M.A., Islam, T., 2013a. Appraisal of SMOS soil moisture at a catchment scale in a temperate maritime climate. J. Hydrol. 498, 292–304.

Srivastava, P.K., Han, D., Ramirez, M.A., Islam, T., 2013b. Machine learning techniques for downscaling SMOS satellite soil moisture using MODIS land surface temperature for hydrological application. Water Resour. Manag. 27, 3127–3144.

Srivastava, P.K., Han, D., Rico-Ramirez, M.A., O'Neill, P., Islam, T., Gupta, M., 2014. Assessment of SMOS soil moisture retrieval parameters using tau-omega algorithms for soil moisture deficit estimation. J. Hydrol. 519, 574–587. http://dx.doi.org/10.1016/j.jhydrol.2014.07.056.

Srivastava, P.K., Han, D., Rico-Ramirez, M.A., O'Neill, P., Islam, T., Gupta, M., Dai, Q., 2015. Performance evaluation of WRF-Noah Land surface model estimated soil moisture for hydrological application: synergistic evaluation using SMOS retrieved soil moisture. J. Hydrol. 529, 200–212. http://dx.doi.org/10.1016/j.jhydrol.2015.07.041.

Sugeno, M., Kang, G.T., 1988. Structure identification of fuzzy model. Fuzzy Set. Syst. 28, 15–33. http://dx.doi.org/10.1016/0165-0114(88)90113-3.

Takagi, T., Sugeno, M., 1985. Fuzzy identification of systems and its applications to modeling and control. IEEE Trans. Syst. Man Cybern. SMC-15, 116–132. http://dx.doi.org/10.1109/tsmc.1985.6313399.

Terzi, Ö., Erol Keskin, M., Dilek Taylan, E., 2006. Estimating evaporation using ANFIS. J. Irrig. Drain. Eng. 132, 503–507.

Wan, Z., Li, Z.-L., 1997. A physics-based algorithm for retrieving land-surface emissivity and temperature from EOS/MODIS data, IEEE Trans. Geosci. Remote Sens., 35 (4), 980–996.

Wan, Z., 2006. MODIS land surface temperature products users' guide. Institute for Computational Earth System Science, University of California, Santa Barbara, CA. http://www.icess.ucsb.edu/modis/LstUsrGuide/usrguide.html.

Wan, Z., Zhang, Y., Zhang, Q., Li, Z.-L., 2004. Quality assessment and validation of the MODIS global land surface temperature. Int. J. Remote Sens. 25, 261–274.

Zadeh, L.A., 1975. Fuzzy logic and approximate reasoning. Synthese 30, 407–428.

Future Challenges in Soil Moisture Retrieval and Applications

Chapter 18

Soil Moisture Retrievals Based on Active and Passive Microwave Data: State-of-the-Art and Operational Applications

J. Muñoz-Sabater[*], A. Al Bitar[†] and L. Brocca[‡]

[*]European Centre for Medium Range Weather Forecasts, Reading, United Kingdom, [†]Centre d'Études Spatiales de la Biosphère, Toulouse, France, [‡]Research Institute for Geo-Hydrological Protection, National Research Council, Perugia, Italy

1 INTRODUCTION

This chapter is organized in five main parts. In the first one, operational soil moisture retrievals based on active C-band data from the advance scatterometer (ASCAT) sensor on board the European meteorological operational program (MetOp) (Figa-Saldana et al., 2002) are considered, and the potential of these data for flood and landslide forecasting is demonstrated. The region of operational application is Italy. The production of soil moisture retrievals from ASCAT data (also in missions such as the Canadian RADARSAT-2 or the first of a series of European Sentinel missions, Sentinel-1) is based on the change detection algorithm. Details about this methodology can be consulted in Wagner et al. (1999, 2013) and Naeimi et al. (2009). This has been one of the first operational retrieval approaches of soil moisture from backscatter information. Recently a generalization of change detection to multiple regression using cumulative distribution function (CDF) transformations was applied to RADARSAT-2 data and validated over Berambadi watershed, South India (Tomer et al., 2015). This part of the chapter focuses on the use of the ASCAT soil moisture product for floods and landslides operational prediction from National and Regional Civil Protection Centres in Italy.

The second part focuses on soil moisture retrievals based on low frequencies passive microwave data. Soil moisture retrievals have been obtained from

Satellite Soil Moisture Retrieval. http://dx.doi.org/10.1016/B978-0-12-803388-3.00018-8

C-band passive microwaves using the advanced microwave scanning radiometer for earth observing system (AMSR-E) sensor data on NASA's aqua satellite (Njoku et al., 2003; Owe et al., 2008). However, L-band is more optimal for soil moisture estimation, because it is less sensitive to roughness and vegetation effects. At these frequencies, the atmosphere is nearly transparent; therefore, both C- and L-band sensors provide all-weather coverage (Eagleman and Lin, 1976; Wang and Choudhury, 1981; Jackson and Schmugge, 1991). L-band acquisition for global soil moisture retrievals has been available since a few weeks after the launch in Nov. 2009 of the Soil Moisture and Ocean Salinity (SMOS) mission of the European Space Agency (ESA) (Kerr et al., 2010), and more recently with the launch of the Soil Moisture Passive and Active (SMAP) (Entekhabi et al., 2010) mission of the National Aeronautics and Space Administration (NASA). Based on these data, Level-3 and Level-4 novel products are starting to emerge. In Section 3, the main principles of the algorithm producing the Level-3 CATDS data sets are explained. The Level-3 SMOS products use the L1B Fourier domain brightness temperatures (TB) as main input, but they use the same forward physical model as ESA Level-2 product (Kerr et al., 2012). Therefore, a very short introduction to the main characteristics of the SMOS Level-2 retrievals is also provided. The potential use of SMOS data is exemplified through two applications for drought and flood prediction.

The third part describes the algorithm of soil moisture estimation for a crucial application, numerical weather prediction (NWP), using both active and microwave remote sensing information. It is expected that the synergetic use of both types of data in land data assimilation systems will support more consistently their influence on forecasts of air temperature and humidity within the boundary layer. Part four presents an overview of the main challenges identified in the delivery of accurate soil moisture data sets based on Earth observation (EO) techniques for operational applications. Finally, part five concentrates in future perspectives aimed at improving modern data sets and their influence in current and future operational applications.

2   STATE-OF-THE-ART SOIL MOISTURE RETRIEVALS AND APPLICATIONS BASED ON ACTIVE MICROWAVE DATA

2.1   Production of H-SAF Soil Moisture Products

Within the "Satellite application facility on support to operational hydrology and water management (H-SAF)" project, three satellite-derived soil moisture products, developed by TU-Wien and the European Center for Medium-Range Weather Forecasts (ECMWF) institutes, are delivered in near-real time: (1) large-scale surface soil moisture by radar scatterometer (H07), (2) small-scale surface soil moisture by radar scatterometer (H08), and (3) profile soil moisture index in the roots region by scatterometer data assimilation (H14).

The algorithm for the ASCAT large-scale soil moisture product (H07) was developed by the Vienna University of Technology (TU Wien) and is from its conception a change detection method. As mentioned above, a full description of the algorithm can be found in Wagner et al. (2013). In the context of EUMETSAT's H-SAF, an approach to disaggregate the 25 km ASCAT surface soil moisture data to 1 km using advanced synthetic aperture radar (ASAR) data acquired by the Envisat satellite has been developed. The method exploits the fact that the temporal dynamics of the soil moisture field is often very similar across a wide range of scales; a phenomenon usually referred to as "temporal stability" (Brocca et al., 2014). This means that a linear model may approximate the relationship between local scale and regional scale measurements. In other words, the small-scale surface soil moisture product (H08) is based on the linear disaggregation of the large-scale product (H07), with the regression coefficients constant in time and derived by the combined analysis of ASCAT and ASAR backscatter measurements. Moreover, an ASCAT root-zone soil moisture profile product (H14) has been developed based on ASCAT surface soil moisture data assimilation into the ECMWF extended Kalman filter land surface data assimilation system (de Rosnay et al., 2013). The retrieved ASCAT root-zone soil moisture is an optimal combination between the modeled first guess, the screen-level temperature and humidity analyses, and the ASCAT-derived surface soil moisture, which is propagated forward in time through the root-zone profile. The ASCAT root-zone soil moisture profile product (H14) is available for four soil layers from surface down to 3 m, with a global daily coverage.

2.2 Description of the Setup for Flood and Landslides Early Warnings

In recent years, the Research Institute for Geo-Hydrological Protection (IRPI-CNR) is running some projects with the National Department of Civil Protection and the Regional Department of Civil Protection of Umbria Region for the exploitation of H-SAF soil moisture products for operational floods and landslides monitoring, and prediction, throughout the Italian territory. Specifically, these projects aim at understanding the real potential of satellite soil moisture products for improving the mitigation of the hydrogeological risk, which represents the major climate-related hazard in Italy.

Before their use within operational systems, satellite soil moisture products need some preprocessing steps to make them suitable for use in hydrological modeling (Brocca et al., 2010). Specifically, the following procedure has been set up for civil protection centers in Italy. First, the three soil moisture products (H07, H08, and H14) are downloaded daily from the H-SAF ftp website and/or the EUMETcast dissemination system. Second, the data are processed to have it ready in a regular grid of ~10 km horizontal resolution, using the nearest neighbor method. Finally, the products are rescaled, using H14 as reference, with predefined linear relationships to match the temporal mean and variance of each product pixel by pixel.

Additionally, in central Italy, an advanced monitoring system has been developed in which H-SAF H07 soil moisture product is compared with both in situ and modeled data in real time (see Fig. 1). The combined use of the three data sources allows a robust and reliable assessment of soil moisture conditions that is fundamental for the prediction of the possible occurrence of floods and landslide in the territory (Ponziani et al., 2012; Massari et al., 2014a).

Two successful examples showing the use of satellite soil moisture data for floods and landslides prediction are described next.

Recently, Massari et al. (2014a, 2015) have developed a simplified continuous rainfall-runoff model (SCRRM) (see Fig. 2A) that exploits at best satellite soil moisture information, and that is highly suitable for operational applications. Indeed, there is no need for continuous and uninterrupted rainfall and temperature time series as in the standard continuous rainfall-runoff models, and SCRRM is flexible and easy to use. Moreover, the low number of parameters that characterizes SCRRM simplifies its applicability on a large scale and in poorly gaged regions. SCRRM was already applied to a small catchment in Greece by Massari et al. (2014b) in the context of a European project addressed to the development of an early warning system for flood and fire risk assessment and management (http://www.flire.eu/en/). Moreover, SCRRM was applied to 35 basins throughout Italy (Fig. 2) (Massari et al., 2015) with two H-SAF soil moisture products (H07 and H14). The results in terms of flood prediction were satisfactory with mean value of the Nash-Sutcliffe efficiency equal to 0.58 and 0.61 when using H14 and H07 as input, respectively. As a consequence, SCRRM is going to be implemented at a national level for flood prediction in Italy by using the near-real time soil moisture products provided through the H-SAF project.

The use of satellite soil moisture data for landslide prediction was successfully demonstrated in Brocca et al. (2012). Specifically, the improvements on the prediction of movements of the Torgiovannetto landslide, located near the famous town of Assisi, were investigated. For the period 2007–2009, the landslide movements were recorded for a series of rainfall events thanks to an extensometers network that was established in 2005. Additionally, meteorological data (rainfall and air temperature) and satellite soil moisture data from H07 product were collected for the same period. Meteorological and soil moisture data were used as input into a multiple linear regression model used to predict the observed landslide movements. As shown in Fig. 3, the use of only rainfall information has low potential for estimating landslide movements with a correlation coefficient, r, equal to 0.219. However, by including satellite soil moisture data in the regression model, the agreement with the observations is significantly higher with $r = 0.821$. On this basis, an operational system was implemented within the Regional Department of Civil Protection of Umbria Region for the monitoring of soil moisture conditions through satellite data (see also Fig. 1) that is currently used for issuing, day-by-day, alerts for the hydrogeological risk.

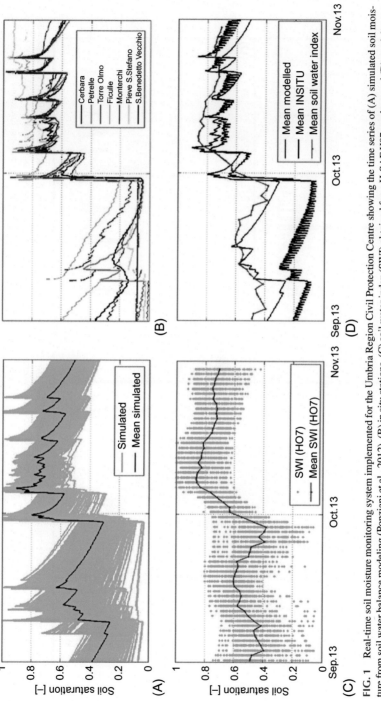

FIG. 1 Real-time soil moisture monitoring system implemented for the Umbria Region Civil Protection Centre showing the time series of (A) simulated soil moisture from soil water balance modeling (Ponziani et al., 2012), (B) in situ stations, (C) soil water index (SWI) obtained from H-SAF H07 product, and (D) spatial mean of modeled, in situ, and satellite data. (D) Provides a real-time assessment of average soil moisture conditions in the territory from the three data sources, thus representing a robust estimate to be used for the forecasting of floods and landslides occurrence.

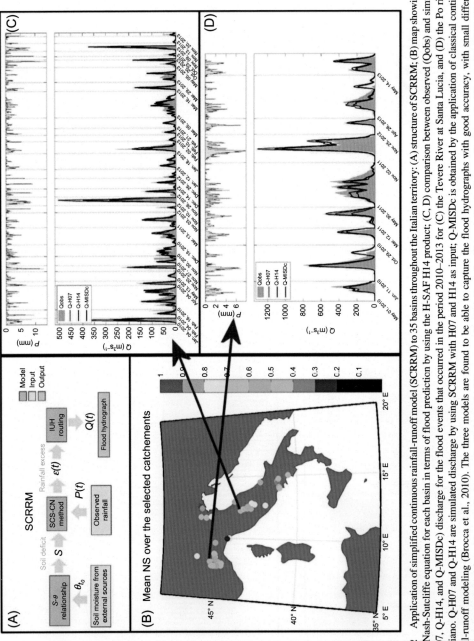

FIG. 2 Application of simplified continuous rainfall-runoff model (SCRRM) to 35 basins throughout the Italian territory: (A) structure of SCRRM; (B) map showing the mean Nash-Sutcliffe equation for each basin in terms of flood prediction by using the H-SAF H14 product; (C, D) comparison between observed (Qobs) and simulated (Q-H07, Q-H14, and Q-MISDc) discharge for the flood events that occurred in the period 2010–2013 for (C) the Tevere River at Santa Lucia, and (D) the Po river at Carigliano. Q-H07 and Q-H14 are simulated discharge by using SCRRM with H07 and H14 as input; Q-MISDc is obtained by the application of classical continuous rainfall-runoff modeling (Brocca et al., 2010). The three models are found to be able to capture the flood hydrographs with good accuracy, with small differences among them. *(Adapted from Massari, C., Brocca, L., Ciabatta, L., Moramarco, T., Gabellani, S., Albergel, C., de Rosnay, P., Puca, S., Wagner, W., 2015. The use of H-SAF soil moisture products for operational hydrology: flood modelling over Italy. Hydrology 2(1), 2–22.)*

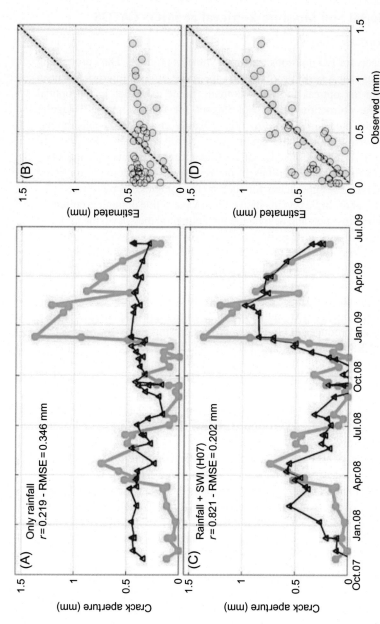

FIG. 3 Comparison between observed (*circles*) and estimated (*triangles*) crack aperture for the Torgiovannetto landslide in central Italy as timeseries (A, C) and scatterplot (B, D). Estimated crack aperture are obtained by using a multiple linear regression model that considers (A, B) only rainfall-related quantities (i.e., total rainfall and maximum rainfall for duration of 1 h), (C, D) rainfall and SWI obtained from H-SAF H07 product (*r*, correlation coefficient; *RMSE*, root mean square error). The improvement due to the use of satellite soil moisture data is evident in the bottom panels with an increase in correlation from 0.219 to 0.821. (*Adapted from Brocca, L., Ponziani, F., Moramarco, T., Melone, F., Berni, N., Wagner, W., 2012. Improving landslide forecasting using ASCAT-derived soil moisture data: a case study of the Torgiovannetto landslide in central Italy. Remote Sens. 4(5), 1232–1244.*)

3 STATE-OF-THE-ART SOIL MOISTURE RETRIEVALS AND APPLICATIONS BASED ON PASSIVE MICROWAVE DATA

3.1 Multiorbit Soil Moisture Retrieval at SMOS CATDS

3.1.1 Theoretical Basis

Level-3 data sets for EO missions are time synthesis products. Data processing of EO missions using multiple revisits of the same area of interest has been widely used in the optical and radar domain. Some examples in the optical domain are the correction of aerosol impact for visible images (Hagolle et al., 2008), the cloud impact correction (Hagolle et al., 2010), and the use of multiple revisits for the land cover classification algorithm (Inglada and Mercier, 2007). With the increase of multitemporal data sets and the open access policy to them, communities in the remote sensing domain are being organized around new approaches for such methodologies. As mentioned previously, the use of multiple revisits in the microwave active domain is common. In the passive microwaves and despite the high potential they have, the majority of operational soil moisture retrieval algorithms are not based on multiple revisits. The main reasons are the wide swath, the coarse resolution, and high revisit time [1 day for AMSR-E (Njoku et al., 2003)], 2–3 days for SMOS (Kerr et al., 2010) and SMAP (Entekhabi et al., 2010).

Because of its novelty, the basic information in the production of the Level-3 data soil moisture retrievals from the SMOS satellite is provided in this section. The generation is produced at CATDS, a postprocessing center for SMOS data initiated by CNES (Centre National d'Etudes Spatiales). The processing chains running at CATDS do not only produce time synthesis products for the mission in a gridded format but they also take advantage of the frequent revisit of SMOS to make a multiorbit, multiparameter retrieval. The Level-3 processor uses as a forward model the same physical model as the Level-2 processor. A short summary of the model features is provided here. The SMOS Level-2 retrieval can be divided into two major components.

The first component is a physical model that computes the TB at the antenna reference frame forced by ancillary data and physical parameters value. The selected physical model for the SMOS mission is the L-band microwave emission of the biosphere (L-MEB) (Wigneron et al., 2007) tau-omega model. A comparison of several models used for soil moisture retrieval from microwave sensors is presented in Miernecki et al. (2014). The main features of the L-MEB physical model implementation into the SMOS operational processors are:

- Single scattering is considered.
- A combined retrieval of soil moisture and vegetation optical thickness in nominal surfaces by use of angular signature information.
- Dual polarization is modeled; full polarization is used only to take into account the Faraday rotation and the geometric rotation to change the reference frame.

- Consideration of antenna patterns (weighting function) change due to the acquisition configuration and mean antenna pattern (ATBD L2). Fig. 4 (from Al Bitar et al., 2012) shows the mean weighting function and its associated cumulative normalized function. The mean weighting function expresses the average contribution with respect to the distance. The cumulative function expresses how much of the signal is contained in a disc of given radius. For example, at ~20 km radius, the 0.5 coefficient corresponds to 3 dB noise level and about 86% of the signal if a homogeneous surface is considered. This corresponds to the nominal resolution of the sensor.
- Surface heterogeneity is considered through aggregated land cover at 4 km. The contributions are then aggregated based on the mean antenna pattern. An area of 125 km by 125 km is considered at each retrieval node, ensuring that the complete surface is covered.
- Dynamic changes in surface heterogeneity are considered through the use of weather data from ECMWF.

Since the mission launch many modifications have been applied to the operational processing model either through improved parameterization (Rahmoune et al., 2014) or through the choice of physical models (Mialon et al., 2015).

The second component is an optimization scheme that minimizes a Bayesian cost function, thus minimizing the differences between the observed TB and the modeled one while retrieving the physical parameter values. Adding to the previous components a series of preprocessing and postprocessing steps makes it possible to filter the input data for undesired effects, like the decrease of quality due to spatial sampling or radio frequency interference (RFI) (Oliva et al., 2012; Richaume et al., 2014).

The physical approach conducted at Level-3 is the same one as for Level-2. In fact at the core of the processing the same implementation for the physical

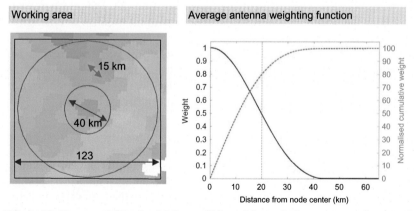

FIG. 4 Working area *(left)* and weighting coefficient of the synthetic antenna as a function of distance.

model is used. The main difference in Level-3 is the use of several orbits to retrieve soil moisture. This has an impact on the postprocessing steps for selecting the orbits and the optimization scheme to retrieve the parameters. Because Level-2 retrieval is a multiparameter retrieval, Level-3 is then a multiorbit, multiparameter retrieval. The reasons that motivate the use of the multiorbit approach are the following:

- The angular sampling of a radiometric accuracy at the border of the swath is reduced, and using a multiorbit approach can help to improve the number of successful retrievals at the border.
- It is expected that the vegetation optical depth change between consecutive retrievals at ascending or descending passes separately is highly correlated. In fact, in the microwaves domain, the vegetation optical depth is mainly correlated to the vegetation water content, which, in turn, is correlated to the leaf area index (LAI).

Other general motivations for Level-3 products are to provide a gridded product, in contrast to swath-based products, more adapted for the scientific community, and to provide products at 25 km sampling more consistent with the sensor nominal resolution.

3.1.2 Orbit Selection

The previous motivations guided the elaboration of the retrieval algorithm. First, the following criteria are applied for the selection of revisits:

- Ascending and descending orbits processing are completely separated.
- TB products are produced. In fact, this is equivalent to a rerun of the last step of the Level-1 processing chain to make the transition from the Fourier domain to the spatial domain based on the new grid.
- TB products are filtered at high altitudes where more than one revisit occurs to select a revisit per day (latitudes above 60°N and S). The selection criterion is the minimum distance from the center of the swath. The criterion is applied for each grid node.
- For each day, a 7-day period is considered for each node. Three revisits are considered independently for each grid node. One for the central date, one for the 3.5 days before, and one for the 3.5 days after. The selection is made based on minimum distance from the swath center (see Fig. 5).

3.1.3 Retrieval Algorithm

As a result of applying the previous steps, observed TB at the antenna reference frame are obtained. A Bayesian cost function that includes multiorbits information is constructed. This is achieved by incorporating in the retrieval approach a temporal autocorrelation function of the retrieved parameters (in this case the vegetation optical depth). The cost function becomes

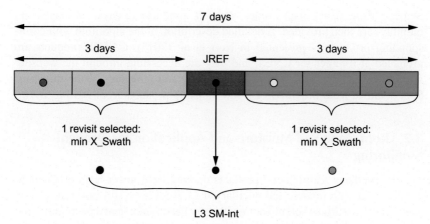

FIG. 5 Selection of revisit orbits for the multiorbit retrieval at SMOS CATDS.

$$\text{Cost} = (\text{TB}_M - \text{TB}_F)^t \cdot [\text{Cov}_{\text{TB}}]^{-1} \cdot (\text{TB}_M - \text{TB}_F)$$
$$+ (P - P_0)^t \cdot [\text{Cov}_P]^{-1} \cdot (P - P_0) \tag{1}$$

where $\text{Cov}_{\text{TB}} - \sigma_{\text{TB}}^2$ is the error covariance matrix of TB data assuming no temporal correlation. For vegetation optical depth the error covariance matrix term (considering temporal autocorrelation and no cross correlation between the different parameters) is defined as

$$\text{Cov}_{P_i} = \sigma_{i0}^2(t_1) \cdot \begin{pmatrix} 1 & \rho_i(t_1,t_2) & \cdots & \rho_i(t_1,t_{NV}) \\ \rho_i(t_1,t_2) & 1 & \cdots & \cdots \\ \cdots & \cdots & 1 & \cdots \\ \rho_i(t_1,t_{NV}) & \cdots & \cdots & 1 \end{pmatrix} \tag{2}$$

and for all other parameters $\text{Cov}_{P_i} = \sigma_{i0}^2(t_n) \cdot I$. The vegetation optical thickness correlation is modeled with a Gaussian autocorrelation function:

$$\text{Correl}_{\text{Tau}}(t_1, t_2) = \text{Corr}_{\max}(t_1, t_2) \cdot \exp\left(-\frac{(t_1 - t_2)^2}{\text{Tc}^2} \right) \tag{3}$$

where

- t_1 and t_2 are the time (expressed in days) corresponding to the vegetation optical thickness retrievals,
- $\text{Corr}_{\max}(t_1, t_2)$ is the maximum amplitude of the correlation function between t_1 and t_2,
- Tc is the characteristic correlation time for the vegetation optical thickness Tau, and
- Tc = 30 days for forests and Tc = 10 days for low vegetation.

Starting from daily maps, time synthesis products (3 days, 10 days, and monthly) are then provided. A detailed description of the algorithm with corresponding products is presented in Kerr et al. (2013). CATDS products are produced at iFremer for CNES, and they are accessible through the website: http://www.catds.fr. Fig. 6 shows the median soil moisture for the month of Sep. obtained from the SMOS L3 soil moisture products from 2010 to 2016.

3.2 Root-Zone Soil Moisture and Application for Drought Monitoring

Level-4 products are high-end products derived from lower-level products in combination with models and ancillary data. Root-zone soil moisture can be accessed from surface soil moisture either by the use of a land data assimilation system or by the use of empirical approaches. Root-zone soil moisture is the primary variable in the monitoring of agricultural drought. At CATDS, a data-driven approach is used to compute the root-zone soil moisture. Wagner et al. (1999) suggested the use of an exponential filter to access root-zone soil moisture from shallow soil moisture. Albergel (2008) provided a sequential formulation of the model and tested it over local sites in the southwest of France. Qiu et al. (2014) showed that surface soil moisture can be considered as a reliant proxy to root-zone soil moisture. At SMOS CATDS Level-4, root-zone soil moisture is provided by applying the exponential filter in combination with the Food and Agriculture Organization (FAO, United Nations) method FAO56, the latter used to compute the transpiration in a deeper soil layer (10 cm to 1 m).

The water availability in the root-zone is computed using a double bucket hydrological model. The model is forced only by surface soil moisture and no precipitation data is used. The main remote sensing data used is the SMOS L3 CATDS retrieved soil moisture. The MODIS NDVI is used to compute vegetation transpiration rates. The other ancillary data sources are operational climate analysis data from the National Centers for Environmental Prediction (NCEP), soil texture from FAO, and surface cover from ECOCLIMAP-II.

3.2.1 From Near-Surface to the 20 cm Layer

The surface layer extending from the 0–20 cm is the place where complex processes like infiltration, runoff, percolation, and evaporation take place. It is possible to parameterise a global model at 25 km resolution taking into consideration all those processes, but the uncertainty associated with the parameters will be high. For this reason a simple water budget model is used (also referred to as exponential filter) to compute the water content in the 0–20 cm layer. The sequential formulation of the exponential filter presented by Albergel (2008) is used. The only parameter needed is the time lag in days that defines the rate of transfer of the water to deeper layers. This simple formulation has been developed for microwave remote sensing data and is well adapted for SMOS data.

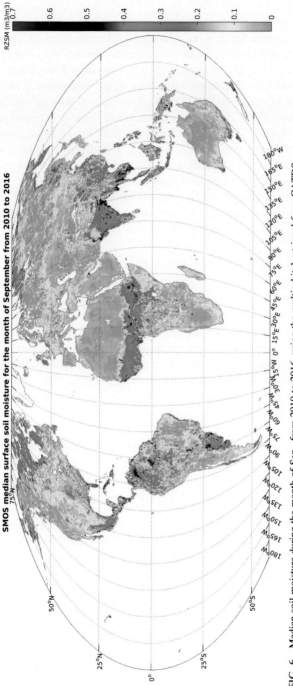

FIG. 6 Median soil moisture during the month of Sep., from 2010 to 2016, using the multiorbital retrieval from CATDS.

3.2.2 From the 20 cm Layer to the Root-Zone Layer

The layer of interest for drought is the 0–200 cm layer. The exponential filter used in the previous layer is not adapted to take into account the vegetation transpiration and it does not take into account capillary rise. For this reason, a budget model based on a linearized Richards equation formulation is selected for the second layer. The transpiration rate of the vegetation is computed by means of the FAO method (Allen, 1998). Two types of data are needed to drive the transpiration: the LAI of the vegetation and climatic data to compute potential leaf transpiration. The NDVI is obtained from the 16-days global MODIS data interpolated over the EASE 25 km grid. The climate data are downloaded through NCEP servers. These data are interpolated to the SMOS product grid and used to compute the potential transpiration.

As mentioned above, the root-zone soil moisture can be used to monitor the agricultural drought. Fig. 7 shows the status of the root-zone soil moisture for Jul. 15, 2012, obtained from SMOS CATDS processing.

3.2.3 Case of the Horn of Africa 2011 Droughts

Fig. 8 shows the root-zone soil moisture index from SMOS for Oct. 2011 over Africa. This index varies between 0 for dry conditions and 1 for wet conditions. The soil moisture is considered as equal to wilting point during dry conditions and to field capacity during wet conditions. It shows the extent of the droughts over the horn of Africa. The wet conditions over the western part of the Sahel are also clearly visible. An extraction of the root-zone soil moisture during the dry season corresponding to the Sahel transition region, between latitudes 9°S and 22°N, is shown at the bottom of Fig. 8. The figure clearly depicts the wet conditions in the interior Niger Delta and in the south of Sudan, which is a highly irrigated agricultural area. This shows the potential of a soil moisture-based index to monitor not only drought due to rainfall deficit but also vegetation stress due to irrigation deficit at regional scale.

4 STATE-OF-THE-ART SOIL MOISTURE ANALYSIS FOR NWP

Soil moisture accurate initialization in NWP models is crucial for good quality forecasts from short to medium ranges (Beljaars et al., 1996; Douville et al., 2000; Drusch and Viterbo, 2007). Due to the long memory of the soil moisture reservoir, its influence can be extended from monthly to subseasonal time scales (Koster et al., 2011; Dirmayer, 2000; Ni-Meister et al., 2005). During many years and due to the lack of satellite data, soil moisture has been constrained with in situ observations of air temperature and humidity from the SYNOP network, preventing in this way model drifts toward unrealistic states of the soil. The ingestion of the soil moisture information contained in screen-level variables has been carried out with a land data assimilation system (LDAS). The core of the ECMWF soil moisture analysis is a point-wise simplified extended

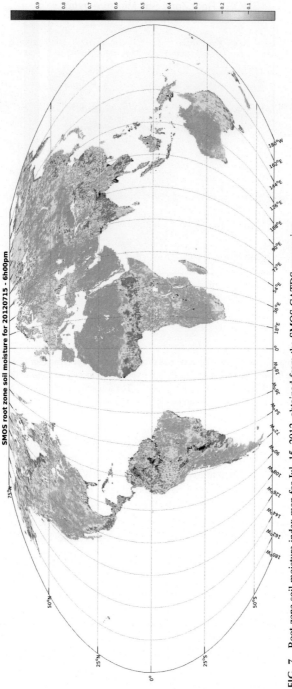

FIG. 7 Root-zone soil moisture index map for Jul. 15, 2012, obtained from the SMOS CATDS processing.

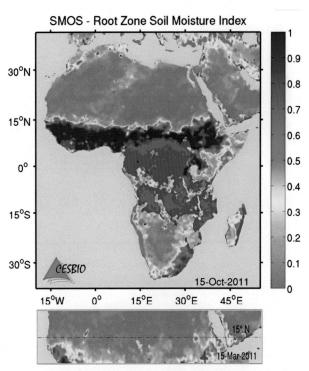

FIG. 8   The top panel shows the root-zone soil moisture index over Africa for Oct. 15, 2011. The droughts over the horn of Africa can be clearly observed. The bottom panel shows the soil moisture conditions over the Sahel region during the dry season. The wet areas of the internal Niger Delta and the South Khartoum agricultural area can be clearly observed.

Kalman filter (SEKF) (Drusch et al., 2009a; de Rosnay et al., 2012), which is a data assimilation scheme particularly suitable for low-dimensional problems, as the ECMWF soil moisture analysis. At each ECMWF 12-h assimilation cycle and for each model grid point the SEKF computes soil moisture increments for the top three soil layers of the hydrology tiled ECMWF scheme of Surface Exchanges over land (HTESSEL) (Balsamo et al., 2009). The objective is producing small corrections of the model forecasted value consistent with the observations. Soil moisture increments ($\Delta \mathbf{sm}$) are computed at analysis time (t_a) based on the misfit between observations ($\mathbf{y}^o$) and the model equivalent of the observations ($H_i(\mathbf{x}^b)$) at the time of the observations (t_i), also called innovation vector. The uncertainty of the forecast soil moisture value and the observations are also accounted for explicitly in the background error covariance matrix $\mathbf{B}$ and in the observation error covariance matrix $\mathbf{R}$, respectively. Soil moisture increments are calculated as

$$\Delta \mathbf{sm}(t_a) = \mathbf{K}_i \left[\mathbf{y}^o(t_i) - H_i(\mathbf{x}^b) \right] \qquad (4)$$

being

$$\mathbf{K}_i = \left[\mathbf{B}^{-1} + \mathbf{H}_i^T \mathbf{R}^{-1} \mathbf{H}^T\right]^{-1} \mathbf{H}_i^T \mathbf{R}^{-1} \tag{5}$$

the Kalman gain, which modulates the amount of correction of the background soil moisture value. H_i is the forward observation operator bringing observations and model equivalents to the same space for comparison purposes. In the case of brightness temperatures, H_i is the radiative transfer model (RTM) that simulates TB at the top of the atmosphere as a function of soil initial states. The ECMWF RTM for low frequencies passive microwaves is the community microwave emission model (Drusch et al., 2009b; de Rosnay et al., 2009; Muñoz-Sabater et al., 2011). The linearized version of H_i is the Jacobian of the observation operator and is represented by $\mathbf{H}_i$. It estimates the sensitivity of the model equivalent of the observations to individual perturbations of the model state vector by a small amount. For each element of the control state vector, a perturbed model integration is required; this being the main driving cost of the SEKF in the operational system. The greater the confidence about the observations, the more weight the assimilated observations will have in the computed soil moisture increments.

The current ECMWF operational system uses 2 m temperature and relative humidity analysis as input for the observation vector (available at synoptic times), and since recent dates soil moisture index observations from the ASCAT active system. The ECMWF operational soil moisture analysis also includes the possibility of using L-band TB from the SMOS passive system (Muñoz-Sabater, 2015), alone or in combination with screen-level variables and ASCAT soil moisture retrievals. The soil moisture background errors of the three top soil layers of the HTESSEL model are currently fixed and static during the assimilation window. The standard deviation of these variables is assumed to be $\sigma_{sm} = 0.01$ m^3 m^{-3}. The operational observation errors in the **R** matrix are also fixed and account for the standard deviation of 2 m temperature, relative humidity, and ASCAT soil moisture observations. Since May 2015, these values were reviewed and set, respectively, to $\sigma_T = 1$ K, $\sigma_{RH} = 4\%$, and $\sigma_{ASCAT} = 0.05$ m^3 m^{-3}. The observation error for SMOS TB is currently being optimized. It has the advantage that is specific to each observation, with a minimum value equivalent to the radiometric accuracy, making the SMOS TB uncertainty more objective than a constant fixed value. Fig. 9 shows an example of accumulated soil moisture increments obtained by assimilating screen-level variables and SMOS TB observations.

The contribution of ASCAT soil moisture retrievals for better initialization of soil moisture in the ECMWF atmospheric system was studied first by Scipal et al. (2008). In their experiment they used a nudging scheme to assimilate ASCAT data; however, compared to the operational optimal interpolation scheme, the use of scatterometer data slightly degraded the forecast scores. More recently, de Rosnay et al. (2012) matched the climatology of ASCAT soil

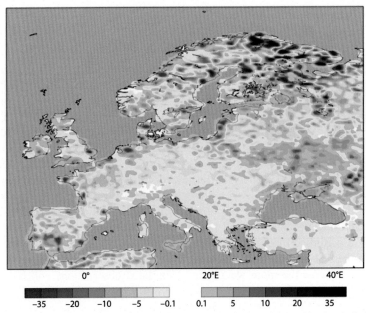

FIG. 9 Accumulated soil moisture increments for Jun. 2010 in mm, after combined assimilation of screen-level variables (T_{2m} and RH_{2m}) and SMOS TB.

moisture data to that of the ECMWF model through CDF matching and used a SEKF to assimilate the scatterometer data. The new soil moisture state did not improve the correlation with ground validation data significantly. However, improvement in the model, in the bias correction methodology, and in the quality of the scatterometer data from MetOp-B has optimized the use of ASCAT soil moisture retrievals for short-range forecasts.

Muñoz-Sabater (2015) has shown that the assimilation of SMOS TB in the ECMWF soil moisture analysis typically reduces the amount of water stored in the soil. The negative sign of the soil moisture increments points in the right direction, as they partially correct for a wet bias associated with ECMWF's land surface scheme. Compared to screen-level variables, the assimilation of SMOS TB produces large corrections of soil moisture, which is a combination of the large sensitivity of SMOS data to soil moisture variations, and the larger weight given to the SMOS observations in the LDAS compared to screen variables. An exhaustive validation against more than 400 in situ stations distributed around the world in summer 2010 showed that SMOS observations are especially informative of the soil moisture content in semiarid climates. However, the information provided by screen-level variables, in combination with SMOS TB, has been demonstrated to be especially valuable in temperate climates. Concerning the implications for the forecast of atmospheric variables, the impact of the new soil moisture state due to the joint assimilation of screen-level variables and

SMOS data showed very little impact in 2 m temperature, air temperature, and relative humidity. Larger impact was observed at specific regions, with mixed signs depending on the geographical location. This result may evidence an over-tuning of the parameters controlling coupling processes at the soil-atmosphere interface.

Despite the preliminary findings above, two aspects should be taken into consideration. First, an optimization of the SMOS observation weight in the land data assimilation system, as well as the revision of other parameters of the soil moisture analysis, will likely produce an updated picture of the real impact of the use of SMOS observations for soil moisture initialization in a weather forecasting model. And second, the combination of SMOS and ASCAT data (as planned to be integrated in the operational system) may solve some of the deficiencies of using only a single type of observation for soil moisture updates.

5 CAVEATS IN THE PREDICTION OF SOIL MOISTURE RETRIEVALS AND ITS APPLICATIONS

The production of accurate soil moisture retrievals based on EO-based techniques is a crucial requirement for their use in operational applications. This chapter has inspected the fundaments of the algorithms producing SMOS soil moisture retrievals, for Level-3 and Level-4, the latter being more adapted to end-user applications. It also has presented the use of state-of-the-art soil moisture retrievals and analyses based on both active and passive microwave data sets for hydrological and meteorological applications. The continuous availability and production of such data sets is essential for operational applications. However, several external factors can hamper their production. One of them is the contamination of the geophysical signal by RFI (Njoku et al., 2005; Oliva et al., 2012; Soldo et al., 2015). This is critical in passive L-band, where the level of observed energy is very low. Indeed, some areas across the globe see the number of retrievals strongly reduced by RFIs, resulting in large gaps in given regions, and as a consequence limiting their use for land applications. Retrievals under dense forests are also another challenging area because of the high attenuation of the soil emission. The lack of long L-band time series has prevented large-scale applications that depend on historical data sets. It is expected that the integration of soil moisture data sets from L-Band observations into the ESA climate change initiative (CCI) soil moisture essential climate variable (ECV), a consistent multisensors soil moisture data set will be available in the near future.

Another issue frequently neglected is the indirect connection existent between soil moisture and the noncontinuous nature of the observations. Soil moisture retrievals may be inconsistent with model simulations, as they may utilize different land surface parameters and background information including soil texture, surface temperature, and vegetation. In fact, soil moisture retrieval involves a prior estimate, frequently from a model. Errors of the prior estimate are often correlated to the background value, and they are commonly not taken into account. These reasons, along with the faster availability of the data, are arguably important reasons to consider the use of direct TB in assimilation schemes for NWP systems (Muñoz-Sabater, 2015), and they have to be carefully considered to produce high-quality soil moisture data sets.

Despite the obstacles mentioned above, satellite soil moisture products have reached a good level of accuracy and maturity, especially those based on longer time series with active observations. However, their use for operational hydrological applications is still very limited. The main difficulty is due to the spatial mismatch between the resolution of current satellite soil moisture products (from $\sim$400 to $\sim$1500 km^2) and the study area under investigation (frequently below 100 km^2). Some authors have already demonstrated that even coarse-resolution soil moisture products can still provide useful information for improving runoff and landslide prediction, even for small to medium basins/areas (e.g., Brocca et al., 2010, 2012). The modeling of land surface physical processes is evolving at a faster pace than the resolution at which satellite retrievals are being improved. This tendency will not change in the years to come, likely decades. Therefore, if satellite soil moisture retrievals are intended for use in the new wave of operational applications demanding high-resolution information of soil variables, this mismatch has to be addressed (Wood et al., 2011). Downscaling coarser remote sensing observations to the model resolution, using a set of fine resolution auxiliary data (Merlin et al., 2010; Piles et al., 2011), is one of the most accepted methods used today. Even some recent studies claim the downscaled approach to be carried out with implicit bias correction (Verhoest et al., 2015). The reverse approach consists of upscaling model outputs to the satellite footprint resolution (Crow et al., 2012). The challenge is even greater if different types of data sets at different spatial resolutions and revisit time are intended to be used simultaneously, as it is for the case explained in Section 6. In this case some compromises may be necessary. For example, Fascetti et al. (2016) compared different sources of soil moisture retrievals with very different characteristics (spatial resolution, observation time, units). The comparison was only possible by means of several assumptions, as the rescaling of the active source of soil moisture retrievals or the spatial collocation through the nearest neighbor approach. They concluded that the way the active data is rescaled has an impact on the comparison results.

The exploitation of spaceborne observations in land applications is also limited by the shallow penetration depth ($\sim$1–5 cm), which is significantly lower than the effective depth at which hydrological or meteorological applications need to extract information from. But even this shallow information can be physically propagated at deeper layers by means of data assimilation methods (Houser et al., 1998; Walker et al., 2001; Muñoz-Sabater et al., 2007, 2008) or by empirical approaches, such as that used in Wagner et al. (1999) and Al Bitar et al. (2013).

The constant presence of biases is another important element hindering the efficient exploitation of soil moisture retrievals for a wide range of applications. Biases are the cause of systematic differences between the statistical moments of the retrievals and the models. One reason is the fact that while satellite instruments provide an averaged measurement of a shallow layer of the surface, models estimate the numerical value of a prognostic variable over a smaller area but representative of a deeper thickness. The use of different modeling and/or observational approaches typically leads to predictions with different systematic relationships to the assumed truth. Simplifications in the model parameterization and in the retrieval algorithm are also sources of biases. Prior to the joint use of soil moisture retrievals and models, biases should be removed. A popular technique used to remove biases consists of matching the statistical moments of two independent distributions, also called cdf-matching. Frequently, the observations climatology is matched to that of a model equivalent. The remaining disagreement between the "unbiased" observations and the model counterpart contains useful information to constrain the time-space evolution of modeled soil moisture. Although this technique is based on the availability of long time series, several authors have found this technique to work even for short records of satellite data (Reichle and Koster 2004). Other strategies of bias correction methods use least squares regression techniques (Crow et al., 2005; Crow and Zhan, 2007) or variance-based techniques as triple collocation analysis (Stoffelen, 1998). However, one should bear in mind that in reality, even if the model and observations are subjected to a bias correction approach, residual biases frequently remain in the systems.

A final important issue to consider is the format and the data volume (Muñoz-Sabater et al., 2012). Indeed, end-users are frequently uncomfortable working with large data sets and with unconventional data formats. Even though it may look like a negligible problem, this is still today one of the major obstacles for the effective use of satellite soil moisture data in several operational domains. Therefore, it is advisable to invest in capacity building for making, on one hand, the satellite products more user-friendly and, on the other hand, the possible users aware of the large potentials that these products can offer.

In the production of continuous, high-quality, and useful soil moisture data sets for operational use, the above are not the only obstacles and caveats to be taken into account, yet they are common to all soil moisture retrievals algorithms and they need to be addressed for each specific application.

6 FUTURE OUTLOOK

Soil moisture data sets based on EO techniques have brought a lot of benefit to a wide range of daily applications such as those shown in this chapter, and many others such as fire risk assessment and water resources management of agricultural water monitoring. The observed benefits should encourage us to carry out more investigations to foster the use of satellite soil moisture data in different communities (hydrology, geomorphology, agriculture, etc.).

With the purpose of improving the accuracy of satellite rainfall products, some approaches using satellite soil moisture data were recently developed (Crow et al., 2009; Pellarin et al., 2013; Brocca et al., 2013; Wanders et al., 2015). Among them, Brocca et al. (2013) proposed a new method for estimating rainfall using soil moisture observations, called SM2RAIN. The method is based on the inversion of the soil water balance equation; that is, it estimates the rainfall by using the change in time of the amount of water stored into the soil, thus considering the "soil as a natural rain gage." SM2RAIN has been applied both on a local (Brocca et al., 2013, 2015) and on a global scale (Brocca et al., 2014) with ground and satellite soil moisture data as input with satisfactory results in terms of rainfall estimation. Moreover, Massari et al. (2014b) and Ciabatta et al. (2015b) found that the correction of rainfall through SM2RAIN provides improvement in flood modeling when compared to the use of rain gage observations only.

On this basis, in the next phase of the H-SAF project, the development of a new precipitation product in near-real time is expected that integrates satellite soil moisture (through SM2RAIN) and precipitation products. Basically, this new product will integrate the "top-down" and "bottom-up" perspectives for rainfall estimation from remote sensing. The retrieval of rainfall in the state-of-the-art products is based on a "top-down" approach; that is, rainfall is obtained through the inversion of the atmospheric signals scattered or emitted by atmospheric hydrometeors. In the new concept on which the SM2RAIN algorithm was developed, rainfall is obtained in a "bottom-up" perspective from the knowledge of the amount of water stored in the soil and measured from remote sensing. The two perspectives are independent and their integration is capable of providing a higher-quality rainfall product that takes advantage of the benefits of both perspectives. First validation results in Italy have shown highly promising results (Ciabatta et al., 2015a) as the integrated product is found to outperform state-of-the-art products and to provide reliable information for hydrological applications (Ciabatta et al., 2015b).

Soil moisture is identified as an ESA ECV for climate change studies. It is thus important to have a long (decadal) and consistent data set of soil moisture obtained from multiple EO missions. One of the outcomes of the ESA CCI on soil moisture is a consistent decadal data set of observed surface soil moisture from microwave remote sensing. The second phase of this program will include SMOS data and, eventually, SMAP data will be included too. The overlap of SMOS and SMAP missions will make possible the continuity of L-band observations for soil moisture. With the emergence of the European Copernicus program [previously known as Global Monitoring for Environment and Security (GMES)], synergistic studies between active (Sentinel-1) and passive (SMOS, SMAP) sensors are expected, providing new surface soil moisture products that make use of the advantages of the different sensors to provide a high-temporal (3 days) and high-spatial (100 m) resolution within recommended accuracy. By associating these products with optical data and crop modeling, new applications related to agricultural management (irrigation, vegetation stress, and others) will be elaborated. A multitude of services is expected to emerge from these activities, accelerating the technological transition from research (and public) to commercial (and private) sectors.

Soil moisture in fully coupled land-atmospheric systems will require special attention because of the enhanced predictability skill coming from the long memory of the soil moisture reservoir. Soil moisture in NWP systems has been optimized to provide sensible and latent heat fluxes, based on the assimilation of atmospheric temperature and moisture errors in LDAS. The risk is that model errors can be accumulated into soil moisture, with negative effects for the realism of the soil water content. However, the availability of observations from new platforms and the irruption of new applications such as carbon flux modeling and river flow forecasts require high-quality soil moisture, as well as other variables of the water cycle as runoff. In this context, the realism of both the surface processes and the numerical value of surface variables are necessary. The improvement of soil moisture in the future will be supported by a more comprehensive modelization of surface processes, such as the explicit processes of vegetation (Boussetta et al., 2013) and the mutual interaction of the water, energy, and carbon cycles. The use of satellite observations (sensitive only to the top cm of the soil) will be enhanced by increasing the soil vertical discretization, currently composed at ECMWF of four layers at 7, 21, 72, and 189 cm. The benefits of adding extra layers will enhance land-atmosphere interaction, and it will extend the memory from the deeper soil layers and the model capability to represent multiple time scales. In the data assimilation side, the flexible structure of the ECMWF SEKF opens a lot of new opportunities, including the exploitation of new satellite surface products, such as SMAP data, and the extension of the SEKF to analyze additional variables, such as soil temperature, snow temperature, and mass and vegetation parameters.

This chapter has explained how the use of scatterometer data from ASCAT and radiometric data from SMOS has the potential to produce better, more realistic soil moisture states not only for weather forecasting but also potentially for other applications such as river flood forecasting or carbon sources estimation. Both types of data have already shown, in certain cases, deficiencies in terms of poor impact in the soil moisture analysis or in air temperature and humidity forecasts. Active and passive systems are complementary in many different ways. An upgraded operational version of the ECMWF surface analysis will combine both types of remote sensing information, along with the traditional use of conventional data, overcoming in this integrated way some of the observed deficiencies and at the same time increasing the efficiency of the LDAS.

ACKNOWLEDGMENTS

The authors acknowledge support from EUMETSAT through the "Satellite Application Facility on Support to Operational Hydrology and Water Management" H-SAF project and the

National Department of Civil Protection of Italy. Also, thanks for the support of ESA for the SMOS project at ECMWF. Finally, acknowledgments to CNES and Ifremer for the SMOS CATDS data sets.

REFERENCES

Al Bitar, A., Leroux, D., Kerr, Y.H., Merlin, O., Richaume, P., Sahoo, A., Wood, E.F., 2012. Evaluation of SMOS soil moisture products over continental US using the SCAN/SNOTEL network. IEEE Trans. Geosci. Remote Sens. 50 (5), 1572–1586.

Al Bitar, A., Kerr, Y., Merlin, O., Cabot, F., Wigneron, J.-P., 2013. Global drought index from SMOS soil moisture. In: IEEE International Geoscience and Remote Sensing Symposium, IGARSS 2013, Melbourne, Australie, 21–26 July.

Albergel, C., Rüdiger, C., Pellarin, T., Calvet, J.-C., Fritz, N., Froissard, F., Suquia, D., Petitpa, A., Piguet, B., Martin, E., 2008. From near-surface to root-zone soil moisture using an exponential filter: an assessment of the method based on in-situ observations and model simulations. Hydrol. Earth Syst. Sci. 12, 1323–1337. http://dx.doi.org/10.5194/hess-12-1323-2008.

Allen, R.G., 1998. Crop Evapotranspiration-Guidelines for Computing Crop Water Requirements-FAO Irrigation and Drainage Paper 56. Food and Agriculture Organization of the United Nations, Rome. No. 6541.

Balsamo, G., Viterbo, P., Beljaars, A., van den Hurk, B., Hirschi, M., Betts, A., Scipal, K., 2009. A revised hydrology for the ECMWF model: verification from field site to terrestrial water storage and impact in the integrated forecast system. J. Hydrometeorol. 10, 623–643.

Beljaars, A.C.M., Viterbo, P., Miller, M., 1996. The anomalous rainfall over the United States during July 1993: sensitivity to land surface parameterization and soil moisture anomalies. J. Hydrometeorol. 124, 362–383.

Boussetta, S., Balsamo, G., Beljaars, A., Agusti-Panareda, A., Calvet, J.-C., Jacobs, C., van den Hurk, B., Viterbo, P., Lafont, S., Dutra, E., Jarlan, L., Balzarolo, M., Papale, D., van der Werf, G., 2013. Natural carbon dioxide exchanges in the ECMWF integrated forecasting system: implementation and offline validation. J. Geophys. Res. 118, 1–24. http://dx.doi.org/10.1002/jgrd.50488.

Brocca, L., Melone, F., Moramarco, T., Wagner, W., Naeimi, V., Bartalis, Z., Hasenauer, S., 2010. Improving runoff prediction through the assimilation of the ASCAT soil moisture product. Hydrol. Earth Syst. Sci. 14, 1881–1893.

Brocca, L., Ponziani, F., Moramarco, T., Melone, F., Berni, N., Wagner, W., 2012. Improving land-slide forecasting using ASCAT-derived soil moisture data: acase study of the Torgiovannetto landslide in central Italy. Remote Sens. 4 (5), 1232–1244.

Brocca, L., Melone, F., Moramarco, T., Wagner, W., 2013. A new method for rainfall estimation through soil moisture observations. Geophys. Res. Lett. 40 (5), 853–858.

Brocca, L., Zucco, G., Mittelbach, H., Moramarco, T., Seneviratne, S.I., 2014. Absolute versus temporal anomaly and percent of saturation soil moisture spatial variability for six networks worldwide. Water Resour. Res. 50 (7), 5560–5576.

Brocca, L., Massari, C., Ciabatta, L., Moramarco, T., Penna, D., Zuecco, G., Pianezzola, L., Borga, M., Matgen, P., Martínez-Fernández, J., 2015. Rainfall estimation from in situ soil moisture observations at several sites in Europe: an evaluation of SM2RAIN algorithm. J. Hydrol. Hydromech. 63 (3), 201–209.

Ciabatta, L., Brocca, L., Massari, C., Moramarco, T., Gabellani, S., Puca, S., Wagner, W., 2015a. Rainfall-runoff modelling by using SM2RAIN-derived and state-of-the-art satellite rainfall products over Italy. J. Appl. Earth Observ. Geoinf. http://dx.doi.org/10.1016/j.jag.2015.10.004.

Ciabatta, L., Brocca, L., Massari, C., Moramarco, T., Puca, S., Rinollo, A., Gabellani, S., Wagner, W., 2015b. Integration of satellite soil moisture and rainfall observations over the Italian territory. J. Hydrometeorol. 16 (3), 1341–1355.

Crow, W.T., Zhan, X., 2007. Continental-scale evaluation of remotely sensed soil moisture products. IEEE Geosci. Remote Sens. Lett. 4, 451–455.

Crow, W.T., Bindlish, R., Jackson, T.J., 2005. The added value of spaceborne passive microwave soil moisture retrievals for forecasting rainfall-runoff partitioning. J. Geophys. Res. 32. http://dx.doi.org/10.1029/2005GL023543.

Crow, W.T., Huffman, G.F., Bindlish, R., Jackson, T.J., 2009. Improving satellite rainfall accumulation estimates using spaceborne soil moisture retrievals. J. Hydrometeorol. 10, 199–212.

Crow, W.T., Berg, A.A., Cosh, M.H., Loew, A., Mohanty, B.P., Panciera, R., de Rosnay, P., Ryu, D., Walker, J.P., 2012. Upscaling sparse ground-based soil moisture observations for the validation of coarse-resolution satellite soil moisture products. Rev. Geophys. 50. http://dx.doi.org/10.1029/2011RG000372.

de Rosnay, P., Drusch, M., Boone, A., Balsamo, G., Decharme, B., Harris, P., Kerr, Y., Pellarin, T., Polcher, J., Wigneron, J.P., 2009. AMMA land surface model intercomparison experiment coupled to the community microwave emission model: ALMIP-MEM. J. Geophys. Res. 114.

de Rosnay, P., Balsamo, G., Albergel, C., Muñoz-Sabater, J., Isaksen, L., 2012. Initialisation of land surface variables for numerical weather prediction. Surv. Geophys. http://dx.doi.org/10.1007/s10712-012-9207-x.

de Rosnay, P., Drusch, M., Vasiljevic, D., Balsamo, G., Albergel, C., Isaksen, L., 2013. A simplified extended Kalman filter for the global operational soil moisture analysis at ECMWF. Q. J. R. Meteorol. Soc. 139 (674), 1199–1213.

Dirmayer, P., 2000. Using a global soil wetness dataset to improve seasonal climate simulation. J. Clim. 13, 2900–2921.

Douville, H., Viterbo, P., Mahfouf, J., Beljaars, A., 2000. Evaluation of optimal interpolation and nudging techniques for soil moisture analysis using FIFE data. Am. Meteorol. Soc. 128, 1733–1756.

Drusch, M., Viterbo, P., 2007. Assimilation of screen-level variables in ECMWF's integrated forecast system: a study on the impact of the forecast quality and analyzed soil moisture. Am. Meteorol. Soc. 135, 300–314.

Drusch, M., de Rosnay, P., Balsamo, G., Andersson, E., Bougeault, P., Viterbo, P., 2009a. Towards a Kalman filter based soil moisture analysis system for the operational ECMWF integrated forecast system. Geophys. Res. Lett. 36. http://dx.doi.org/10.1029/2009GL037716.

Drusch, M., Holmes, T., de Rosnay, P., Balsamo, G., 2009b. Comparing ERA-40 based L-band brightness temperatures with Skylab observations: a calibration/validation study using the community microwave emission model. J. Hydrometeorol. 10, 213–226.

Eagleman, J.R., Lin, W.C., 1976. Remote sensing of soil moisture by a 21 cm passive radiometer. J. Geophys. Res. 81, 3660–3666.

Entekhabi, D., Njoku, E.G., O'Neill, P.E., Kellogg, K.H., Crow, W.T., Edelstein, W.N., Entin, J.K., Goodman, S.D., Jackson, T.J., Johnson, J., Kimball, J., Piepmeier, J.R., Koster, R.D., Martin, N., McDonald, K.C., Moghaddam, M., Moran, S., Reichle, R., Shi, J.C., Spencer, M.W., Thurman, S.W., Tsang, L., Van Zyl, J., 2010. The soil moisture active passive (SMAP) mission. Proc. IEEE 98, 704–716.

Fascetti, F., Pierdicca, N., Pulvirenti, L., Crapolicchio, R., Muñoz-Sabater, J., 2016. A comparison of ASCAT and SMOS soil moisture retrievals over Europe and Northern Africa from 2010 to 2013. Int. J. Appl. Earth Observ. 45 (Part B), 135–142.

Figa-Saldana, J., Wilson, J.J.W., Attema, E., Gelsthorpe, R., Drinkwater, M.R., Stoffelen, M.R.A., 2002. The advanced scatterometer (ASCAT) on the meteorological operational (MetOp) platform: a follow on for European wind scatterometers. Can. J. Remote Sens. 28, 404–412.

Hagolle, O., Dedieu, G., Mougenot, B., Debaecker, V., Duchemin, B., Meygret, A., 2008. Correction of aerosol effects on multi-temporal images acquired with constant viewing angles: application to Formosat-2 images. Remote Sens. Environ. 112 (4), 1689–1701.

Hagolle, O., Huc, M., Pascual, D.V., Dedieu, G., 2010. A multi-temporal method for cloud detection, applied to FORMOSAT-2, VENμS, LANDSAT and SENTINEL-2 images. Remote Sens. Environ. 114 (8), 1747–1755.

Houser, P.R., Shuttleworth, W.J., Famiglietti, J.S., Gupta, H.V., Syed, K.H., Goodrich, D.C., 1998. Integration of soil moisture remote sensing and hydrologic modelling using data assimilation. Water Resour. Res. 34, 3405–3420.

Inglada, J., Mercier, G., 2007. A new statistical similarity measure for change detection in multi-temporal SAR images and its extension to multiscale change analysis. IEEE Trans. Geosci. Remote Sens. 45 (5), 1432–1445.

Jackson, T.J., Schmugge, T.J., 1991. Vegetation effects on the microwave emission from soils. Remote Sens. Environ. 36, 203–212.

Kerr, Y.H., Waldteufel, P., Wigneron, J.-P., Delwart, S., Cabot, F., Boutin, J., Escorihuela, M., Font, J., Reul, N., Gruhier, C., Juglea, S.E., Drinkwater, M.R., Hahne, A., Martin-Neira, M., Mecklenburg, S., 2010. The SMOS mission: new tool for monitoring key elements of the global water cycle. Proc. IEEE 98 (5), 666–687.

Kerr, Y.H., Waldteufel, P., Richaume, P., Wigneron, J.P., Ferrazzoli, P., Mahmoodi, A., Al Bitar, A., Cabot, F., Gruhier, C., Juglea, S.E., Leroux, D., Mialon, A., Delwart, S., 2012. The SMOS soil moisture retrieval algorithm. IEEE Trans. Geosci. Remote Sens. 50 (5), 1384–1403.

Kerr, Y., Jacquette, E., Al Bitar, A., Cabot, F., Mialon, A., Richaume, P., Quesney, A., Berthon, L., Wigneron, J.-P., 2013. CATDS SMOS L3 Soil Moisture Retrieval Processor: Algorithm Theoretical Baseline Document (ATBD). CESBIO, Toulouse.

Koster, R., Yamada, T.J., Balsamo, G., Berg, A., Boisscric, M., Dirmeyer, P., Doblas-Reyes, F., Drewitt, G., Gordon, C., Guo, Z., et al., 2011. The second phase of the global land-atmosphere coupling experiment: soil moisture contribution to subseasonal forecast skill. J. Hydrometeorol. 12, 805–822.

Massari, C., Brocca, L., Barbetta, S., Papathanasiou, C., Mimikou, M., Moramarco, T., 2014a. Using globally available soil moisture indicators for flood modelling in Mediterranean catchments. Hydrol. Earth Syst. Sci. 18, 839–853.

Massari, C., Brocca, L., Moramarco, T., Tramblay, Y., Didon Lescot, J.-F., 2014b. Potential of soil moisture observations in flood modelling: estimating initial conditions and correcting rainfall. Adv. Water Resour. 74, 44–53.

Massari, C., Brocca, L., Ciabatta, L., Moramarco, T., Gabellani, S., Albergel, C., de Rosnay, P., Puca, S., Wagner, W., 2015. The use of H-SAF soil moisture products for operational hydrology: flood modelling over Italy. Hydrology 2 (1), 2–22.

Merlin, O., Al Bitar, A., Walker, J., Kerr, Y., 2010. An improved algorithm for disaggregating microwave-derived soil moisture based 880 on red, near-infrared and thermal-infrared data. Remote Sens. Environ. 114, 2305–2316.

Mialon, A., Richaume, P., Leroux, D., Bircher, S., Al Bitar, A., Pellarin, T., … Kerr, Y.H., 2015. Comparison of Dobson and Mironov dielectric models in the SMOS soil moisture retrieval algorithm. IEEE Trans. Geosci. Remote Sens. 53 (6), 3084–3094.

Miernecki, M., Wigneron, J.P., Lopez-Baeza, E., Kerr, Y., De Jeu, R., De Lannoy, G.J., … Richaume, P., 2014. Comparison of SMOS and SMAP soil moisture retrieval approaches using tower-based radiometer data over a vineyard field. Remote Sens. Environ. 154, 89–101.

Muñoz Sabater, J., Fouilloux, A., de Rosnay, P., 2012. Technical implementation of SMOS data in the ECMWF integrated forecasting system. IEEE Geosc. Remote Sens. Lett. 9 (2), 252–256. http://dx.doi.org/10.1109/LGRS.2011.2164777.

Muñoz-Sabater, J., 2015. Incorporation of microwave passive brightness temperatures in the ECMWF soil moisture analysis. Remote Sens. 7 (5), 5758–5784. http://dx.doi.org/10.3390/rs70505758.

Muñoz-Sabater, J.M., Jarlan, L., Calvet, J.-C., Bouyssel, F., de Rosnay, P., 2007. From near-surface to root-zone soil moisture using different assimilation techniques. J. Hydrometeorol. 8, 194–206.

Muñoz-Sabater, J., Rudiger, C., Calvet, J.C., Fritz, N., Jarlan, L., Kerr, Y., 2008. Joint assimilation of surface soil moisture and LAI observations into a land surface model. Agric. For. Meteor. 148, 1362–1373. http://dx.doi.org/10.1016/j.agrformet.2008.04.003.

Muñoz-Sabater, J., de Rosnay, P., Balsamo, G., 2011. Sensitivity of L-band NWP forward modelling to soil roughness. Int. J. Remote Sens. 32, 5607–5620.

Naeimi, V., Scipal, K., Bartalis, Z., Hasenauer, S., Wagner, W., 2009. An improved soil moisture retrieval algorithm for ERS and METOP scatterometer observations. IEEE Trans. Geosci. Remote Sens. 47 (7), 1999–2013.

Ni-Meister, W., Walker, J., Houser, P., 2005. Soil moisture initialization for climate prediction: characterization of model and observation errors. J. Geophys. Res. 110. http://dx.doi.org/10.1029/2004JD005745.

Njoku, E.G., Jackson, T.J., Lakshmi, V., Chan, T.K., Nghiem, S.V., 2003. Soil moisture retrieval from AMSR-E. IEEE Trans. Geosci. Remote Sens. 41 (2), 215–229.

Njoku, E.G., Ashcroft, P., Chan, T.K., Li, L., 2005. Global survey and statistics of radio-frequency interference in AMSR-E land observations. IEEE Trans. Geosci. Remote Sens. 43 (5), 938–947.

Oliva, R., Daganzo, E., Kerr, Y.H., Mecklenburg, S., Nieto, S., Richaume, P., Gruhier, C., 2012. SMOS radio frequency interference scenario: status and actions taken to improve the RFI environment in the 1400–1427-MHz passive band. IEEE Trans. Geosci. Remote Sens. 50 (5), 1427–1439.

Owe, M., De Jeu, R.A.M., Holmes, T.R.H., 2008. Multi-sensor historical climatology of satellite-derived global land surface moisture. J. Geophys. Res. 113, F01002. http://dx.doi.org/10.29/2007JF000769.

Pellarin, T., Louvet, S., Gruhier, C., Quantin, G., Legout, C., 2013. A simple and effective method for correcting soil moisture and precipitation estimates using AMSR-E measurements. Remote Sens. Environ. 136, 28–36.

Piles, M., Camps, A., Vall-llossera, M., Corbella, I., Panciera, R., Rudiger, C., Kerr, Y.H., Walker, J., 2011. Downscaling SMOS-derived soil moisture using MODIS visible/infrared data. IEEE Trans. Geosci. Remote Sens. 49 (9), 3156–3166.

Ponziani, F., Pandolfo, C., Stelluti, M., Berni, N., Brocca, L., Moramarco, T., 2012. Assessment of rainfall thresholds and soil moisture modeling for operational hydrogeological risk prevention in the Umbria region (central Italy). Landslides 9 (2), 229–237.

Qiu, J., Crow, W.T., Nearing, G.S., Mo, X., Liu, S., 2014. The impact of vertical measurement depth on the information content of soil moisture times series data. Geophys. Res. Lett. 41 (14), 4997–5004.

Rahmoune, R., Ferrazzoli, P., Singh, Y.K., Kerr, Y.H., Richaume, P., Al Bitar, A., 2014. SMOS retrieval results over forests: comparisons with independent measurements. IEEE J. Select. Top. Appl. Earth Observ. Remote Sens. 7 (9), 3858–3866.

Reichle, R., Koster, R., 2004. Bias reduction in short records of satellite soil moisture. Geophys. Res. Lett. 31. http://dx.doi.org/10.1029/2004GL020938.

Richaume, P., Soldo, Y., Anterrieu, E., Khazaal, A., Bircher, S., Mialon, A., Al Bitar, A., Rodriguez-Fernandez, N., Cabot, F., Kerr, Y., Mahmoodi, A., 2014. RFI in SMOS measurements: update on detection, localization, mitigation techniques and preliminary quantified impacts on soil moisture products. In: EEE International Geoscience and Remote Sensing Symposium (IGARSS), July 13–18, 2014. pp. 223–226.

Scipal, K., Drusch, M., Wagner, W., 2008. Assimilation of an ERS scatterometer-derived soil moisture index in the ECMWF numerical weather prediction system. Adv. Water Resour. 31, 1101–1112.

Soldo, Y., Cabot, F., Khazaal, A., Miernecki, M., Slominska, E., Fieuzal, R., Kerr, Y.H., 2015. Localization of RFI sources for the SMOS mission: a means for assessing SMOS pointing performances. IEEE J. Select. Top. Appl. Earth Observ. Remote Sens. 8 (2), 617–627.

Stoffelen, A., 1998. Toward the true near-surface wind speed: error modeling and calibration using triple collocation. J. Geophys. Res. 103 (C4), 7755–7766.

Tomer, S.K., Al Bitar, A., Sekhar, M., Zribi, M., Bandyopadhyay, S., Sreelash, K., Sharma, A., Corgne, S., Kerr, Y., 2015. Retrieval and multi-scale validation of soil moisture from multi-temporal SAR data in a semi-arid tropical region. Remote Sens. 7 (6), 8128–8153.

Verhoest, N.E.C., van den Berg, M.J., Martens, B., Lievens, H., Wood, E.F., Pan, M., Kerr, Y.H., Al Bitar, A., Tomer, S.K., Drusch, M., Vernieuwe, H., De Baets, B., Walker, J.P., Dumedah, G., Pauwels, V.R.N., 2015. Copula-based downscaling of coarse-scale soil moisture observations with implicit bias correction. IEEE Trans. Geosci. Remote Sens. 53 (6), 3507–3521. http://dx.doi.org/10.1109/TGRS.2014.2378913.

Wagner, W., Lemoine, G., Rott, H., 1999. A method for estimating soil moisture from ERS scatterometer and soil data. Remote Sens. Environ. 70, 191–207.

Wagner, W., Hahn, S., Kidd, R., Melzer, T., Bartalis, Z., Hasenauer, S., Figa, J., de Rosnay, P., Jann, A., Schneider, S., Komma, J., Kubu, G., Brugger, K., Aubrecht, C., Zuger, J., Gangkofner, U., Kienberger, S., Brocca, L., Wang, Y., Bloeschl, G., Eitzinger, J., Steinnocher, K., Zeil, P., Rubel, F., 2013. The ASCAT soil moisture product: a review of its specifications, validation results, and emerging applications. Metcorol. Z. 22 (1), 5–33.

Walker, J.P., Willgoose, G.R., Kalma, J.D., 2001. Onedimensional soil moisture profile retrieval by assimilation of near-surface measurements: a simplified soil moisture model and field application. J. Hydrometeorol. 2, 356–373.

Wanders, N., Pan, M., Wood, E.F., 2015. Correction of real-time satellite precipitation with multi-sensor satellite observations of land surface variables. Remote Sens. Environ. 160, 206–221.

Wang, J.R., Choudhury, B.J., 1981. Remote sensing of soil moisture content over bare field at 1.4 GHz frequency. J. Geophys. Res. 86, 5277–5282.

Wigneron, J.P., Kerr, Y., Waldteufel, P., Saleh, K., Escorihuela, M.J., Richaume, P., Ferrazzoli, P., de Rosnay, P., Gurney, R., Calvet, J.C., Grant, J.P., Guglielmetti, M., Hornbuckle, B., Mätzler, C., Pellarin, T., Schwank, M., 2007. L-band microwave emission of the biosphere (L-MEB) model: description and calibration against experimental data sets over crop fields. Remote Sens. Environ. 107 (4), 639–655.

Wood, E.F., Roundy, J.K., Troy, T.J., van Beek, L.P.H., Bierkens, M.F.P., Blyth, E., de Roo, A., Döll, P., Ek, M., Famiglietti, J., Gochis, D., van de Giesen, N., Houser, P., Jaffë, P.R., Kollet, S., Lehner, B., Lettenmaier, D.P., Peters-Lidard, C., Sivapalan, M., Sheffield, J., Wade, A., Whitehead, P., 2011. Hyperresolution global land surface modeling: meeting a grand challenge for monitoring earth's terrestrial water. Water Resour. Res. 47, W05301.

Chapter 19

Emerging and Potential Future Applications of Satellite-Based Soil Moisture Products

E. Tebbs[*], F. Gerard[†], A. Petrie[†] and E. De Witte[‡]

[*]King's College London, Department of Geography, Strand, London, United Kingdom, [†]Centre for Ecology & Hydrology, Wallingford, United Kingdom, [‡]Airbus Defence and Space Ltd, Stevenage, United Kingdom

1 INTRODUCTION

Soil moisture (SM), a key variable controlling the fluxes of water, carbon, and energy between the atmosphere and the land surface, is firmly established as one of the most critical priorities for Earth Observation (EO). The Group on Earth Observations (GEO) has identified SM as the second-most important EO parameter after precipitation, because it addressed so many user needs (Group on Earth Observations, 2012); it was identified as a priority for all GEO societal benefit areas, including agriculture, disasters, health, biodiversity, ecosystems, water, climate, energy, and weather. SM has also been recognized as an Essential Climate Variable by the Global Climate Observing System (GCOS, 2010).

Over the last few decades, huge progress has been made in the remote sensing of SM from satellite platforms, and, as a result, there have been considerable advances in the application of this technology for studying Earth systems processes (Petropoulos et al., 2015). Further technological advances are expected in the coming years, including improvements in the spatial resolution, revisit frequency, and accuracy of SM observations (Ochsner et al., 2013; Piles et al., in press). These developments are expected to enhance our knowledge of SM, in terms of its temporal and spatial variability, and so will contribute to our scientific understanding of surface and atmospheric processes and dynamics. This will also create potential for extending the current applications of satellite-based SM data and developing applications in new discipline areas. For example, SM data has huge operational potential for flood prediction, drought early warning, and agricultural management. Considerable progress has already been made in the application of SM data within the fields of numerical weather prediction,

Satellite Soil Moisture Retrieval. http://dx.doi.org/10.1016/B978-0-12-803388-3.00019-X

climate science, and hydrological modelling. However, comprehensive integration of satellite-based SM measurements in these disciplines is still a work in progress, and there are also many more fields where SM information has potential to be usefully applied (Ochsner et al., 2013).

A key limitation of current satellite SM products is that they do not always match the user requirements in terms of accuracy, temporal and spatial resolution, penetration depth, or a combination of the above. New SM products are becoming available that seek to overcome these limitations by combining several EO datasets into a product that can be used in an operational context (Petropoulos et al., 2015). The first dedicated satellite for soil moisture i.e. SMOS launched in 2009 by ESA has been providing data at 15 km for 6 years, followed by the NASA SMAP satellite, launched in Jan. 2015, retrieves information from an L-band passive radiometer (1.41 GHz) with the aim of delivering SM products with improved accuracy and spatial resolutions and coverage (daily and global) (Entekhabi et al., 2010). These improved capabilities will extend the range of viable applications for satellite SM data. In the longer term, new EO missions need to better address the needs of the user community. Hence, there has been a move towards involving users from an early stage of the mission development so that they can help to define the requirements (e. g., the user advisory group for the SMAP mission; Moran et al., 2010). The involvement of key stakeholders early on in the missions will help to raise awareness of the current capabilities within the user community. It will also promote the development of much needed new models and tools which incorporate satellite SM measurements.

2 POTENTIAL FUTURE APPLICATIONS OF SM

SM data has the potential to be used in a wide range of applications across the ecological and hydrological sciences. This section gives an overview of the applications identified from a survey carried out at the UK's Centre for Ecology and Hydrology (CEH). The research at CEH cuts across disciplines and focusses on land and freshwater ecosystems and their interaction with the atmosphere. The institute has an extensive programme of long-term monitoring, analysis, and modelling to deliver UK and global environmental data, providing early warnings of change and management solutions for the land and freshwaters. The survey consisted of interviews with a number of CEH research staff, chosen to represent the science areas and applications within CEH that are expected to benefit from spatial SM data. Questions focussed on establishing, in each case, the importance of a satellite SM product, the anticipated improvements, the specific product requirements, and the barriers to uptake. The applications identified from this review are summarized in Table 1. In the subsequent sections, each of these applications are discussed in more detail and additional information from the literature has been included.

TABLE 1 Summary of Potential Applications for a Satellite-Based Soil Moisture Product Identified by CEH Scientists

Discipline	Potential Applications	Temporal Requirements	Spatial Requirements	Barriers to Uptake
Carbon sequestration	Investigate the impact of SM on vegetation productivity	Monthly	1–100 m	Models need to be adapted to improve sensitivity to SM
Biodiversity	Investigate response of species to climate change and extreme conditions (e.g., drought)	Ideally weekly	100 m resolution	Models need to be adapted to incorporate SM
	Understand species dynamics and produce improved species distribution maps			
Habitat condition monitoring	Assess response of vegetation community to changes in SM—e.g., for heathland	Seasonal and annual means sufficient	10–100 m	Microwave signal may have difficulty penetrating the vegetation
Drought	Improve drought prediction models through better representation of spatial variability in SM	Daily data useful for hydrological modelling	Spatial scale of hydrological models (1 km^2)	Shallow penetration depth of EO SM[a]. Use of a two-layer model would help[b]
	Estimate Wetness Index (WI) as a drought indicator			
Agricultural management	Determine when SM level right for sowing or operating machinery	Daily	Plot to field scale	Estimating SMD of the root-zone require deeper layer SM estimates[b]
	Improving irrigation efficiency			
	Estimate soil moisture deficit (SMD)			
Flooding	Help to constrain hydrological models to improve flood forecasts	Daily	Meters to several kilometers	Depth sampled by EO typically top 1 cm, ideally need top 10–15 cm or more
	Improve surface water flood modelling by looking at soil saturation			

Continued

TABLE 1 Summary of Potential Applications for a Satellite-Based Soil Moisture Product Identified by CEH Scientists—cont'd

Discipline	Potential Applications	Temporal Requirements	Spatial Requirements	Barriers to Uptake
Land surface modelling	Improve representation of SM heterogeneity in models Identify where soil-physics in the models needs changing Model validation	Daily	Catchment scale (1 km grid squares)	Need surface and subsurface SM
Landscape epidemiology (Health)	Model impact of future climate scenarios on pathogens Assess risks Relevant to range of pathogens, including plant fungal infections and vector-borne diseases, particularly of midges and mosquitoes	Daily-weekly	100 m–1 km	Ideally need soil moisture to >10 cm depth. Health issues in connection with SM are not high on the agenda
Greenhouse gas emissions	Assess accuracy of GHG emissions estimates, particularly N_2O and methane Improve national emissions estimates for submission to UNFCCC	*It is not yet clear whether spatial or temporal should be prioritized*		Emission inventory guidelines currently do not use variable SM in the calculations
Reservoir management	Near real-time monitoring of reservoirs Decision support—e.g., inform managers when to open spillways to reduce flooding	Daily	Meters to several kilometers[c]	Low latency required so managers know predictions at least a day in advance
Contaminants	Improve representation of runoff in pollution event modelling	Daily	10 m–1 km	Requires near-real time data

[a]Blyth et al. (1993).
[b]Ragab (1995).
[c]Doubkova et al. (2009).

2.1 Agriculture

Crop growth is regulated, among other factors, by SM. Consequently, SM estimates from satellites are a valuable indicator for agricultural monitoring, which can be used to predict future crop yields. This information allows decision makers to identify shortfalls and take appropriate action (e.g., changing import-export strategies) to ensure national economic security. Several studies have looked at the feasibility of using satellite SM to monitor yields. For example, the Temperature Vegetation Dryness Index (TVDI), derived from MODIS Land Surface Temperature (LST) and Enhanced Vegetation Index, has been used to estimate surface SM availability, and hence predict yields, for soybean and wheat crops in the Argentine Pampas (Holzman et al., 2014). TVDI showed a strong correlation with SM and was able to accurately predict crop yields 1–3 months before harvest. Microwave-based satellite SM estimates have shown similar potential; the assimilation of SM data from the Advanced Microwave Scanning Radiometer-EOS (AMSR-E) into a crop model improved simulated crop yields, particularly under very dry conditions (Ines et al., 2013). However, AMSR-E was insensitive to SM under dense vegetation. The new SMOS mission is designed for high accuracy even under dense vegetation, thereby allowing effective SM measurements during the critical period of crop growth; hence, it has potential to improve crop yield forecasts (Ines et al., 2013). Yield estimates ideally require root-zone SM, but remote sensing is only able to measure SM in the top few centimeters; therefore, methods have been developed which use surface SM values to predict subsurface SM (e.g., two layers model; Ragab, 1995).

Extreme climatic events, such as drought and flooding, can have a serious impact on agricultural productivity. EO SM data can be used to predict these events and thereby provide valuable information on potential food insecure regions to policy makers and NGOs. For example, Champagne et al. (2015) found that SM data from the Soil Moisture Ocean Salinity (SMOS) mission was a good real-time indicator of climate-related agricultural risk. SMOS was able to identify periods of drought and excess water that led to crop losses. In the particular case of drought, EO-based SM data shows potential for improving drought prediction models, which have, up to now, largely been based on soil water balance models forced by global precipitation data.

Satellite-based SM products have huge operational potential for the efficient management of farming systems. For this reason, the future SAOCOM-1 (Satellites for Observation and Communications) Argentinian SAR mission has better agricultural decision-making as primary mission goal. If SM information can be provided at the farm scale, it could be used to identify when and where moisture levels are appropriate for using machinery or for sowing seeds. For irrigation management, SM observations are required at the field and subfield scale (tens to hundreds of meters). Fieuzal et al. (2011) developed a method for retrieving topsoil moisture with high-resolution

(30 m) Advanced Synthetic Aperture Radar (ASAR) data. ASAR data was combined with Formosat-2 NDVI imagery (8 m) and a vegetation model to correct for variation in biomass water content. Using this method, Fieuzal et al. (2011) were able to detect irrigated areas throughout the whole growing season at the field scale (~5 ha). This presents an attractive operational approach; however, it requires large volumes of optical and microwave imagery. Hence, there is potential to use satellite-based SM data for improving irrigation efficiency and contributing to "irrigation decision support systems." However, in order to be economically viable for farmers the cost of accessing an EO SM product would have to be smaller than the increase in profit directly resulting from the improved irrigation efficiency. Furthermore, the information would need to be provided in a format that integrates seamlessly with the farmers' workflow and which facilitates decision-making. Developing such a product will likely require contributions from both private enterprise and the scientific community.

Rather than direct SM measurements, an alternative method for irrigation scheduling is by monitoring the soil moisture deficit (SMD). In the United Kingdom, the Met Office (http://www.metoffice.gov.uk/) currently provides weekly and monthly modelled SMD data to the agricultural sector and water resource planners at cost. The SMD estimates are produced as part of the Meteorological Office Rainfall and Evaporation Calculation System (MORECS), which provides near real-time assessments of rainfall, evaporation, and SM at a 40 km grid spacing based on meteorological stations across the United Kingdom (Hough and Jones, 1997; Clark, 2002). There is scope for improving the accuracy of the Met Office SMD product by incorporating SM data from EO. For example, Srivastava et al. (2013) showed that combining SMOS SM data with a land surface model (LSM) improved SMD prediction.

2.2   Biodiversity and Habitat Condition Monitoring

EO-based SM is a useful indicator for monitoring remote, water-dependent habitats. For example, Jacome et al. (2013) demonstrated the potential of high-resolution (5.2 × 7.6 m) SAR data, from RADARSAT-2, for monitoring SM as an indicator of wetland dynamics and condition. SM is also a potential biodiversity indicator; in situ SM has been shown to correlate with species richness for some habitat types (Litaor et al., 2008). Sass et al. (2012) found that volumetric SM estimates based on SAR data, from ERS-1 and -2, exhibited a strong correlation with vascular-plant species richness in the area around the Sundance Provincial Park, Alberta, Canada. Based on this relationship, they developed a predictive model to map vascular-plant species richness and to identify where the park boundary could be expanded to improve its ecological value. Hence, satellite SM data provides a valuable source of information for conservation and, in future, there is potential to use it for the management of protected areas, including for defining ecologically relevant boundaries.

In addition to being a valuable indicator of habitat condition, satellite SM data has the potential to improve our understanding of environmental conditions driving species dynamics, beyond those that are commonly considered, such as temperature and rainfall (Turner et al., 2003). For example, Oliver et al. (in press) used spatiotemporal estimates of "aridity" around hundreds of butterfly monitoring sites, based on Met Office temperature and rainfall data, to investigate the interacting effects of climate change and habitat fragmentation on drought-sensitive butterflies. For further study of these interactions, EO SM data would be an improvement because it provides a more direct estimate of SM in the landscape, which is likely to be a key factor driving variations in butterfly host-plant quality. SM data can also be used to investigate ecosystem resilience and to examine the response of plant and animal communities to extreme events such as droughts. Stampoulis et al. (2014) used microwave remote sensing observations to investigate the resilience of SM and vegetation water content to prolonged dry spells across different vegetation types; their study shed light on the role of hydrologic regime in determining biotic composition and pattern. As this is an emerging application area, very few studies have used EO-based SM data to investigate these relationships. Hence, there is huge potential to use satellite SM estimates in future to investigate the response of vegetation communities to changes in SM. For example, CEH researchers suggested that SM data could be used to examine the impact of heathland drying on biodiversity, since some heath species are very sensitive to soil water conditions (Gloaguen, 1987). However, this will be challenging because the presence of woody vegetation such as heath can influence the accuracy of SM measurements (Petropoulos et al., 2015).

2.3   Carbon Sequestration

SM is crucial in regulating the distribution and productivity of vegetation in both agricultural and natural systems. Consequently, SM plays an important role in controlling terrestrial carbon uptake (Chen et al., 2014). Satellite observations can be used to study spatial and temporal variations in the relationship between SM and vegetation, and, as a result, it can improve our understanding of the terrestrial carbon cycle. At a continental scale, there is a high degree of spatial correlation between SM and vegetation cover, which can be clearly observed in satellite imagery (Fig. 1), while at a regional to local scale the relationship can be more complicated. For example, a case study over mainland Australia (Chen et al., 2014) found a significant positive correlation between NDVI and satellite-based SM, with NDVI lagging 1 month behind SM. However, the study found strong regional variations in the temporal characteristics, with vegetation in drier areas being more sensitive to SM. These observations are useful for understanding the impact of SM on vegetation at various temporal

AMSRE_LAND3-NETCDF.002 AMSR–E Aqua Daily L3 Surface Soil Moisture [g/cm³]
(01Jan2003 – 31Dec2010)

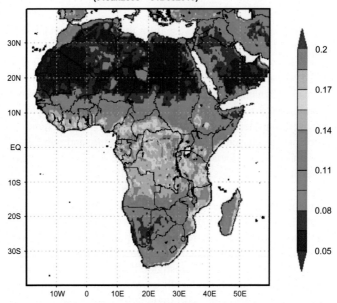

MOD13C2.005 Normalized Difference Vegetation Index (NDVI) 5.8km [none]
(Jan2003 – Dec2010)

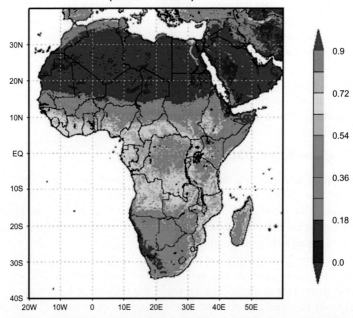

FIG. 1 Maps of AMSR-E/Aqua Daily L3 Surface SM (Njoku, 2004; top) and MODIS NDVI (bottom), showing the high degree of spatial correlation between the two datasets at the continental scale. It also shows some smaller scale regional patterns that are present in the SM image but not in the NDVI image and vice versa, highlighting the different information contained in the two datasets. (Credit: *Produced by Emma Tebbs using the Giovanni online data system, developed and maintained by the NASA GES DISC.*)

scales, and the results are useful for improving coupled vegetation-climate models.

The incorporation of remotely sensed SM data into an ecosystem carbon flux model has been shown to improve both the spatial pattern and the magnitude of carbon flux estimates (Verstraeten et al., 2010). CEH researchers expect that a flood event followed by persistently high SMs could impact on soil aeration and thus crop and grassland carbon cycling and productivity. To investigate this, SM data from satellites could be integrated with a dynamic plant-soil model, which estimates carbon flux and net primary productivity (the N14C model; Tipping et al., 2012). The particular model would first need to be made more responsive to SM, through the introduction of soil aeration. In addition, integrating the plant-soil model with a hydrological model would deliver dissolved organic carbon, a measure of water quality, and provide the potential for quantifying aquatic carbon sinks.

2.4 Land Surface Modelling

SM is an important variable in land surface modelling because it influences the surface energy balance, hydrological processes, and biogeochemical cycles. Satellite SM data can also be incorporated into the models to improve the spatial representation of SM heterogeneity. For example, LSMs have difficulty representing landscape-scale SM patterns due to irrigation. SM data from advanced microwave satellites such as SMOS, SMAP, ASCAT, AMSRE/2 are able to capture irrigation signals. Furthermore, the assimilation of satellite SM (e.g., from AMSR-E) in LSMs can improve SM dynamics and lead to improved CO_2 and energy flux predictions (Rebel et al., 2012).

EO-based SM can also be used for the validation of LSMs. For example, The Joint UK Land Environment Simulator (JULES) is a community LSM which is used as a stand-alone model or as part of the Met Office Unified Model for weather forecasting and climate prediction. Sub-grid cell variability in SM is modelled within JULES using simple statistical relationships (Best et al., 2011) and the availability of accurate EO-based SM measurements would allow validation of these relationships and evaluation of how the statistical parameters vary with landscape properties. The SM data could be used to answer questions such as "How do we model within grid cell soil moisture variability given a mean amount of rainfall?"; "How quickly does each grid cell drain after a rainfall event?"; and "Where is the moisture going and why?". Hence, a SM product would be extremely valuable in terms of validating the soil water transport equations in the model, particularly the effects of heterogeneity of soils and SM.

2.5 Hydrology

SM controls the partitioning of precipitation into runoff and infiltration, and as such it is an important hydrological parameter. The potential for incorporating satellite SM data in the hydrological sciences has already been demonstrated, for applications such as flood prediction and water resources management (Wagner et al., 2007). Currently, most hydrological models still calculate SM from a water balance of rainfall and potential evaporation. An EO SM product is unlikely to fully replace these calculations, because the penetration depth of the satellite is not sufficient to provide a direct measurement of total deep layer SM. However, EO SM data can be used as an input to help constrain these models by providing better representation of spatial heterogeneity in SM. For example, assimilating SM estimates from the ASCAT (C-band radar) satellite sensors into rainfall-runoff models has been proven to improve flood forecasting (Brocca et al., 2012a). For surface water flooding, again the application of EO is limited by its low-sampling depth (typically top 1 cm), but the data could be useful for giving an indication of topsoil saturation. Satellite observations of saturated soil at the surface can give improved warnings of surface water flooding susceptibility. Fleiss et al. (2011) used the ASCAT 25 km Soil Water Index (SWI) product to examine the precursor conditions to the severe 2010 floods in Pakistan. They found that positive SWI anomalies preceded the floods and could potentially be used as part of an early flood warning system in the future. Additionally an operational flood forecasting system is under practice using SMOS data (http://www.cesbio.ups-tlse.fr/SMOS_blog/?tag=floods). Since topsoil saturation is an indicator of runoff, SM data could also be used for reservoir management. Near real-time monitoring would be possible if daily SM data were available with low-latency times (Piles et al., in press). The information could be used for decision-support systems, for example to tell managers when to open spillways to reduce flooding. Access to remotely sensed SM data over large spatial areas would provide an additional calibration point for large-scale spatially distributed water resource model and thus would improve simulations.

2.6 Nutrient Transport and the Fate of Contaminants

Knowledge of SM allows us to predict the amount of surface runoff that can be expected for a given amount and intensity of rainfall. Surface runoff and erosion are key vectors for the transport of nutrients (nitrate, phosphate) and other chemicals (pesticides, herbicides, metals) to surface waters in agricultural areas (e.g., Sharpley et al., 2001). Management and reduction of this transport depends on understanding (i) how and why runoff and erosion occur, and (ii) the influence of spatial and temporal variability in SM, at scales from subfield to whole catchment, on the generation of runoff and erosion. The CEH large-scale spatially distributed water resource model (GWAVA; Meigh and Tate, 2002) has a water quality module that is crucial for predicting potential concentrations of contaminants (e.g., nanoparticles and microplastics) and nutrients transported from land into water courses through runoff or remaining in the soils through mainly infiltration. Improved estimates of SM would give better predictions of the fate of contaminants and nutrients. The availability of SM data in near real-time would enable temporally relevant water quality simulations which would be significant improvement on the currently generated distributions of concentration based on historic data.

2.7   Landscape Epidemiology

SM is an important variable influencing the spread of many plant pathogens, such as soil borne fungal diseases, and vector-borne human and animal diseases (e.g., malaria carried by the mosquito and Bluetongue disease of cattle carried by biting midges). SM data from satellites has the potential to help us understand, and so predict, the role SM has on changes in the abundance of pathogens. The data could be incorporated into models that are used to develop scenarios and risk maps; but this is dependent on ensuring the biological mechanisms of the correlations are well understood in the first place. Currently, NDVI and rainfall data are used as a SM proxy in many models, for example for mapping the risk of West Nile virus (Brownstein et al., 2002). Berger et al. (2013) used TVDI, from MODIS LST and NDVI, to determine surface moisture conditions as a way of monitoring blacklegged tick habitat and human health risk in southern New England. Incorporating satellite-based SM data may help refine the model predictions and the improved models might then support projections under future climate scenarios. For example, the data could be used to investigate the impact of projected warming and future drought events on pathogen or virus abundance and habitat suitability (Hay et al., 2002).

2.8   Greenhouse Gas Emissions

SM influences the fluxes of greenhouse gases between the land and the atmosphere. Currently, modelled GHG emissions estimates, produced for the UK's National Atmospheric Emissions Inventory (NAEI) for submission to UNFCCC, do not incorporate interannual variability in environmental conditions. Improved knowledge of SM variability, both spatially and temporally, would improve the accuracy of the emissions estimates and would allow better estimation of the large uncertainties associated with GHG emissions. SM data would be particularly useful for determining nitrous oxide emissions, because the critical factors are SM and nitrogen application rate. Higher SM also encourages methane production (e.g., paddy rice fields).

2.9   Natural Hazards

SM data can be used to identify precursor conditions for water-related hazards including flooding, droughts, wildfires, and landslides. Hence, SM is a useful proxy for predicting natural hazards because it can identify early changes in the environment before the impacts are felt. Near real-time satellite SM data has huge potential to act as an early warning for various natural hazards. The prospect of using SM data for predicting drought and floods has already been discussed; therefore, this section focuses on the forecasting of wildfires and landslides, which are closely related to drought and floods respectively.

Wildfire risk is often closely associated with the occurrence of drought. Live fuel moisture content (LFMC), a key variable for determining fire risk, has been shown to correlate strongly with in situ SM (Qi et al., 2012). Hence, there is potential to use remotely sensed SM estimates to improve regional scale fire forecasts. However, satellites can only retrieve near-surface SM which may not be ideal for estimating LFMC (Qi et al., 2012). Optical satellite indices that are sensitive to soil and vegetation moisture content, such as the normalized multiband drought index (NMDI), have been used successfully for active fire detection (Wang et al., 2008). Microwave products also show potential; Aubrecht et al. (2011) found that the ASCAT SWI showed a strong spatial correlation with fire occurrence. Near real-time SM data has been used for operational fire monitoring; the Barcelona fire prevention service used high-resolution (1 km) SM maps for the Iberian Peninsula, produced using a combination of SMOS and MODIS data, as an indicator of fire risk (Piles et al., 2013). Downscaled SM data has also been incorporated into decision-support systems for post-fire rehabilitation planning, and there are plans to incorporate data from the new SMAP mission into the same system (Weber et al., 2013).

SM data can also be used to improve predictions of the spatial and temporal occurrence of rainfall-triggered landslides. Areas of high-SM content, identified from RADARSAT-1 imagery, have been shown to agree well with the locations of historical landslide events in Kuala Lumpur (Hassaballa et al., 2014). Furthermore, Brocca et al. (2012b) found that the inclusion of ASACT-derived SM in a multiple linear regression model improved the prediction of landslide movement for a rock slope in central Italy (correlation coefficient, $r = 0.40$ for rainfall only and $r = 0.85$ for rainfall + ASCAT SWI). Hence, there is potential to use EO SM data to produce landscape risk maps and to improve the accuracy of early warning systems.

3   SM FOR ADDRESSING FOOD AND WATER SECURITY IN AFRICA

The African continent is in urgent need of accurate information on its freshwater resources, which are under huge pressure due to climate change and rapid population growth. Africa has a human population of 1.2 billion, which is expected to double over the next 35 years to reach a 2.4 billion by 2050; by which time, Africa will represent around 25% of the global population (UNICEF, 2014). Adding to the scarcity of Africa's water resources are the effects of climate change. The Intergovernmental Panel on Climate Change (IPCC) predicts major changes in precipitation patterns—that will be felt more in Africa than anywhere else—and an increase in extreme flood and drought events (Fig. 2;

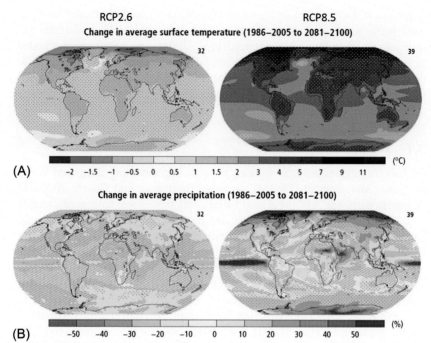

FIG. 2   Change in average surface temperature (a) and change in average precipitation (b) based on multi-model mean projections for 2081–2100 relative to 1986–2005 under the RCP2.6 (left) and RCP8.5 (right) scenarios. The number of models used to calculate the multi-model mean is indicated in the upper right corner of each panel. Stippling (i.e., dots) shows regions where the projected change is large compared to natural internal variability and where at least 90% of models agree on the sign of change. Hatching (i.e., diagonal lines) shows regions where the projected change is less than one standard deviation of the natural internal variability. This figure is taken from the 2014 IPCC report (IPCC, 2014; Figure SPM.7).

IPCC, 2014). Temperature changes due to climate change will result in very large reductions in crop yield, with cereal yield reductions up to 30% in Africa as early as 2080 (Parry et al., 2004). This will have a huge impact on the high proportion of Africa's population that live in rural areas and depend mainly on rain-fed agriculture for their livelihoods. It is clear that sustainable water management and food security are two of the most important challenges facing Africa. EO-based SM measurements have the potential to contribute to tackling these issues by providing improved hydrological information.

The ESA's Tiger initiative was established to assist African countries in exploiting EO for inventorying, monitoring, and assessing water resources (Fernandez-Prieto and Palazzo, 2007). As part of this initiative, the SHARE (Soil moisture for Hydrometeorological Applications) project was set up to address the needs of the hydrological community by producing a medium resolution (1 km) SM product using ASAR Global Mode (GM) data, with twice weekly measurement frequency (Doubkova et al., 2009). The SHARE SM products were produced for the Southern Africa Development Community, and

Australia, and distributed freely on an operational basis. This dataset has been used for a wide range of applications including crop yield modelling, and drought and flood forecasting (Bartsch et al., 2009). The ASAR mission ended in Apr. 2012, when the Envisat satellite failed; however, there is potential to develop a global operational system based on data from Sentinel-1A (launched in 2014) and 1B (scheduled for launch in 2016).

3.1 Water Management

Globally, countries are now working towards implementing Integrated Water Resource Management (IWRM) strategies (Stålnacke and Gooch, 2010). In Africa, in particular, IWRM is vital for ensuring that people have access to safe water and adequate sanitation, and it is also important for safeguarding food security by promoting the efficient use of available water resources, for example for irrigation. In most African countries, information on water resources is sparse and unreliable, which presents major challenges for implementing IWRM (Fernandez-Prieto and Palazzo, 2007). EO-based SM data can provide valuable information about water resources and therefore has potential to contribute to effective water governance in Africa. For example, Bartsch et al. (2008) demonstrated that the ESA Tiger SHARE SM product can clearly be related to river discharge measurements in the upper Okavango basin, southwestern Africa, and, furthermore, it can be used to monitor the dynamics of wet and inundated areas. Hence, this study illustrated the huge operational potential of microwave SM products for managing water resources in Africa. In arid and semiarid regions, the management of groundwater resources is particularly important and this requires accurate knowledge of the rate of recharge. Groundwater recharge can be estimated by combining information from a variety of sources, including remote sensing, in a comprehensive groundwater model (Kinzelbach et al., 2002). Utilizing satellite-derived SM in these models provides information on the spatial variability, which point observations fail to capture.

Knowing the spatial and temporal evolution of surface SM is equally important for flood forecasting. Flash floods in dry areas are increasingly common in Africa, and the number of flood-related fatalities and associated economic losses have increased dramatically (Di Baldassarre et al., 2010). The humanitarian impacts of flooding in Africa are frequently reported on the Integrated Regional Information Networks (IRIN) website (http://www.irinnews.org/). Flash floods develop at space and time scales that conventional observation systems are not able to monitor. As a result, there is a lack of basic understanding of the spatial and temporal variability of these key hydrological parameters which are needed to drive the hydrometeorological models used by water resource authorities to predict floods. The combination of comprehensive groundwater models and near real-time SM measurements from satellites is required to improve understanding and forecasting of these flood events.

3.2 Food Security

As human population grows in Africa, crop yields will fail to keep up with demand unless urgent action is taken. An agricultural revolution based on the principles of better soil management is needed to improve productivity (Glatzel et al., 2014). Degraded soils retain less moisture and, hence, biomass productivity is reduced. SM also has a large influence on the microbial communities (Bell et al., 2009) that are pivotal in maintaining soil quality and vegetation productivity. The monitoring and evaluation of SM conditions will help to target interventions for soil protection and restoration in Africa. Furthermore, there is potential to expand irrigated areas substantially in Sub-Saharan Africa (You et al., 2010). Realizing this potential is essential for future food security. Making the right decisions on where, when, and how much to irrigate is critical for both water use efficiency and crop health. Over- or under-irrigating is likely when SM is not monitored. Current commercial irrigation support tools rely on point observations collected using SM probes (e.g., http://www.lindsayafricagb.com/soil-moisture-monitoring). Space-based SM observations have huge potential for SM monitoring irrigation in less intensive farming systems, where ground-based measurements are not collected (Moran et al., 2010).

In the horn of Africa, drought is a recurring phenomenon which poses significant threats to food security, particularly in regions of rain-fed agriculture. Systematic EO-based SM measurements have the potential to vastly improve predictions of drought and famine. The data will complement existing drought analysis tools that are based on precipitation anomaly, hydrological models, or vegetation indices (Anderson et al., 2012). A key aim of the SMAP mission is to contribute to the predictions of the Famine Early Warning Systems Network (http://www.fews.net/), which has, up to now, primarily been based on weather forecasts. Anderson et al. (2012) showed that remote SM-based drought monitoring systems, based on microwave, thermal, or modelled SM estimates, are robust across the different climatic and ecological zones of East Africa. The remotely measured SM data can be used as an indicator for the onset, evolution, and severity of drought events, and for anticipating the impacts on human-ecological systems. Microwave SM data can also contribute to improved global preharvest yield forecasts (Bolten et al., 2010), which are important for food security.

4 CURRENT LIMITATIONS AND FUTURE DEVELOPMENTS

4.1 Sensors and Algorithms

The spatial resolution of SM observations is a major constraint for many applications. Due to the scale of the processes being studied, or the resolution of existing models, SM data with resolutions of 10–100 m is required for many purposes. For example, hydrological models typically operate at a scale of meters to several kilometers (Doubkova et al., 2009). SAR systems have been used in many studies due to their high-spatial resolution (Blyth et al., 1993;

Jacome et al., 2013); however, they lack the spatial coverage required for frequent global monitoring. New sensors such as SMOS, SMAP and Sentinel-1 offer improved spatial and temporal resolutions (Wagner et al., 2009). For studies which make use of historic SM data, spatial resolution will still be an issue. In these cases, a number of downscaling approaches have been developed which can be applied to SM observations (Chakrabarti et al., 2014; Merlin et al., 2008; Piles et al., 2011). There is also potential for exploiting synergies between different satellite sensors to produce products with higher spatial and temporal resolution and/or improved accuracy (Ochsner et al., 2013; Petropoulos et al., 2015).

The shallow depths sampled by satellites are not ideal for many applications which require root-zone SM estimates (Piles et al., in press). Hence, further work is needed to investigate the relationship between root-zone and surface SM (Ragab, 1992, 1995), particularly for vegetated pixels (Holzman et al., 2014). Future sensors, such as SMAP, will sample greater depths (0–5 cm) allowing them to make measurements which are more representative of SM beyond the surface skin layer. The ability to detect SM accurately under vegetation is important for many applications; in these cases, SAR systems provide a good option because they are better suited to penetrating vegetation (Sass et al., 2012). Some studies also require long-term time series of SM and approaches have been developed for combining data from different satellite sensors to produce a single SM archive; for example, the ESA Climate Change Initiative dataset (http://www.esa-soilmoisture-cci.org/). Owe et al. (2008) have also developed a multisensor historical SM dataset which used a consistent retrieval approach. However, the sensors used have different technical specifications and radiometric characteristics, so the users should exercise care when interpreting the dataset.

4.2 Integration With In Situ Measurements

Previously, in situ sensors were only capable of making point observations of SM, but recently a new approach has been developed, the cosmic-ray soil moisture observing system (COSMOS), which enables area-averaged measurements that sample a circular area with a diameter of $\sim$400 m (Zreda et al., 2012). These new in situ observations, which sample greater depths (typically tens of centimeters) than satellites, are attractive for a range of applications (hydrology, GHG modelling, etc.), where previously remote sensing was the only way to obtain spatially representative information. However, limitations of the COSMOS sensors include limited coverage and a reduced range under wet conditions. In the United Kingdom, CEH is establishing the first UK COSMOS network which will provide real-time SM data from a range of sites across the United Kingdom (http://cosmos.ceh.ac.uk). Other COSMOS networks exist in Germany and the United States. COSMOS has been proposed as a means of validating EO products and vice versa. Some ground observation networks are already being used for calibrating and validating satellite-based SM retrievals, for example FLUXNET and the International Soil Moisture Network (ISMN; Dorigo et al., 2011). This points us towards another possibility which is the integrated use of satellite and in situ SM measurements. To this end, GEO Water Societal Benefit Area recommended a "complimentary global system of space-based and in-situ observational networks" as they believe that this would be key to "driving more accurate hydrological and water resource management models" (Group on Earth Observations, 2010). SM is hard to measure with traditional in situ networks alone with appropriate spatial and temporal resolution and coverage. A combination of state-of-the-art space infrastructure and new in situ measurement technologies may offer a solution.

5 CONCLUSIONS

As we have seen throughout this chapter, satellite-based SM observations have potential to be used in a wide range of application areas across the hydrological and ecological sciences. The level of progress made in incorporating satellite SM measurements varies across the disciplines. In some areas, such as land surface modelling and hydrology, SM is already being incorporated into models with a great deal of success. In other areas, there is still need to develop or adapt models to utilize SM data as an input. Within these exciting emerging application areas, including landscape epidemiology and habitat condition monitoring, there have been some initial demonstration studies which show potential for further work. Their implementation in the future is dependent on the availability of high accuracy and high-spatial resolution SM data.

There is particular potential, and need, for operationalizing the application of EO-based SM in the areas of agriculture and natural hazards monitoring. The transition from research demonstration projects to full operational systems is an important challenge for the future. New satellite missions are required to provide enhanced operational capabilities. While the SMAP mission takes us a step closer to this, it is not viable in the long term as a fully operational system due to its high cost and short lifespan. SMAP is a research and operational demonstration mission which aims to inform the development of future fully operational SM missions (Moran et al., 2010). Future systems will need to be devised that offer the spatial resolution typically associated with radar systems but at much higher temporal resolutions and overcoming the issues related to the accuracy of the SM retrieval. To place these systems in an operational use, they will also need to be designed with a low-cost approach in mind. The need for operationalizing SM products is particularly great for developing countries and water-stressed regions, found across much of the African continent. Here the potential gains from satellite-based SM data are high and the data has the potential to make positive changes to peoples' lives by enhancing sustainable water management and food security. The SHARE project has already demonstrated an operational system in Africa and Australia based on ASAR data (Doubkova et al., 2009) and there is potential to build on this framework, and incorporated data from new sensors, in future.

The provision of an operational SM product with high temporal and spatial resolutions, and enhanced retrieval accuracy, is needed in order to cover a much wider range of applications for which SM information is required. Furthermore, to improve the uptake of SM data it will be important to make it easily accessible for users and to provide data with short latency times. Users should be involved from an early stage of future SM mission design and value-added products should be produced for some applications. There is also a need for integrating EO-based SM data with existing tools and models. New methods will need to be developed, including assimilation techniques, to promote the effective use of the new SM products. Overall, the development of new EO sensors, retrieval algorithms and models which integrate SM data, will support a diversity of application areas and improve our understanding of the role of SM in driving important Earth surface processes.

ACKNOWLEDGMENTS

Thanks to all CEH staff who took part in interviews: Simon Smart and Ed Rowe (carbon sequestration); Helen Roy, Tom Oliver, John Redhead (Biological Records Centre); Jamie Hannaford, Christel Prudhomme, Ragab Ragab (drought, agriculture, irrigation, hydrology, reservoirs); Steve Lofts, Claus Svendsen (contaminants); Doug Clark, Eleanor Blyth (land surface modelling/JULES); Bob Moore, Vicky Bell (flood forecasting); Steven White, Beth Purse, Susan Withenshaw (pathogens); Garry Hayman, Pete Levy, Ulli Dragosits (greenhouse gasses). Visualizations used in Fig. 1 in this chapter were produced with the Giovanni online data system, developed and maintained by the NASA GES DISC.

REFERENCES

Anderson, W.B., Zaitchik, B.F., Hain, C.R., Anderson, M.C., Yilmaz, M.T., Mecikalski, J., Schultz, L., 2012. Towards an integrated soil moisture drought monitor for East Africa. Hydrol. Earth Syst. Sci. 16, 2893–2913. http://doi.org/10.5194/hess-16-2893-2012.

Aubrecht, C., Elvidge, C.D., Baugh, K.E., Hahn, S., 2011. Identification of wildfire precursor conditions: linking satellite based fire and soil moisture data. In: Computational Vision and Medical Image Processing: VipIMAGE 2011. CRC Press, Boca Raton, pp. 347–353.

Bartsch, A., Doubkova, M., Pathe, C., Sabel, D., Wagner, W., Wolski, P., 2008. River flow & wetland monitoring with ENVISAT ASAR Global Mode in the Okavango basin and delta. In: Proceedings of the Second IASTED Africa Conference Water Resource Management (AfricaWRM 2008), Gaborone, Botswana, pp. 152–156.

Bartsch, A., Doubkova, M., Wagner, W., 2009. ENVISAT ASAR GM soil moisture for applications in Africa and Australia. In: ESA Conference on Earth Observation and Water Cycle Science, pp. 18–20.

Bell, C.W., Acosta-Martinez, V., McIntyre, N.E., Cox, S., Tissue, D.T., Zak, J.C., 2009. Linking microbial community structure and function to seasonal differences in soil moisture and temperature in a Chihuahuan Desert grassland. Microb. Ecol. 58, 827–842. http://dx.doi.org/10.1007/s00248-009-9529-5.

Berger, K.A., Wang, Y., Mather, T.N., 2013. MODIS-derived land surface moisture conditions for monitoring blacklegged tick habitat in southern New England. Int. J. Remote Sens. 34 (1), 73–85. http://dx.doi.org/10.1080/01431161.2012.705447.

Best, M.J., Pryor, M., Clark, D.B., Rooney, G.G., Essery, R.L.H., Menard, C.B., Harding, R.J., et al., 2011. The Joint UK Land Environment Simulator (JULES), model description. Part 1: energy and water fluxes. Geosci. Model Dev. 4, 677–699. http://dx.doi.org/10.5194/gmdd-4-641-2011.

Blyth, K., Biggin, D.S., Ragab, R., 1993. ERS-1 SAR for monitoring soil moisture and river flooding. In: Proc. 2nd ERS-1 Symposium 'Space at the Service of Our Environment', Hamburg, Germany, Oct. 1993. European Space Agency Publication Sp. 361, Nordwijk, The Netherlands, vol. 2, pp. 839–844.

Bolten, J.D., Crow, W.T., Zhan, X., Jackson, T.J., Reynolds, C.A., 2010. Evaluating the utility of remotely sensed soil moisture retrievals for operational agricultural drought monitoring. IEEE J. Sel. Top. Appl. Earth Obs. Remote Sens. 3 (1), 57–66. http://dx.doi.org/10.1109/JSTARS.2009.2037163.

Brocca, L., Moramarco, T., Melone, F., Wagner, W., Hasenauer, S., Hahn, S., 2012a. Assimilation of surface- and root-zone ASCAT soil moisture products into rainfall-runoff modeling. IEEE Trans. Geosci. Remote Sens. 50, 2542–2555. http://dx.doi.org/10.1109/TGRS.2011.2177468.

Brocca, L., Ponziani, F., Moramarco, T., Melone, F., Berni, N., Wagner, W., 2012b. Improving landslide forecasting using ASCAT-derived soil moisture data: A case study of the torgiovannetto landslide in central Italy. Remote Sens. 4, 1232–1244. http://doi.org/10.3390/rs4051232.

Brownstein, J.S., Rosen, H., Purdy, D., Miller, J.R., Merlino, M., Mostashari, F., Fish, D., 2002. Spatial analysis of West Nile virus: rapid risk assessment of an introduced vector-borne zoonosis. Vector Borne Zoonotic Dis. 2 (3), 157–164. http://dx.doi.org/10.1089/15303660260613729.

Chakrabarti, S., Bongiovanni, T., Judge, J., Zotarelli, L., Bayer, C., 2014. Assimilation of SMOS soil moisture for quantifying drought impacts on crop yield in agricultural regions. IEEE J. Sel. Top. Appl. Earth Obs. Remote Sens. 7 (9), 3867–3879.

Champagne, C., Davidson, A., Cherneski, P., L'Heureux, J., Hadwen, T., 2015. Monitoring agricultural risk in Canada using L-band passive microwave soil moisture from SMOS. J. Hydrometeorol. 16 (1), 5–18. http://dx.doi.org/10.1175/JHM-D-14-0039.1.

Chen, T., de Jeu, R.a.M., Liu, Y.Y., van der Werf, G.R., Dolman, A.J., 2014. Using satellite based soil moisture to quantify the water driven variability in NDVI: a case study over mainland Australia. Remote Sens. Environ. 140, 330–338. http://dx.doi.org/10.1016/j.rse.2013.08.022.

Clark, C., 2002. Measured and estimated evaporation and soil moisture deficit for growers and the water industry. Meteorol. Appl. 9, 85–93. http://dx.doi.org/10.1017/S1350482702001093.

Di Baldassarre, G., Montanari, A., Lins, H., Koutsoyiannis, D., Brandimarte, L., Blschl, G., 2010. Flood fatalities in Africa: from diagnosis to mitigation. Geophys. Res. Lett. 37 (22), 2–6. http://dx.doi.org/10.1029/2010GL045467.

Dorigo, W.A., Wagner, W., Hohensinn, R., Hahn, S., Paulik, C., Xaver, A., … Jackson, T., 2011. The International Soil Moisture Network: a data hosting facility for global in situ soil moisture measurements. Hydrol. Earth Syst. Sci. 15, 1675–1698. http://dx.doi.org/10.5194/hess-15-1675-2011.

Doubkova, M., Bartsch, A., Pathe, C., Sabel, D., Wagner, W., 2009. The medium resolution soil moisture dataset: overview of the SHARE ESA DUE TIGER project. In: International Geoscience and Remote Sensing Symposium (IGARSS), pp. 116–119. http://dx.doi.org/10.1109/IGARSS.2009.5416930.

Entekhabi, D., Njoku, E.G., O'Neill, P.E., Kellogg, K.H., Crow, W.T., Edelstein, W.N., Van Zyl, J., et al., 2010. The soil moisture active passive (SMAP) mission. Proc. IEEE 98 (5), 704–716. http://dx.doi.org/10.1109/JPROC.2010.2043918.

Fernandez-Prieto, D., Palazzo, F., 2007. The role of Earth observation in improving water governance in Africa: ESA's TIGER initiative. Hydrogeol. J. 15, 101–104. http://dx.doi.org/10.1007/s10040-006-0118-0.

Fieuzal, R., Duchemin, B., Jarlan, L., Zribi, M., Baup, F., Merlin, O., Garatuza-Payan, J., et al., 2011. Combined use of optical and radar satellite data for the monitoring of irrigation and soil moisture of wheat crops. Hydrol. Earth Syst. Sci. 15, 1117–1129. http://dx.doi.org/10.5194/hess-15-1117-2011.

Fleiss, M., Kienberger, S., Aubrecht, C., Kidd, R., Zeil, P., 2011. Mapping the 2010 Pakistan floods and its impact on human life—a post-disaster assessment of socio-economic indicators. In: Geoinformation for Disaster Management (GI4DM), Antalya, Turkey.

GCOS, 2010. Implementation Plan for the Global Observing System for Climate in Support of the UNFCC (2010 Update). (Online). Available, http://www.wmo.int/pages/prog/gcos/Publications/gcos-138.pdf (accessed 03.07.15.) GCOS-138, 180 pp.

Glatzel, K., Conway, G., Alpert, E., Brittain, S., 2014. No ordinary matter: conserving, restoring and enhancing Africa's soils. A Montpellier panel report, December 2014.

Gloaguen, J.C., 1987. On the water relations of four heath species. Vegetatio 70, 29–32.

Group on Earth Observations, 2010. Critical Earth observations priorities: water societal benefit area. Task US-09-01a.

Group on Earth Observations, 2012. Task US-09-01a: critical Earth observation priorities. Final report (second ed.).

Hassaballa, A.A., Althuwaynee, O.F., Pradhan, B., 2014. Extraction of soil moisture from RADARSAT-1 and its role in the formation of the 6 December 2008 landslide at Bukit Antarabangsa, Kuala Lumpur. Arab. J. Geosci. 7 (7), 2831–2840. http://doi.org/10.1007/s12517-013-0990-6.

Hay, S.I., Randolph, S.E., Rogers, D.J., 2002. Remote sensing and geographic information systems in epidemiology. Emerg. Infect. Dis. 8 (4), 448–449. http://www.ncbi.nlm.nih.gov/pmc/articles/PMC2730234/.

Holzman, M.E., Rivas, R., Piccolo, M.C., 2014. Estimating soil moisture and the relationship with crop yield using surface temperature and vegetation index. Int. J. Appl. Earth Obs. Geoinf. 28, 181–192. http://dx.doi.org/10.1016/j.jag.2013.12.006.

Hough, M.N., Jones, R.J.A., 1997. The United Kingdom Meteorological Office rainfall and evaporation calculation system: MORECS version 2.0—an overview. Hydrol. Earth Syst. Sci. 1 (2), 227–239. http://dx.doi.org/10.5194/hess-1-227-1997.

Ines, A.V.M., Das, N.N., Hansen, J.W., Njoku, E.G., 2013. Assimilation of remotely sensed soil moisture and vegetation with a crop simulation model for maize yield prediction. Remote Sens. Environ. 138, 149–164. http://dx.doi.org/10.1016/j.rse.2013.07.018.

IPCC, 2014. Climate Change 2014: Synthesis Report. In: Core Writing Team, Pachauri, R.K., Meyer, L.A. (Eds.), Contribution of Working Groups I, II and III to the Fifth Assessment Report of the Intergovernmental Panel on Climate Change. IPCC, Geneva, Switzerland. 151 pp.

Jacome, A., Bernier, M., Chokmani, K., Gauthier, Y., Poulin, J., De Sève, D., 2013. Monitoring volumetric surface soil moisture content at the La Grande basin boreal wetland by radar multi polarization data. Remote Sens. 5, 4919–4941. http://doi.org/10.3390/rs5104919.

Kinzelbach, W., Aeschbach, W., Alberich, C., Goni, I.B., Beyerle, U., Brunner, P., Chiang, W.-H., Rueedi, J., Zoellmann, K., 2002. A survey of methods for groundwater recharge in arid and semi-arid regions. In: Early Warning and Assessment Report Series, UNEP/DEWA/RS.02-2United Nations Environment Programme, Nairobi, Kenya. ISBN 92-80702131-3.

Kumar, S.V., Peters-Lidard, C.D., Santanello, J.a., Reichle, R.H., Draper, C.S., Koster, R.D., Jasinski, M.F., et al., 2015. Evaluating the utility of satellite soil moisture retrievals over irrigated areas and the ability of land data assimilation methods to correct for unmodeled processes. Hydrol. Earth Syst. Sci. Discuss. 12, 5967–6009. http://dx.doi.org/10.5194/hessd-12-5967-2015.

Litaor, M.I., Williams, M., Seastedt, T.R., 2008. Topographic controls on snow distribution, soil moisture, and species diversity of herbaceous alpine vegetation, Netwot Ridge, Colorado. J. Geophys. Res. 113, 1–10. http://dx.doi.org/10.1029/2007JG000419.

Meigh, J.R., Tate, E.L., 2002. The Gwava model—development of a global-scale methodology to assess the combined impact of climate and land use changes. In: EGS General Assembly Conference Abstracts (Vol. 27). p. 1276.

Merlin, O., Walker, J.P., Chehbouni, A., Kerr, Y., 2008. Towards deterministic downscaling of SMOS soil moisture using MODIS derived soil evaporative efficiency. Rem. Sens. Environ. 112(10), pp. 3935–3946.

Moran, S., Entekhabi, D., Njoku, E., O'Neill, P., 2010. Report of the 1st SMAP applications workshop. JPL Publication 10-05 report, pp. 1–36.

Njoku, E.G., 2004. AMSR-E/Aqua Daily L3 Surface Soil Moisture, Interpretive Parameters, & QC EASE-Grids. Version 2, NASA National Snow and Ice Data Center Distributed Active Archive Center, Boulder, CO. http://dx.doi.org/10.5067/AMSR-E/AE_LAND3.002.

Ochsner, T.E., Cosh, M.H., Cuenca, R.H., Dorigo, W.A., Draper, C.S., Hagimoto, Y., Zreda, M., et al., 2013. State of the art in large-scale soil moisture monitoring. Soil Sci. Soc. Am. J. 77 (6), 1888–1919. http://dx.doi.org/10.2136/sssaj2013.03.0093.

Oliver, T.H., Marshall, H.H., Morecroft, M.D., Brereton, T.M., Prudhomme, C., Huntingford, C., 2015. Interacting effects of climate change and habitat fragmentation on drought-sensitive butterflies. Nat. Clim. Change.

Owe, M., de Jeu, R., Holmes, T., 2008. Multisensor historical climatology of satellite-derived global land surface moisture. J. Geophys. Res. Earth Surf. 113, 1–17. http://doi.org/10.1029/2007JF000769.

Parry, M.L., Rosenzweig, C., Iglesias, A., Livermore, M., Fischer, G., 2004. Effects of climate change on global food production under SRES emissions and socio-economic scenarios. Global Environ. Change 14, 53–67. http://dx.doi.org/10.1016/j.gloenvcha.2003.10.008.

Petropoulos, G.P., Ireland, G., Barrett, B., 2015. Surface soil moisture retrievals from remote sensing: current status, products & future trends. Phys. Chem. Earth, Parts A/B/C 44. http://dx.doi.org/10.1016/j.pce.2015.02.009.

Piles, M., Camps, A., Vall-Llossera, M., Corbella, I., Panciera, R., Rüdiger, C., Kerr, Y.H., Walker, J., 2011. Downscaling SMOS-derived soil moisture using MODIS visible/infrared data. Geosci. Remote Sens., IEEE Trans. on, 49(9), pp. 3156–3166.

Piles, M., Vall-Llossera, M., Camps, A., Sanchez, N., Martinez-Fernandez, J., Martinez, J., Riera, R., et al., 2013. On the synergy of SMOS and Terra/Aqua MODIS: high resolution soil moisture maps in near real-time. In: International Geoscience and Remote Sensing Symposium (IGARSS), pp. 3423–3426. http://dx.doi.org/10.1109/IGARSS.2013.6723564.

Piles, M., Petropoulos, G.P., Ireland, G., Sanchez, N., in press. A novel method to retrieve soil moisture at high spatio-temporal resolution based on the synergy of SMOS and MSG SEVIRI observations. Rem. Sens. Environ.

Qi, Y., Dennison, P.E., Spencer, J., Riaño, D., 2012. Monitoring live fuel moisture using soil moisture and remote sensing proxies. Fire Ecol. 8 (3), 71–87. http://dx.doi.org/10.4996/fireecology.0803071.

Ragab, R., 1992. Assessment of the relationship between remotely sensed topsoil moisture content and profile moisture content. In: Eley, F.J., Granger, R., Martin, L. (Eds.), Soil Moisture Modelling and Monitoring for Regional Planning.Workshop Held at The National Hydrology Research Centre, Environment Canada, Saskatoon, Saskatchewan, Canada, March 9–10, 1992. NHRI Symposium No. 9 Proceedings, pp. 141–153.

Ragab, R., 1995. Towards a continuous operational system to estimate the root zone soil moisture from intermittent remotely sensed surface moisture. J. Hydrol. 173, 1–25.

Rebel, K.T., De Jeu, R.A.M., Ciais, P., Viovy, N., Piao, S.L., Kiely, G., Dolman, A.J., 2012. A global analysis of soil moisture derived from satellite observations and a land surface model. Hydrol. Earth Syst. Sci. 16, 833–847. http://dx.doi.org/10.5194/hess-16-833-2012.

Sass, G.Z., Wheatley, M., Aldred, D.A., Gould, A.J., Creed, I.F., 2012. Defining protected area boundaries based on vascular-plant species richness using hydrological information derived from archived satellite imagery. Biol. Conserv. 147, 143–152. http://dx.doi.org/10.1016/j.biocon.2011.12.025.

Sharpley, A.N., McDowell, R.W., Kleinman, P.J.A., 2001. Phosphorus loss from land to water: integrating agricultural and environmental management. Plant Soil 237, 287–307. http://dx.doi.org/10.1023/A:1013335814593.

Srivastava, P.K., Han, D., Rico-Ramirez, M.A., Al-Shrafany, D., Islam, T., 2013. Data fusion techniques for improving soil moisture deficit using SMOS satellite and WRF-NOAH land surface model. Water Resour. Manage. 27, 5069–5087. http://dx.doi.org/10.1007/s11269-013-0452-7.

Stålnacke, P., Gooch, G.D., 2010. Integrated water resources management. Irrig. Drain. Syst. 24, 155–159. http://dx.doi.org/10.1007/s10795-010-9106-6.

Stampoulis, D., Andreadis, K., Granger, S.L., Fisher, J.B., Behrangi, A., Das, N.N., Turk, J., 2014. Quantifying the resilience of vegetation and soil moisture during dry spells using satellite remote sensing. In: AGU Fall Meeting Abstracts.

Tipping, E., Rowe, E.C., Evans, C.D., Mills, R.T.E., Emmett, B.A., Chaplow, J.S., Hall, J.R., 2012. N14C: a plant–soil nitrogen and carbon cycling model to simulate terrestrial ecosystem responses to atmospheric nitrogen deposition. Ecol. Modell. 247, 11–26. http://dx.doi.org/10.1016/j.ecolmodel.2012.08.002.

Turner, W., Spector, S., Gardiner, N., Fladeland, M., Sterling, E., Steininger, M., 2003. Remote sensing for biodiversity science and conservation. Trends Ecol. Evol. 18 (6), 306–314. http://dx.doi.org/10.1016/S0169-5347(03)00070-3.

UNICEF, 2014. Generation 2030 Africa.

Verstraeten, W.W., Veroustraete, F., Wagner, W., Roey, T., Heyns, W., Verbeiren, S., Feyen, J., 2010. Remotely sensed soil moisture integration in an ecosystem carbon flux model. The spatial implication. Clim. Change 103 (1–2), 117–136. http://dx.doi.org/10.1007/s10584-010-9920-8.

Wagner, W., Blöschl, G., Pampaloni, P., Calvet, J.-C., Bizzarri, B., Wigneron, J.-P., Kerr, Y., 2007. Operational readiness of microwave remote sensing of soil moisture for hydrologic applications. Nord. Hydrol. http://dx.doi.org/10.2166/nh.2007.029.

Wagner, W., Sabel, D., Doubkova, M., Bartsch, A., Pathe, C., 2009. The potential of Sentinel-1 for monitoring soil moisture with a high spatial resolution at global scale. ESA Special Publications SP-674.

Wang, L., Qu, J.J., Hao, X., 2008. Forest fire detection using the normalized multi-band drought index (NMDI) with satellite measurements. Agr. Forest Meteorol. 148, 1767–1776. http://dx.doi.org/10.1016/j.agrformet.2008.06.005.

Weber, K., Schnase, J.L., Carroll, M., Brown, M.E., Gill, R., Haskett, G., Gardner, T., 2013. RECOVER—an automated burned area emergency response decision support system for post-fire rehabilitation management of Savanna ecosystems in the Western US. In: AGU Fall Meeting Abstracts, vol. 1, p. 1543.

You, L., Ringler, C., Nelson, G., Wood-sichra, U., Robertson, R., Wood, S., Sun, Y., et al., 2010. What is the irrigation potential for Africa? A combined biophysical and socioeconomic approach. IFPRI discussion paper 00993, June 2010.

Zreda, M., Shuttleworth, W.J., Zeng, X., Zweck, C., Desilets, D., Franz, T., Rosolem, R., 2012. COSMOS: the cosmic-ray soil moisture observing system. Hydrol. Earth Syst. Sci. 16, 4079–4099. http://dx.doi.org/10.5194/hess-16-4079-2012.

Index

Note: Page numbers followed by *f* indicate figures and *t* indicate tables.

Printed in the United States
By Bookmasters